Great Lakes

Lessons in Participatory Governance

Great Lakes

Lessons in Participatory Governance

Editors

Velma I. Grover
Adjunct Professor
Faculty of Environmental Studies
York University
Toronto, ON
Canada

Gail Krantzberg
Professor and Director
Centre for Engineering and Public Policy
School of Engineering Practice
McMaster University
Hamilton, ON
Canada

CRC Press
Taylor & Francis Group
Boca Raton London New York

CRC Press is an imprint of the
Taylor & Francis Group, an **informa** business
A SCIENCE PUBLISHERS BOOK

CRC Press
Taylor & Francis Group
6000 Broken Sound Parkway NW, Suite 300
Boca Raton, FL 33487-2742

Cover illustration: Courtesy of United States Environmental Protection Agency Great Lakes National Program office

Printed in the United States of America on acid-free paper
Version Date: 20120327

International Standard Book Number: 978-1-57808-769-3 (Hardback)

Library of Congress Cataloging-in-Publication Data

Great Lakes : lessons in participatory governance / editors, Velma I.
Grover, Gail Krantzberg.
p. cm.
Includes bibliographical references and index.
ISBN 978-1-57808-769-3 (hardcover : alk. paper)
1. Environmental protection--Great Lakes (North America) 2. Restoration ecology--Great Lakes (North America) 3. Environmental policy--North America. 4. Political participation--Great Lakes Region(North America) 5. Water quality management--Great Lakes (NorthAmerica) 6. Great Lakes (North America)--Environmental conditions. 7. Environmental protection--Great Lakes (Africa) 8. Political participation--Great Lakes Region (Africa) 9. Water quality management--Great Lakes (Africa) 10. Great Lakes (Africa)--Environmental conditions. I. Grover, Velma I. II. Krantzberg, Gail.
TD171.3.G72G72 2012
333.91'630977--dc23

2011038889

Visit the Taylor & Francis Web site at
http://www.taylorandfrancis.com

CRC Press Web site at
http://www.crcpress.com

Science Publishers Web site at
http://www.scipub.net

Preface

Great Lakes in both North America and Africa constitute a large chunk of planet's surface freshwater. However, increased pollution, decreasing water quantity, over-fishing, urban sprawl and climate change are threatening the sustainability of the Great Lakes. What happens to the Great Lakes has significance not only for the riparian countries but also the surrounding regions and the global community as a whole, as the waters and fisheries of these lakes provide enormous economic and social benefits for the people of surrounding regions.

Great Lakes have seen a fair bit of pollution and contamination with industrial and agricultural growth along its shores. Land-use changes have been observed in the catchments of the Great Lakes, mainly in response to population increase and consequent demand for resources. Clearing of land for agriculture (and large scale farming), lumbering, and human settlements result in increased surface run-off and increased sediment and nutrient input in the lakes. In some cases like Lake Erie there have been major challenges with eutrophication and invasive species, and Lake Victoria has faced its own challenges in terms of invasive species. This industrial and agricultural build-up along the shores of lakes has led to accumulation of toxins in the Great Lakes. [1]Initially with industrialization and large scale farming, there were few regulations to prevent discharge of effluents in water bodies and to protect water quality and as a result toxin loading to the lakes was in some cases severe. Most of these toxins came from either agricultural runoff (contaminated with fertilizers, herbicides or other chemicals used in farms) or from industrial discharges (since lakes provide easy access for transportation of raw materials and finished goods a lot of industries sprung along lake side but they also just discharged untreated effluent in the lakes). With new regulations and programs such as Remedial Action Plans (RAPs) there has been significant progress in the control of point-source pollution (where polluters can be directly identified), however, non-point source pollution (where the sources are diffuse or even unknown) remains much more difficult to control or regulate.

RAPS are being developed and implemented at Areas of Concern (43, now 39 because some four Areas of Concern have completed implementation and

[1]USEPA, http://www.epa.gov/glnpo/atlas/glat-ch4.html

have been delisted). The Areas of Concern are locations throughout the Great Lakes basin ecosystem where environmental quality is particularly compromised. According to United States and Canada Great Lakes Water Quality Agreement of 1987, RAPs are to embody a systematic and comprehensive ecosystem approach to restoring and protecting beneficial uses (the ability of fish, wildlife and humans to thrive) in the Areas of Concern. The Agreement calls for the federal governments, in cooperation with state and provincial governments, to ensure the public is consulted throughout the development and implementation of the RAPs. While costly clean up efforts remain, there are notable advances in remediation and prevention programs. Essential elements that characterize successful initiatives include true participatory decision making, a clearly articulated and shared vision, and focused and deliberate leadership.

Most of the RAPs at geographic Areas of Concern were based on ecosystem management principles and involved stakeholder participation in both identifying issues of concern, desired future regeneration states, and remedial interventions. The focus of this edited volume is on these participatory governance regimes in environmental management in general and in Great Lakes in particular. Editors of the book believe that there is a wealth of experience in this area which is not published, especially in scholarly literature. The book brings an up-to-date information and thinking on remediation and regeneration around the lakes. The book documents such experiences from North American and African lake systems. The book concludes with transferable finding about revitalizing the ecosystem integrity of the lakes, which can be replicated in other watersheds of the world.

Velma I. Grover
Gail Krantzberg

Hamilton, ON, Canada
March, 2012

Contents

SECTION 5: INDUSTRIAL LEADERSHIP

SECTION 6: CONCLUSIONS

List of Contributors

Abdel-Fattah Sommer

92 Melbourne St., Hamilton, Ontario, L8P2A6, Canada.

E-mail: abdelfs@mcmaster.ca

Boughton Lori

Pennsylvania Department of Environmental Protection, Northwest Regional Office, 230 Chestnut Street, Meadville, PA 16335, USA.

E-mail: lboughton@state.pa.us

Crane Thomas R.

Great Lakes Commission, Eisenhower Corporate Park, 2805 S. Industrial Hwy., Suite 100, Ann Arbor, MI 48104-6791, USA.

E-mail: tcrane@glc.org

Cruickshank Abigail

24048 109th Avenue, Maple Ridge, British Colombia V2W 124, Canada.

E-mail: abcrvicks@gmail.com

Fox Bonnie

Conservation Ontario, 120 Bayview Parkway, Box 11, Newmarket, Ontario, Canada, L3Y 4W3.

E-mail: bfox@conservationontario.ca

Gaden Marc

Great Lakes Fishery Commission, 2100 Commonwealth Blvd, Ste. 100, Ann Arbor, MI 48105, USA.

E-mail: marc@glfc.org

Goddard Chris

Great Lakes Fishery Commission, 2100 Commonwealth Blvd. Ste. 100, Ann Arbor, MI 48105, USA.

E-mail: cgoddard@glfc.org

Grover Velma I.

916-981 Main St. W, Hamilton, ON, L8S 1A8, Canada.

E-mail: velmaigrover@yahoo.com

Hall John D.

Hamilton Harbour Remedial Action Plan, 867 Lakeshore Road, Burlington, Ontario, L7R 4A6, Canada.

E-mail: John.hall@ec.gc.ca

James Phil

Credit Valley Conservation Authority, 1255 Old Derry Road, Missisauga, ON L5N 6R4, Canada.

E-mail: pjames@creditvalleyca.ca

Krantzberg Gail

Centre for Engineering and Public Policy in the School of Engineering Practice, McMaster University, ITB 109, 1280 Main St, W, Hamilton, ON L8S 4L7, Canada.

E-mail: krantz@mcmaster.ca

Kubursi Atif

Econometric Research Ltd. and McMaster University, 1280 Main St, W, Hamilton, ON L8S 4L7, Canada.

E-mail: Kubursi@mcmaster.ca

Kuper George H.

Council of Great Lakes Industries, 3600 Green Court, Suite 710, Ann Arbor, Michigan 48105-1570, USA.

E-mail: ghk@cgli.org

Kusmierczyk Irek W.

Department of Political Science, Vanderbilt University, VU Station B#351817, Nashville, Tennessee 37235, USA.

E-mail: i.kusmierczyk@vanderbilt.edu

Lawrence Patrick L.

Department of Geography and Planning, University of Toledo, Toledo, Ohio 43606, USA.

E-mail: patrick.lawrence@utoledo.edu

Lawrence Ted

Great Lakes Fishery Commission, 2100 Commonwealth Blvd, Ste. 100, Ann Arbor, MI 48105, USA.

E-mail: ted@glfc.org

Lukes Robb

Credit Valley Conservation Authority, 1255 Old Derry Road Missisauga, ON L5N 6R4, Canada.

E-mail: rlukes@creditvalleyca.ca

McLaughlin Chris

Walter G. Booth School of Engineering Practice, McMaster University, Hamilton, Ontario, Canada, L8S 4K1.

E-mail: chris.mclaughlin@mcmaster.ca

Meek Sonya

Toronto and Region Conservation Authority, 5 Shoreham Drive, Downsview, Ontario, Canada, M3N 1S4.

E-mail: smeek@trca.on.ca

Mkumbo Oliva C.

Lake Victoria Fisheries Organization, Plot No. 7B/7E Busoga Square, Belle Avenue, P.O. Box 1625, Jinja 256, Uganda.

E-mail:

Muldoon Paul

Environmental Lawyer.

E-mail: Paul.muldoon@sympatico.ca

O'Connor Kristin M.

Hamilton Harbour Remedial Action Plan, 867 Lakeshore Road, Burlington, Ontario, L7R 4A6.

E-mail: Kristin.oconnor@ec.gc.ca

Okurut Tom O.

Lake Victoria Basin Commission Secretariat, P.O. Box 1510-40100,

Kisumu, Kenya.

E-mail: lvbc@lvbcsec.org

Othero Doreen M.

Lake Victoria Basin Commission Secretariat, P.O. Box 1510-40100, Kisumu, Kenya.

E-mail: lvbc@lvbcsec.org

Patterson Tracy

Freeman and Associates, Suite 325, 701 Rossland RD East Whitby, ON L1N 9K3.

E-mail: tfp@sympatico.ca

Shaw Ian A.C.

Arcelor Mittal Dofasco, 1330 Burlington St. East, Hamilton, Ont. L8N 3J5, Canada.

Vanderlinden Kyle

Credit Valley Conservation Authority, 1255 Old Derry Road, Missisauga, ON L5N 6R4, Canada.

E-mail: kvanderlinden@creditvalleyca.ca

Zimmer Christine

Credit Valley Conservation Authority, 1255 Old Derry Road, Missisauga, Ont. L5N 6R4, Canada.

E-mail: czimmer@creditvalleyca.ca

Section 1

Setting the Scene

CHAPTER 1

Introduction and Roadmap of the Book

Velma I. Grover

916-981 Main St. W, Hamilton, ON, L8S 1AB, Canada
E-mail: velmaigrover@yahoo.com

INTRODUCTION

Lakes attract people, industry, and commerce, because lakes (and rivers) provide an excellent mode of transportation. They are thus facing competitive exploitation and link to livelihoods of large populations and high levels of economic activity, such as fishing and related industries that are an important source of gross domestic product (GDP), in addition to tourism, which can amount to billions of dollars. Lakes are also important in enhancing aquatic biodiversity, providing ecological services, generating hydro-power, providing a source of food, a source of drinking water supply, and creating areas for people to enjoy the aesthetic beauty and recreational activities that large lakes offer. More than 33 million people inhabit the Great Lakes Basin, including about a third of Canada's population and 15% of United States' population. Hence it is important to maintain the quality of lakes to support eco-system services. Since the Great Lakes in North America are trans-boundary, instruments and institutions are needed for shared water governance. The history of these institutions can be traced back to 1909.

Historically, one of the biggest challenges facing management of Great Lakes has been the presence of toxic contaminants. The accumulation of toxins in the Great Lakes is largely the result of industrial and agricultural build-up along the shores of the lakes over the past hundred years. With early industrialization and farming, there were few, if any, regulatory requirements protecting lake quality and, as such, many toxins were added to the lake systems. During the 20th century, many new chemicals were used in agricultural production, which were later discovered to have severe effects on the ecosystem. Mineral-based industries were situated on the lakes because of the ease with which materials could be transported. An unfortunate side effect of those locations has been that wastes often have been intentionally and accidentally dumped into the lakes. While there has been significant progress in the control of such point-source pollution (where polluters can be directly identified), non-point source pollution (where the sources are diffused or even unknown) remains much more difficult to control or regulate. Present day threats now extend to alien invasive species and climate change.

Governance institutions were established in the Great Lakes region, in response to high-profile crisis situations. The Boundary Waters Treaty of 1909 evolved in response to water quantity disputes involving hydropower development and water quality problems causing human diseases and deaths that peaked around 1900. As eutrophication problems worsened in the 1950s and 1960s, the Great Lakes Water Quality Agreement (GLWQA) of 1972 was attached to the Treaty as a permanent reference. The Agreement has since been revised in 1978 and 1987 to address other environmental issues. It was in 1987 that amendments were made to strengthen the programs, practices and technologies mentioned in the 1978 amendment to increase the accountability for their implementation, including setting up a timeline for implementation. At this time annexes were added to address atmospheric deposition of toxic pollutants, contaminated sediment, groundwater and non-point sources of pollution. Forty two areas around the Great Lakes were identified as Areas of Concern (AOC), these areas were heavily degraded and required restoration plans and programs to regenerate eco-system services. At this point Annexes were also added to incorporate the development and implementation of Remedial Action Plans for AOC and Lake-wide Management Plans to control critical pollutants. Remedial Action Plans (RAPs) for Great Lakes Areas of Concern (AOCs) are perhaps the best example of community-based environmental protection in existence. Through the collaboration between public and private institutions, the RAPs apply a watershed approach to ecosystem regeneration and protection, as they progress towards the re-establishment of beneficial uses.

This book describes the actions taken at AOCs through RAPs with an aim to highlight the successes and failures of different sites. An attempt is made to draw lessons learnt from a few sites. The book has five sections. The first section of the book gives a general introduction about the institutions established for managing water quality/quantity, followed by a general introduction to Water Quality Agreements and Remedial Action Plan. Section 2 first focuses on the principles of Integrated Water Resources Planning and Implementation in general and then gives case studies on how they are applied in Ontario. Section 3 discusses some case studies of RAP implementation plans around the Great Lakes, while Section 4 focuses on the African Great Lakes and brings out some comparative analysis of the Great Lakes in both continents. Section 5 of the book looks at the Industrial Leadership in improving water quality in the Great Lakes region, and the last section is the concluding section.

Section 1: Setting the Scene

The first section of the book focuses on the general introduction to the principles involved in the remedial action plan such as participatory governance, innovations in decision making and the role of municipalities.

The next chapter in this section, "Great Lakes – Great Responsibilities: History of and Lessons in Participatory Governance" by Tom Crane looks at the history and role of governments and the institutions of the Great Lakes – St. Lawrence River basin in the governance of the Great Lakes. The chapter explores how water has historically influenced and still influences the role and relationships of governments at different levels spanning over two countries. The author also explains the main treaties, conventions, agreements and other legal mechanisms which have shaped governance and institutions in the region. Major institutions such as International Joint Commission, Great Lakes Fishery Commission are briefly described with an emphasis on how these organizations fit into the governance framework of the region as well as how these institutions have evolved as a result of significant events and changes that have occurred; both to the ecosystem and the response of governments to these changes.

This is followed by a chapter on "Governance in the Great Lakes: A Regime in Transition" in which Paul Muldoon discusses the participatory governance in the Great Lakes in the ecological as well as institutional context. This chapter provides a brief review of the Great Lakes governance regime under the GLWQA followed by a discussion on some key themes or lessons learned that could be useful at present (2011) when the GLWQA is being renegotiated to include new threats and revamp or update governance regimes.

The next chapter by Chris McLaughlin, "Innovations in Decision Making", focuses on the potential of adaptive management and decision analysis, to "expand and enrich" adaptive Great Lakes governance. The author describes the principles and fundamental character of these two techniques as a means to develop an institutional framework to help structure participatory decision making and adaptive governance in the Great Lakes that more directly links science to policy and management activity to program outcomes.

One of the most important players in RAP is the local government, however its role in trans-boundary management and the decision making process is not clearly defined. Irek W. Kusmierczyk in "Transnational Municipal Networks of American and Canadian Local Governments in the Context of Bilateral Environmental Relations: The Emergence of a European Phenomenon in the

Great Lakes Basin", discusses the increasing capacity of local government in trans-boundary environmental management (essentially meaning the degree to which actors can participate in, and influence, governance or decision-making processes). The focus of the chapter is mainly on the transformation of the role of local government in environmental policymaking in the Great Lakes Basin and specifically on "the Great Lakes and St. Lawrence Cities Initiative" which is the only example of a *transnational municipal network* operating across the border between Canada and the United States.

To sustain the implementation of appropriate policies and programs, it is important to know the costs/benefits associated with them. Although, restoration, enhancement and protection of the watersheds and waterfront are essential to provide "green infrastructure", amenities and services (including eco-system services provided by clean environment) that can attract industrial, commercial and residential intensification to developed areas, they come with a cost. Atif Kubursi presents an analysis of the costs and benefits of such restorations and RAPs in the Great Lakes Region in the next chapter of the book "The Economic Benefits of Remedial Action Plans in Ontario".

Section 2: Principles of Integrated Water Resources Management

Section Two of the book focuses on the Principles of Integrated Water Resources Planning and Implementation in general and more specifically with case studies on how they are applied in Ontario. The section begins with a general overview of Integrated Water Resources Management by Abigail Cruickshank and Velma I Grover. The chapter describes the basic principles of IWRM, development of IWRM, how it can be implemented and the hurdles in implementation of IWRM in general.

The next chapter "Lessons Learned from Implementing Low Impact Development in the Credit River Watershed" by Christine Zimmer, Phil James, Kyle Vanderlinden, Robb Lukes, and Tracy Patterson. Monitoring results in the Credit River valley show that the health of the watershed declines as one moves from the largely rural Upper Watershed to the highly urbanized Lower Watershed. To prevent further deterioration an immediate strategic shift in current land use practices is needed. The strategy involves adopting Low Impact Development (LID) principles and practices in both developing areas and retrofitting existing urban areas. A key component to implementing LID is delivering an effective pollution prevention program to prevent pollutants from contaminating surface and groundwater supplies. The authors rightly point out that adoption of LID involves storm water practices on private property, hence engaging the public is a critical component of wide scale implementation. Unfortunately to date environmental campaigns have been unsuccessful in engaging and changing behaviour. This chapter discusses how marketing and outreach initiatives, informed by sound market research and a fulsome understanding of the target demographic(s), is an effective approach to securing uptake and desired long term behaviour change within the intended population. Informational- and educational-based campaigns have not achieved the desired results because they fail to address the intrinsic motivations of the target audience(s).

Implementation actions for the protection and enhancement of the Great Lakes and St. Lawrence ecosystem usually take place at the local scale and need local partners. This requires a framework that links larger Great Lakes/ St. Lawrence objectives for restoration and development to local priorities that would also engage local partners and citizens. Bonnie Fox and Sonya Meek in "Conservation Ontario's Coordinating Mandate to Promote an Integrated Watershed Management Approach to Protection of the Great Lakes and St. Lawrence Ecosystem" explain that this general framework is lacking in Ontario but existing Conservation Authorities' Integrated Watershed Management approach could be more fully utilized to advance objectives. The chapter describes Conservation Ontario, Conservation Authorities and their integrated watershed management approach. It then discusses the benefits of integrated watershed management to the Great Lakes/St. Lawrence, the lessons learned from its application to date, and illustrates leading edge efforts in the Rouge River watershed. The authors also describe the benefits of using integrated watershed management approach across the Great Lakes and St. Lawrence Basins and makes recommendations for moving forward.

Section 3: Case Studies

As highlighted by the discussion so far, restoration of lakes is important and it needs the participation of numerous stakeholders. Through unique stories of different RAP experiences, the third section of the book describes how restoration of different water-fronts and eco-systems depend on the nature and causes of local degradation, local partners and the roadmap built by different stakeholders involved. This makes all the RAPs unique and they all have different stories of successes and failures. The section begins with a historical perspective presented by Gail Krantzberg in "North American Remedial Action Plans: a Primer". The author discusses the Great Lakes Quality Agreement and its amendments, in particular Annex 2 which focuses on RAPs. In addition to the history the author also describes the lessons learnt from the past on what has worked, what has not and why, and the role of these lessons learnt informs of a revision to Annex 2, and enhances place-based engagement and implementation for the Great Lakes regeneration. The chapter also highlights the ecosystem approach that is one of the important ingredients for success of the RAP process.

The first case study of RAPs presented in this section is a success story of how Collingwood Harbour was the first AOC to be delisted. In "First Off the List: The Collingwood Harbour Story" Gail Krantzberg begins with a discussion of the issues and problems that existed in Collingwood and how the community came together under the leadership of RAP coordinator and Public Advisory Committee Chair to work towards improving the harbour and the opportunities that came following its improvement. As pointed out by the author governments need to play an important role in building and nurturing community capacity to carry out the complex stakeholder processes. The author concludes that as learnt from the experiences of Collingwood case study: for success of any RAP there are a few central pre-requisites such as: participation of the appropriate actors; development of mutually agreed upon decision making process; common visions and objectives; local ownership political support; public participation and continued funding.

Hamilton is the "steel capital" of Canada, and has had unique challenges to its harbour front, communities and fish habitat from the raw sewage going into Lake Ontario from the industries as well as municipal waste water. "Remedial Action Plan Case Study: Participatory Governance Used in Hamilton Harbour" by John D. Hall and Kristin M. O'Connor describes these issues surrounding Hamilton harbour from the historical perspectives as the city and industries developed in early 1900s followed by the environmental movement in the city and establishment of stakeholder groups and processes for the RAP. The chapter also discusses one of the great challenges, i.e. presence of certain chemicals in Lake Ontario, that is still preventing Hamilton from being delisted as an AOC. The authors then list a number of "lessons learned" from the Hamilton Harbour RAP experience under different categories: contextual lessons learnt (don't lose the original goals, learn from the past success and failures, create a Remedial Action Plan); process lessons learnt (get all the stakeholders at the table, have a good facilitator and well defined roles, have a science based writing team and task groups to explain complex scientific issues, value of senior level representation, good communication plan, and motivated people to keep the plan going); implementation lessons learned (implementing structure is needed, well defined work plan is important, accountability is an essential component for a RAP program to succeed, technical teams and advisory committees are a valuable part of the RAP process, and project advisory groups are needed for proper implementation of the plan); and communication lessons learned (communication with the public and engaging politicians).

The Maumee AOC faced issues similar to those of other areas of concerns: contaminated sediment; nutrient loadings; wetland loss; aquatic habitat degradation; bacterial pollution; aged and degraded combined water and wastewater infrastructure leading to sewer overflows; eutrophication; and non-source pollutants from rural runoff. Patrick L. Lawrence in "As the Water Flows: Community Based Decision-Making and Participatory Planning for the Maumee Area of Concern, Ohio" examines the history, planning framework, successes and challenges of the RAP planning process followed by reflections upon the fundamental barriers and opportunities present in the community, importance of leadership and innovation, and the driving forces needed to achieve improvements necessary to return the rivers and streams to be fishable, drinkable, and swimmable waters for the community. The chapter also gives key evolutionary steps or approaches to community engagement and involvement via a local RAP Advisory Committee and conclusions drawn as to the elements of each step followed by a discussion on changes that occurred as progress continued and the current organizational arrangements developed as a response to varied challenges. The author then discusses the opportunities that emerged during the history of the Maumee RAP and for its continued future success. One of the main challenges faced, as concluded by the author, is to keep all the stakeholders engaged in a long process of restoration.

The main issue in Presque Isle Bay, situated on Lake Erie, was that much of the wastewater from the City of Erie's industrial and domestic sources was discharged directly into the bay or its tributary streams treating them as sewers rather than natural resources. The outbreak of typhoid in the early 1990s compelled the city to construct its wastewater treatment plant. Although the plant has undergone many upgrades since then, yet problems remained as a result of discharges to the bay and lake from combined sewer overflows, unregulated industrial releases, and the treatment plant itself. During heavy storm events, raw sewage flowed directly into the bay, drawing the attention of bay-front businesses and raising concerns among local citizens. However, the main reason for Presque Isle Bay to become the 43rd Area of Concern (AOC) was the report by a local fisherman about the presence of external sores and lesions on brown bullhead catfish caught in the Presque Isle Bay. Lori Boughton in "Remedial Action Plan: A Case Study of the Presque Isle Bay Area of Concern" emphasizes that public involvement is a key to the Great Lakes environmental movement. The success story of Presque Isle Bay's illustrates how regional collaborations and participatory decision-making necessitated by local environmental and economic issues can work successfully. One of the important lessons learnt is that there should be realistic expectations on the part of members and the organization implementing the RAP. One of the biggest tasks in implementation is to identify organizations that will be involved and then define their roles in both decision making and implementation.

It has been a challenge at most places of the areas of concern to get delisted and quite a few of them are still not delisted. Climate change is exacerbating the problem. Sommer Abdel Fattah in her chapter, "Restoring Beneficial Uses in Great Lakes Areas of Concern under the Threats of a Changing Climate" has highlighted how climate change threatens the resilience of beneficial uses and poses challenges for the long term integrity of restored ecosystem services.

Section 4: Comparative Analysis of the Great Lakes

Section Four focuses on African Great Lakes and brings out some comparative analysis of the Great Lakes in both continents. In the first chapter in this section, "A Holistic Approach to Natural Resource Management: A Case of Lake Victoria Basin" Tom O. Okurut and Doreen M. Othero discuss the challenges related to the largest freshwater lake, Lake Victoria, in Africa. In order to realize maximum benefits from the lake and its basin, a paradigm shift in its management, from the Traditional Chiefdoms to a Holistic Approach was gradually introduced starting from 1997. As described by the authors, the adoption of a holistic approach for management of Lake Victoria Basin was driven mainly by: (i) increases in population density (ii) environmental decline; (iii) a multiplicity of stakeholders with different interests, and (iv) duplication of effort in the projects and programs. In order to deal with these threats, a specialized institution of the East African Community (EAC); the Lake Victoria Basin Commission was established in July 2005 to coordinate holistically the initiatives by Governments, Civil Society Organizations and the private sector. National Focal Points were established at the national level in each Partner State to support the Commission. In the chapter, the authors describe how these management approaches have yielded positive impacts in areas of conflict prevention and resolution, investment, projects development, harmonization of relevant laws, networking and sharing of information and research. The experiences from the EAC activities in the Lake Victoria and its Basin indicate that the management and coordination of a trans-boundary Lake System requires a holistic approach anchored on a committed and focused management body to coordinate all initiatives therein with a view to building synergies among different stakeholders.

In "Top-Down and Bottom-Up Approaches in the Management of the Laurentian Great Lakes and Lake Victoria Fisheries: A Comparison of Two Shared Water Bodies" Marc Gaden, Oliva C. Mkumbo, Ted Lawrence, and Chris Goddard give a comparative analysis of fisheries management in the Great Lakes in Africa and North America with a focus on the primary structures and institutions established to facilitate inter-jurisdictional cooperation, collaborative processes, or co-management. As described in the chapter, fishery management authority in both regions is diffused, with two nations, eight states, the province of Ontario, and U.S. tribes involved in Great Lakes fishery management and three partner states—Kenya, Tanzania, and Uganda—and scores of local entities called "Beach Management Units" involved in Lake Victoria fishery management.

The chapter also discusses how fishery management in both regions involves both "top-down" and "bottom-up" elements.

Section 5: Industrial Leadership

Section Five is a forward looking piece where authors look at the Industrial Leadership in improving water quality in the Great Lakes region. The section begins with a contribution from George H. Kuper, "Industry and Governance in the Great Lakes Basins". Industries were attracted around Great Lakes area because of the abundance of raw material including natural resources and water but with no regulations on wastewater discharge untreated industrial and municipal water was discharged directly into the lakes. However, as discussed in the chapter, now these industries are working with ministries to set up standards for effluent discharge in the lakes and have also formed the bi-national Council of Great Lakes Industries. The author also describes other initiatives such as: participation in meetings to showcase the measures taken by industry towards implementation of sustainability pursuits; participation of industry on the IJC's bi-national Virtual Elimination Task Force followed by becoming an advocate of Task Force findings and inviting governments and environmental groups to form multi-stakeholder forums on specific issues; and sponsoring a comprehensive assessment of the state of the Lakes ecosystem to contribute to the understanding of the status and needs for restoring and protecting the Great Lakes Ecosystem. The chapter also describes the industry's role in chemicals management, water management, Great Lakes restoration, Great Lakes Water Quality Agreement, and the challenges the industries are facing in the region (the challenges include increased regulation, regional policy initiatives, pressures from global competition and weak economic environments including invasive species). The chapter concludes with the future of industry as a stakeholder.

This is followed by a chapter by Ian A.C. Shaw, "Hamilton Harbour's Regeneration – An Industrial Perspective" which discusses specifically the role of ArcelorMittal Dofasco, a steel mill in Hamilton, as an active participant in the clean-up and restoration of Hamilton Harbour. As discussed by Shaw since 1989 AM Dofasco has reduced total process effluent discharges to Hamilton Harbour by 99%. In the future it is expected that water will become an even more critical issue for the industry driving the need for further innovation in water efficiency, conservation and restoration efforts of the Great Lakes.

Section 6: Conclusions

With rapid industrialization and mass scale agriculture with pesticides and herbicides we have lost sight of how much we are degrading the Great Lakes and its eco-system. With a new vision to have some control over this degradation the areas of concerns were identified and some progress has been made in the quality of lakes in North America. We hope that the lessons learnt documented in this book can help raise awareness around the world to protect our precious water resources. It can be concluded from all the chapters that the key to success is involving all the key stakeholders from the very beginning in a meaningful way, have a well defined work plan and the right partners. It is also important to have a right mix of funding, innovation and leadership (facilitator) to lead the process. Accountability, transparency, good communication plan and celebrating every small success along the way to delisting are also some of the essential components of successful RAP implementation. The ongoing challenge is to keep all the stakeholders motivated and engaged in the long process, and to have continued progress at the areas of concern in implementation of RAPs through creative solutions to complex problems, partnerships and commitment for funding.

CHAPTER 2

Great Lakes – Great Responsibilities: History of and Lessons in Participatory Governance

Thomas R. Crane

Great Lakes Commission, Eisenhower Corporate Park,
2805 S. Industrial Hwy., Suite 100, Ann Arbor, MI 48104-6791
E-mail: tcrane@glc.org

INTRODUCTION

This chapter will look at the formation, history and role of governments and institutions in the water rich region of the Great Lakes – St. Lawrence River Basin. As we examine the development of the region, one thing becomes clear; water is a unifying factor that has influenced and continues to influence the role and relationships of government at many levels and spanning two countries. As we enter deeper into the 21st century, it will be interesting to see how the region's system of governance evolves to meet the challenges of a changing world where environmental protection, resource management and economic development are requiring more than ever, a regional if not a global approach to management.

This chapter will also examine the main treaties, conventions, agreements and other legal mechanisms which have shaped governance and institutions in the region. Major institutions will be briefly described with an emphasis on how these organizations fit into the governance framework of the region as well as how these institutions have evolved as a result of significant events and changes that have occurred; both the changes to the ecosystem and the response of governments to those changes.

This chapter will not discuss in detail the programs for implementation that have arisen from the various treaties, conventions and agreements between and within the two countries of Canada and the U.S. Other chapters of this book will examine significant programs such as the Remedial Action Plan (RAP) program born out of the Great Lakes Water Quality Agreement amendments of 1987.

BACKGROUND ON THE GREAT LAKES

The five Great Lakes of North America and their connecting channels make up the largest freshwater system in the world. The Great Lakes have a total water surface area of nearly 95,000 square miles. They contain approximately 24.6 quadrillion liters of water (6.5 quadrillion gallons) which is 18-20% of the world's supply of fresh surface water and 95% of the fresh surface water in the U.S. Spread even over the contiguous U.S., the Great Lakes would flood the land with 8 to 10 feet of water and the water contained in the Great Lakes would cover the combined continents of North America, South America and Africa to a depth of over one foot. Thousands of streams and rivers drain more than 521,000 square kilometers (201,000 square miles) of the basin and feed directly into the Great Lakes (Government of Canada and U.S. Environmental Protection Agency 1995).

With these facts in mind, it is no wonder that the Great Lakes are considered the world's greatest freshwater resource. Along with the St. Lawrence River, the Great Lakes have played and continue to exert a profound influence in the establishment, advancement and sustainment of the regional and national economies of the U.S. and Canada. The unique geographical, ecological and climatological characteristics of the Great Lakes and the land area making up the drainage basin have shaped the socio-economic heritage of the region well before European exploration and settlement began in the late 16th century. Since, the 19th century, after most of the early political and physical barriers to settlement had been resolved the Great Lakes – St. Lawrence River region became home to many generations of people as a place to live, work, worship and educate their children and an equally attractive place for many businesses and industries to locate (Colborn et al. 1990).

From the moment that the glaciers that formed the Great Lakes began to recede almost 14,000 years ago, the freshwater seas called the Great Lakes have influenced every aspect of life and culture in the region. The region is blessed with huge forests, large wilderness areas that support abundant wildlife populations, rich agricultural lands, extensive mineral deposits and water resources that support numerous water-based industries and a world-class sport fishery (Gauthier and Manninen 1999).

By virtue of the expansiveness and complexities of the Great Lakes – St. Lawrence River system one might expect that the management of the Great Lakes – St. Lawrence River resources demand a complex multi-jurisdictional approach. Indeed, those physical, ecological, socio-economic and cultural attributes are rivaled by the number and diversity of governments and institutions with some role to play in the management of Great Lakes – St. Lawrence River air, land and water resources (Donahue 1987).

The current institutional structure for governance in the Great Lakes – St. Lawrence River region is both large and complex. Two federal governments, eight U.S. states, two Canadian provinces, several region-wide organizations, approximately 120 Native American/First Nation authorities, dozens of nongovernmental organizations representing the region's business, industrial, environmental and conservation interests and hundreds of local government jurisdictions and agencies have some legal authority or responsibility for matters pertaining to resource or ecosystem management in the Basin (Hildebrand et al. 2002).

Individually and collectively, these entities contribute to the governance and management structure of the region and therefore influence the Great Lakes – St. Lawrence River decision making at some level.

HISTORY OF BINATIONAL COOPERATION AND GOVERNANCE IN THE GREAT LAKES

The evolution of Great Lakes – St. Lawrence River institutions for governance has been a fascinating study and to a large extent has been influenced by U.S.–Canada relations occurring over several hundred years. While the Treaty Between the United States and Great Britain Relating to Boundary Waters, and Questions Arising Between the United States and Canada (Boundary Waters Treaty of 1909) and the Great Lakes Water Quality Agreement of 1972 (revised in 1978 and 1987) would become the cornerstones of renowned policy innovations in environmental and resource management, the need for a binational approach to govern the conservation and development of the region's water resources was beginning to take shape much earlier in the history of the two countries.

Before the signing of the Boundary Waters Treaty in 1909, prior treaties and agreements had been entered into by the sovereign governments of the times. The treaty constitutes the most potent international legal device that two or more countries can employ to address a common issue. In the U.S., treaty power is designated expressly to the President, contingent upon the advice and consent of the Senate. Similarly in Canada, treaties require the approval of the Prime Minister with advice and consent from Parliament (Botts and Muldoon 2005).

The Jay Treaty of 1794 established peaceful trade between the U.S. and Great Britain. It was innovative in that it agreed that disputes over wartime debts and the American–Canadian boundary were to be sent to arbitration—one of the first major uses of arbitration in diplomatic history. Later the Rush-Bagot Pact and Convention of 1818 focused specifically on the Great Lakes and was an agreement between the U.S. and Great Britain to eliminate their naval fleets in the Great Lakes and Lake Champlain following the war of 1812 (Donahue 1987). Subsequent decades in the 19th century saw the signing of a series of treaties and agreements that addressed mutual navigation rights and related issues.

Boundary Waters Treaty of 1909

At the turn of the 20th century, there was an increased recognition of the need to approach water resources management and development in shared Canadian–U.S. river basins in a manner consistent with the general interests of the two countries rather than on an issue-by-issue basis. This recognition, led to the signing of the Boundary Waters Treaty of 1909. The Boundary Waters Treaty addressed the need for a more sophisticated arrangement to tackle development-induced water management problems and it addressed the need for the establishment of an international body (i.e., the International Joint Commission), to settle transboundary water disputes (Botts and Muldoon 2005). It is worth mentioning here some of the other factors of the times that influenced the signing of the Boundary Waters Treaty of 1909.

First, the issue specific approach was becoming unacceptable in light of development-induced water management problems that reached critical levels around the time of the signing of the Treaty. Some of these management problems included water quality problems in Lakes Michigan associated with the development of the Chicago Sanitary and Ship Canal (i.e., the Chicago Diversion), the lack of a comprehensive plan for the development of hydropower for the generation of electricity along the St. Marys and Niagara Rivers, irrigation disputes west of the Great Lakes in the St. Marys and Milk River basins and country specific infrastructure projects and improvements to support commercial navigation in the Great Lakes.

Second, the two countries possessed no water management rules or regulations to help them solve these problems and no institutional means for developing and coordinating joint plans and applying the rules once developed.

Third, diplomatic needs were also evolving, away from issue specific approaches for managing shared resources to a more comprehensive approach that required the development of rules and regulations as well as an international agency to apply them (Donahue 1987).

The Boundary Waters Treaty has stood the test of time and is far-reaching in scope. The treaty provides the principles for Canada and the U.S. to follow in using the shared waters between the two countries. For instance, both countries must agree to any project that would change the natural levels or flows of boundary waters. The treaty also states that waters shall not be polluted on either side of the boundary to the injury of health or property on the other side. The principles of the treaty, as relevant now as they were 100 years ago, helped pave the way for other governance models later in the 20th century.

CONTRIBUTION OF THE BOUNDARY WATERS TREATY TO GREAT LAKES GOVERNANCE

The contributions of the Boundary Waters Treaty to Great Lakes – St. Lawrence River governance are many. The Boundary Waters Treaty was ahead of its time in that it focused attention on natural (ecosystem) boundaries rather than political boundaries. It was also innovative in that it recognized the environmental, economic and cultural value of the shared resources of the Great Lakes – St. Lawrence River. And equally important, it was an early example recognizing the importance of binational cooperation for management of shared resources.

Institutional Framework for Regional Governance

The Boundary Waters Treaty of 1909 was a landmark agreement in promoting U.S–Canada cooperation. But it also greatly influenced the evolution of the institutional structure for governance which will be discussed below.

As mentioned, the Boundary Waters Treaty created the International Joint Commission (IJC) and paved the way for other Great Lakes-specific initiatives such as the Convention on Great Lakes Fisheries, which established the binational Great Lakes Fisheries Commission (GLFC), and the Great Lakes Basin Compact which established the Great Lakes Commission (GLC). At the time these organizations were formed in the mid-1950s, there was a recognized need for regional organizations with a watershed focus in order to address serious threats to the ecosystem and to provide leadership and advice to governments regarding opportunities to expand the regional economy primarily through the promotion of tourism, recreation and Great Lakes shipping (Donahue 1987). Later on, other important regional organizations were formed, such as the Council of Great Lakes Governors the Great Lakes and St. Lawrence Cities Initiative, the Council of Great Lakes Industries, the Alliance for the Great Lakes, Great Lakes United and the Healing Our Waters Coalition among others. These organizations were born out of the need to bring other leaders and constituents formally into the Great Lakes governance arena and will be discussed later in the chapter.

International Joint Commission

The Boundary Waters Treaty under Article VI established the IJC, which is governed by three members (commissioners) from each country appointed by the President (U.S.) and the Prime Minister (Canada). The ongoing work of the IJC helps to fulfil the treaty's purpose of preventing disputes as well as resolving them (Botts and Muldoon 2005).

The IJC is a permanent binational body created to prevent disputes related to usage of the boundary waters between the two countries and to settle questions and disputes arising across the entire common frontier (Donahue 1987).

The IJC provides the framework for binational cooperation on issues relating to water and air pollution and the regulation of water levels and flows. Within this broad mandate the IJC has several functions:

First, using quasi-judicial authority, the IJC approves or disapproves applications from governments, companies or institutions for construction projects that would affect the levels and flows or boundary and transboundary waters.

Second, the IJC has investigative powers. Using what is called a "reference" (a request from both governments), the IJC investigates specific situations and makes recommendations to the governments on how to address problems and issues of common interest between the two countries. When a reference is given, the investigation is carried out by a board of experts which prepares a report and makes non-binding recommendations for action to the two countries.

Third, the IJC has arbitral power under Article X of the Treaty making the IJC an important body in dispute resolution.

Fourth, the IJC also monitors compliance with Orders of Approval under Article VIII of the Treaty for structures along the boundary waters. In addition to the Great Lakes – St. Lawrence River system, the IJC assists governments in managing other waters along the shared border (Botts and Muldoon 2005).

The Boundary Waters Treaty created the IJC to be independent of federal governments, although each country established a section office with permanent staff in the national capitals of Washington and Ottawa. For technical information and policy advice when conducting its investigations, the IJC has relied up advisory boards and special committees.

Boards are usually made up of staff members from government agencies from both the U.S. and Canada who have the appropriate authority and expertise to carry out their responsibilities of the particular study that they are involved in. Prior to the Great Lakes Water Quality Agreement, these boards were generally of two types. One is a permanent control board for hydrologic management related to the operation of control structures that regulate levels and flows. The second is a special temporary board usually established for a specific purpose, such as the conduct of a reference (Botts and Muldoon 2005).

Great Lakes Water Quality Agreement

The role of the IJC changed significantly after the establishment of the Great Lakes Water Quality Agreement. First signed in 1972, and last amended in 1987, the Agreement affirms the rights and obligations of Canada and the U.S. under the Boundary Waters Treaty of 1909 in particular their mutual obligation not to pollute boundary waters. The Agreement is considered a model of international cooperation and is the flag ship example that has influenced U.S. and Canada relationships for the Great Lakes and laid a foundation for binational resource management for the protection and enhancement of the Great Lakes for more than a generation.

Agreements, like treaties and conventions, are important vehicles for international resource management and participatory governance. The Great Lakes Water Quality Agreement will be discussed here as it relates to the evolving role of the IJC and later with regard to its influence on institutions in the region.

After the signing of the first Agreement, much of the work of the IJC in the Great Lakes shifted to assisting governments in achieving their goal of cleaning up the lakes and preventing further pollution in the system.

To carry out its responsibilities under the Agreement, the IJC established a Great Lakes Regional Office in Windsor, Ontario (with binational staff) and was given an evaluation role to provide accountability for achievement of goals under the Agreement as well as the responsibility to provide a public information service for the programs (Botts and Muldoon 2005).

In addition, the Water Quality Agreement created the provision for ongoing advisory boards to provide periodic reports and make recommendations to the IJC regarding the status of accomplishments under the Agreement. Three of the most important of these boards are:

1. The Water Quality Board (WQB) which is the primary advisory board to the IJC under the Agreement. WQB membership includes heads of provincial and state environmental protection agencies with co-chairmanship provided by the Ontario office of Environment Canada and the U.S. EPA office in Chicago.
2. The Science Advisory Board (SAB) advises the IJC on science related matters under the Agreement. In was originally called the Research Advisory Board until 1978 when it became the SAB. It was first an advisory body to the WQB; later, it became a direct advisor to the IJC itself (Botts and Muldoon 2005). The SAB is responsible for developing recommendations on all matters related to research and the development of scientific knowledge pertinent to Great Lakes water quality. The SAB includes managers of Great Lakes research programs and other recognized experts in certain fields of science and research.

The membership has included social scientists and representatives of industry and nongovernmental organizations.

3. The Council of Great Lakes Research Managers was established by the IJC in 1984 and given board status in 1994. The purpose of the Council is to enhance the ability of the IJC to provide effective leadership, guidance, support and evaluation of Great Lakes research as it applies to the Agreement.

EMERGENCE OF REGIONAL GREAT LAKES GOVERNMENTAL INSTITUTIONS

The Post-World War II era brought significant changes to the Great Lakes – St. Lawrence River region. At the time, the region's population was continuing to grow, industrial expansion was occurring, dominated by heavy industry and manufacturing. In addition, large infrastructure projects were being planned and built, including the construction of the Mackinac Bridge, connecting Michigan's lower and upper peninsulas, as well as construction of numerous locks and dams along the St. Lawrence River that created the international Great Lakes – St. Lawrence Seaway. The completion of this project opened up the Great Lakes to the Atlantic Ocean (through the St. Lawrence River) creating expansion opportunities for Great Lakes – St. Lawrence maritime transportation industry. At this same time, public and scientific concern about Great Lakes water quality was being voiced in both countries (Botts and Muldoon 2005). This awareness which ushered in the "environmental era" of the early 1960s also created an institutional response that would influence Great Lakes – St. Lawrence River participatory governance for many decades to come.

Great Lakes Fishery Commission
(http://www.glfc.org/)

The U.S. and Canada again exercised their sovereign powers to form another important regional institution. The Convention on Great Lakes Fisheries was entered into by the U.S. and Canada in 1955 after many decades of worsening water quality problems and the decline in Great Lakes fisheries, which had begun as early as the 1830s in the lower Great Lakes, reached catastrophic levels with the introduction and spread of the parasitic sea lamprey in the 1930s.

Under the Convention on Great Lakes Fisheries, the Great Lakes Fishery Commission (GLFC) was formed. According to its mission statement, the GLFC has two major responsibilities: 1) to formulate a coordinated research program between the U.S. and Canada the goals of which are to identify ways to nurture the maximum sustained productivity of Great Lakes fish stocks and, based on that research, to recommend specific management initiatives to the governments; and 2) to formulate and implement a program to eradicate or minimize sea lamprey populations in the Great Lakes.

The GLFC is made up of eight Commissioners (four appointed from each the U.S. and Canada) and one U.S. Alternate Commissioner. Commissioners of the U.S. are appointed by the President for six-year terms. Commissioners of Canada are appointed by, and serve at the pleasure of, the Privy Council Office (PCO) which provides consultation and collaboration with international and domestic sources inside and outside Canada government (including the provinces and territories).

The GLFC's research programs are shaped and defined by the Strategic Vision of the GLFC for the First Decade of the New Millennium, established fish community objectives, and a Joint Strategic Plan for Management of Great Lakes Fisheries. To achieve its objectives, the GLFC uses a comprehensive advisory body made up of the Board of Technical Experts, the Sea Lamprey Integration Committee, the Sea Lamprey Research Board, the Council of Lake Committees, lake committees and their technical committees, the Law Enforcement Committee, and the Great Lakes Fish Health Committee (Great Lakes Fishery Commission 2006).

The GLFC, as primarily a single-issue entity has had a significant impact on Great Lakes – St. Lawrence River participatory governance. This is due in part to its successful committee structure which encourages participation from a variety of different government groups and other stakeholders. In addition, the GLFC has been a leader in sharing information and working with Congress and other legislative groups to ensure that funding and programs are available to support ecosystem protection and restoration.

Great Lakes Commission
(http://www.glc.org/)

At the same time that the Great Lakes Fishery Commission was established, the Great Lakes states established an organization to represent their interests and to work collaboratively on basin-wide ecosystem management issues. This organization is the Great Lakes Commission (GLC), an interstate compact agency founded in state and federal law and comprised of state agency officials, legislators and governors' appointees from each of the Great Lakes states. Although legally a U.S. institution, the provinces of Ontario and Québec participate with the GLC through a declaration of partnership which was signed in 1999.

The GLC was formed through the passage of P.L. 90-419, passed by the U.S. Congress in 1968 providing consent and establishment of the Great Lakes Basin Compact (hereafter Compact) entered into by the States of Illinois, Indiana,

Michigan, Minnesota, New York, Ohio, Pennsylvania and Wisconsin through enactment of state specific legislation between 1955 and 1963 (Great Lakes Commission 1968).

A compact is a legal mechanism for two of more states to enter into an agreement of common interest. Compacts hold a unique place in U.S. law in part of because of their broad reach and in part because of the status of the parties in relationship to one another. States have quasi-sovereign status (i.e., they are not mere political subdivisions of the federal government) and they have the ability to enter into agreements with other states to help control current and future state actions both individually and collectively (Broun et al. 2006).

The writers of the Compact understood the complexity of the issues surrounding the Great Lakes ecosystem at the time and so by design, the Compact was drafted to allow the GLC to work on and advise the states on literally any issue surrounding the Great Lakes environment and economy.

Article I of the Great Lakes Basin Compact establishes the purpose of the compact to be accomplished through means of joint and cooperative action between the states. Article I (1) calls for the GLC "to promote the orderly, integrated, and comprehensive development, use, and conservation of the water resources of the Great Lakes Basin" (Great Lakes Commission 1968).

The GLC works closely with governors and state legislators; members of the U.S. Congress from the Great Lakes States; municipal, state, provincial and federal agencies; interstate organizations; private sector firms and associations; universities; colleges and individual citizens.

The GLC shares a resource management philosophy that emphasizes stewardship, sustainability and the need for strong partnerships between the states, provinces and the federal governments.

The GLC pursues its mandate via three functions: information sharing among its member states and provinces, coordination of state and provincial positions on regional issues, and advocacy of those positions (mostly on the U.S. side) on which the states and provinces agree.

The GLC has also from time to time, facilitated the development of regional agreements between its member states and other partners.

For example, the GLC working with the International Association of Great Lakes and St. Lawrence River Mayors and the International Association of Great Lakes Ports executed the signing of an important maritime transportation agreement called the "Declaration of Indiana" on 15 May 1991 in Merrillville, Indiana.

The Declaration, signed by 18 leading maritime associations representing ports, shipowners, labour, pilots and other sectors of the maritime transportation community, listed a series of 10 principles to promote the full potential of the Great Lakes – St. Lawrence Seaway. In addition to the guiding principles, the Declaration included an action agenda with 19 recommendations in the areas of government policy, competiveness and cost containment, marketing, infrastructure needs and environmental protection. The Declaration was an important agreement because it presented the maritime transportation industry with a shared vision, a clear set of priorities and a plan to accomplish them (Great Lakes Commission 1992).

The GLC also coordinated the development, signing and public release of an agreement called the "Ecosystem Charter for the Great Lakes – St. Lawrence River Basin". The Ecosystem Charter presented a vision for sustainability for the Great Lakes – St. Lawrence River. It set forth a series of principles, findings and actions in areas related to rights and responsibilities, ecological integrity and diversity, sustainable communities, institutional relations, public information and education and citizen participation. The final Ecosystem Charter language reflected input from scores of policymakers, resource managers and opinion leaders who contributed language at the many roundtables that were convened during the drafting period or during the public comment period when the final draft was reviewed. More than 30 groups and individuals signed the Charter at the signing ceremony in Detroit in conjunction with the GLC's 1994 annual meeting. By the end of 1994 more than 100 signatories had been secured (Great Lakes Commission 1995).

The Ecosystem Charter was an innovative approach and process that brought together numerous important stakeholders from both the U.S. and Canada to develop an agenda for sustainability for the region and the resource. In the end, it was less than a complete success because it was a non-binding agreement and because there was not a single entity or agency responsible for follow through and implementation. In essence it faded away from the public and region's consciousness because it suffered from the old adage, "if it's everyone's it's no one's".

The GLC has had a long a productive history and has contributed to the Great Lakes – St. Lawrence governance regime through its partnerships and involvement with the states and provinces. Recently, the GLC has also broadened its partnerships through a formal Observer program where partner organizations – including U.S. and Canadian federal, regional and tribal governments and nongovernmental organizations – participate extensively in GLC activities.

One downside is that the GLC is directed by and therefore dependent on the political jurisdictions (e.g. the states and provinces) that are its members. So the GLC's effectiveness, or lack thereof, can be a function of the political will and leanings of the jurisdictions of the times (Donahue 1987). The jurisdictions may (and in this author's opinion have in the past) spurned regional cooperation when their domestic issues are of more immediate concern.

OTHER ORGANIZATIONS INFLUENCING GREAT LAKES – St. LAWRENCE RIVER PARTICIPATORY GOVERNANCE

The 1960s and 70s brought additional significant changes to Great Lakes – St. Lawrence River governance as widespread public attention was brought to bear on water quality and pollution issues that were plaguing the Great Lakes. Within the Great Lakes – St. Lawrence River region on both sides of the border, the citizen movement exploded with demonstrations and political campaigns spawning new environmental groups and garnering media attention to the issues of the day.

A series of events that occurred in the late 1960s, including unsightly algal blooms along the shores of Lake Erie and Lake Ontario, a massive alewife die off in Lake Michigan and the infamous scene of the Cuyahoga River catching fire, prompted new action for environmental protection on the part of governments in both Canada and the U.S. In 1970, both the U.S. Environmental Protection Agency and the Canadian Department of Environment (now Environment Canada) were formed (Botts and Muldoon 2005).

Within this backdrop of public awareness and greater scientific evidence of the damaging effects of pollution to the water quality of the Great Lakes, a joint U.S.–Canada working group was formed to study the need for binational action to clean up the Great Lakes. Several years of study and two years of intense negotiation led to the signing of the Great Lakes Water Quality Agreement on April 15, 1972. In Canada, even prior to the signing of the Great Lakes Water Quality Agreement, a ban on phosphates in detergents was enacted as part of the 1970 Canada Water Act. Canadian scientists at that time were confident that phosphorus was the limiting (most crucial) nutrient contributing to accelerated eutrophication in the Great Lakes. Ontario through its implementation of a phosphate reduction and removal of phosphorus in its sewage treatment systems convinced the U.S. to follow the same course of action. Also, the Canada Water Act authorized the establishment of federal-provincial agreements to address water quality and resource management priorities leading to the negotiation of the first Canada-Ontario Agreement (COA) which specifically designated the responsibilities of Ontario and the federal government regarding the Great Lakes Water Quality Agreement (Botts and Muldoon 2005).

As mentioned earlier, the Agreement is a formal commitment on the part of the U.S. and Canada to resolve a wide range of water quality problems facing the Great Lakes and St. Lawrence River. The issues in 1972 were, and in many cases still are, critical to the economies, culture, public health and quality of life of not only the Great Lakes – St. Lawrence River region but to the entire U.S. and Canada.

Both countries have made considerable progress in reversing the impacts of pollution to the physical, chemical and biological integrity of the Great Lakes – St. Lawrence River ecosystem. This progress has occurred as the result of constant and tireless efforts not only on the part of government but by the scores of stakeholders that have been involved in the Great Lakes Water Quality Agreement from the beginning (Botts and Muldoon 2005).

Tremendous strides have been made to clean up the lakes and protect them from further pollution, and governments at all levels have invested billions of dollars in these efforts. Industries have made significant contributions by changing production processes, the products that are produced, and by cleaning up contaminated areas. Municipalities, often supported by other levels of government, have upgraded sewage and water treatment facilities across the basin. And community and environmental groups have worked energetically to monitor progress and improve the environmental condition of the Great Lakes – St. Lawrence River system (Botts and Muldoon 2005).

The Great Lakes Water Quality Agreement was last amended by protocol on November 18, 1987. From the time of the signing of the first Agreement to the time of the renegotiated Agreement, the region also experienced additional changes and an emergence of important groups and institutions that would become important players in participatory governance in the Great Lakes – St. Lawrence River region.

Binational Executive Committee (http://www.binational.net/)

The Binational Executive Committee (BEC) was the chief new binational institution that was created by parties outside of the IJC framework under the 1987 GLWQA Protocol. The BEC is composed of senior-level representatives of Canadian and U.S. federal, state, provincial, and tribal agencies who are accountable for delivering major programs and activities that respond to the terms of the Great Lakes Water Quality Agreement. Several NGOs have been given observer status as well. The BEC co-chairs have generally been the Ontario regional director for Environment Canada and the Director of U.S. EPA's Great Lakes National Program Office (GLNPO) (International Joint Commission 2005).

The BEC generally meets twice a year or as required to do the following things: 1) set priorities and strategic direction for binational programming in the basin; 2) coordinate binational programs and activities; 3) respond to new and emerging issues on the Great Lakes often through the creation of working groups or committees; 4) evaluate progress under the Agreement; and 5) provide advice, comment or other input for the preparation of various binational reports and presentations.

While the BEC has an important binational mission, many observers feel that it has suffered through the years because it has had little public involvement and recently, due to lack of funding within state, provincial and federal governments, agency partners have been less reliable participants with the BEC. The power of the BEC is limited due to its lack of binational decision-making authority and this has also contributed to an overall decline in participation from senior level policy makers and agency heads. In addition, because U.S. EPA and

Environment Canada seem to interact little outside of the BEC framework (e.g., without input from the IJC or other governmental oversight mechanisms or diplomatic channels) many fear that binational relationships within the region are beginning to erode (Botts and Muldoon 2005).

It should be noted that an inherent strength of the Water Quality Agreement is that it provides a venue for government to government interaction and negotiation where the common interests of the Great Lakes supercedes the separate interest of the two countries. Changes in the relationships between U.S. EPA, Environment Canada and the IJC (which no longer retains its governmental oversight role) along with other factors, such as the U.S. country-specific focus under both the Great Lakes Regional Collaboration Restoration Strategy and the Great Lakes Restoration Initiative (to be discussed later) seem to be undermining the binational cooperation that has developed over the 30-plus years of experience with the Water Quality Agreement (Botts and Muldoon 2005). This should be a concern to all who value the successful binational process that recognizes the ecosystem first rather than country-specific priorities.

Nongovernmental Organizations

The role of nongovernmental organizations in participatory governance in the Great Lakes – St. Lawrence River Basin is substantial and has been increasing in importance over the past 40-plus years. These organizations have been historically under-represented in Great Lakes – St. Lawrence River Basin governance but recently their involvement has been crucial to the success of the various initiatives, agreements and programs that have been developed and put in place on both sides of the border.

It would be impractical to try to list and describe the scores of important non-governmental and quasi-governmental organizations that have made positive contributions to Great Lakes – St. Lawrence River governance in this chapter. However, there are numerous ones that are worth noting due to their history, their binational nature, their longevity and their specific mission related to the Great Lakes – St. Lawrence River region.

Council of Great Lakes Governors (http://cglg.org/)

The governors of Illinois, Indiana, Michigan, Minnesota, Ohio, and Wisconsin created the Council of Great Lakes Governors in 1983. The original goal was to create a non-partisan forum to develop regional agreements on issues of common concern. In 1989, the Governors of New York and Pennsylvania joined the Council. In more recent years, the Premiers of Ontario and Québec have joined with the Governors as associate members of the Council. Since its inception, the Council has led important regional efforts to protect the environment and accelerate economic growth.

A major focus for the Council both in its early years and continuing to the present, is the promotion of coordinated regional water resources management policies and programs. In 1985, the governors worked through the Council to create the first regional water management agreement among the Great Lakes States, Ontario and Québec; the Great Lakes Charter. This historic agreement created a mechanism for regional collaboration, a guide for protective measures to be taken by each jurisdiction and information-sharing protocols to improve water resources management decision-making. It also marked the beginning of a long and fruitful partnership with the Premiers of Ontario and Québec. Twenty years after the signing of the Charter, the states and provinces passed the Great Lakes – St. Lawrence River Basin Water Resources Compact and Agreement which detail how the States and Provinces will manage and protect the waters of the Great Lakes – St. Lawrence River Basin in the 21st century and provide the framework for each State and Province to enact laws to protect the basin from increased withdrawals, diversions and consumptive uses of Great Lakes water. The Compact and Agreement also acknowledge the needs of the states and provinces to balance economic development, social development and environmental protection as interdependent and mutually reinforcing pillars of regional sustainable development (Great Lakes – St. Lawrence River Water Resources Regional Body 2005).

Interestingly, the legal mechanism for entering into this water management agreement for the eight Great Lakes states is another interstate compact titled the Great Lakes Basin Water Resources Compact. A more detailed discussion of this compact and the companion agreement for Ontario and Québec occurs later in the chapter.

In 1986, the governors signed the Toxic Substances Control Agreement which formed the Great Lakes Protection Fund in 1989. The Protection Fund was the first multi-state foundation dedicated to improving the environment. The Toxic Substances Control Agreement also led, indirectly, to the establishment of the Great Lakes Water Quality Initiative in the early 1990s. The U.S. EPA-led initiative set the regulatory structure for new water quality regulations in the U.S. The governors coordinated their efforts through the Council when working with the U.S. EPA to ensure that new state water quality regulations were flexible, effective and compatible with the new economy that was growing in the Great Lakes – St. Lawrence River region.

The Council of Great Lakes Governors has established itself as a leading institution on Great Lakes issues and has recently become more active in the Great Lakes policy arena. The Council's strength stems from the fact that the organization represents the view of the Great Lakes Governors. This is a subtle but important difference from the GLC which represents the views of the Great Lakes States. The GLC's structure and governing process is broader and generally involves numerous state and provincial stakeholders. Consensus usually occurs through a resolution process where policy positions are drafted, discussed and passed at the semiannual meetings of the organization. The Council has less bureaucracy and therefore can often develop consensus on issues more quickly. Also, the numerous regional agreements that have been developed by the governors through the Council over the years have been extremely effective in galvanizing other agencies and institutions around the issue that the particular agreement addresses.

The Great Lakes – St. Lawrence River Mayors and the Great Lakes and St. Lawrence Cities Initiative (http://www.glslcities.org/)

Beginning in the mid-1980s the mayors of the region's cities started to become more interested and involved in Great Lakes – St. Lawrence River issues and became important players in the regional governance arena. In 1986, the mayors and representatives of the hundreds of waterfront communities along the Great Lakes and St. Lawrence River first established the International Association of Great Lakes and St. Lawrence Mayors with co-secretariat support provided by the GLC and the St. Lawrence Economic Development Council (SODES). In 1987, the first annual International Great Lakes – St. Lawrence River Mayors conference was held to provide a forum for discussion and cooperation for the promotion and protection of the Great Lakes – St. Lawrence River system. The International Great Lakes – St. Lawrence Mayors conference was convened annually until 2003.

In 2003, the Great Lakes Cities Initiative was formed as a binational coalition of mayors and other local officials to actively work with federal, state, and provincial governments to advance the protection and restoration of the Great Lakes and the St. Lawrence River. The Great Lakes Cities Initiative was created to provide a more formal and continual presence for mayors and other local officials to actively participate in Great Lakes – St. Lawrence River issues relating to governance, economics, and science.

In May 2005, the International Association of Great Lakes–St. Lawrence Cities Mayors and the Great Lakes Cities Initiative formally merged, bringing together the two organizations as the current Great Lakes and St. Lawrence Cities Initiative.

Prior to the establishment of the Great Lakes and St. Lawrence Cities Initiative, the focus of the region's mayors was primarily in the area of economic development, tourism and promotion of the Great Lakes –St. Lawrence Seaway. The Great Lakes and St. Lawrence Cities Initiative seems to have less of a focus on economic development, regional tourism and water-based transportation but places a stronger emphasis on environmental protection with a broader interest in Great Lakes – St. Lawrence River restoration and sustainability issues. Some of the Great Lakes and St. Lawrence Cities Initiative's current programs include promotion of water conservation, a green cities initiative, a pharmaceutical initiative focusing on the collection and disposal of pharmaceuticals and a Great Lakes 21st Century Waterways study to modernize the Chicago waterway system and protect the region from the introduction and spread of Asian Carp.

(For more information and elaboration on the role and accomplishments of the Great Lakes and St. Lawrence Cities Initiative, see Chapter 10.

Canadian Environmental Law Association (http://www.cela.ca/)

The Canadian Environmental Law Association (CELA) is a non-profit, public interest organization established in 1970 to protect the environment and to advocate environmental law reforms. Funded by Legal Aid Ontario, CELA is part of a system of community-based legal clinics across Ontario and is one of 17 which offer services in specialized areas of the law.

CELA, while specifically a Canada-based organization, has been extremely active with binational issues and agreements related to the work of the IJC, the Great Lakes Water Quality Agreement review and the development and passage of the Great Lakes – St. Lawrence River Basin Sustainable Water Resources Agreement, among others.

Pollution Probe (http://www.pollutionprobe.org/)

Another Canadian-based group formed during the environmental movement of the 1960s and early 1970s was Pollution Probe. Pollution Probe is a charitable-based organization that works with government, industry and business, non-profit organizations and private citizens to help define issues and find solutions to problems through research, education and advocacy.

Like CELA, Pollution Probe has worked binationally on water quality issues related to the work of the IJC and has emphasized sustainability in areas related to renewable energy, climate change and water management.

Nature Québec (http://www.naturequebec.org/)

Nature Québec is a not for profit organization representing individuals and organizations working on conservation issues and with an emphasis on the promotion of sustainable development.

Working on ecosystems and biodiversity protection, Nature Québec subscribes to the following principles: 1) To maintain the ecological processes essential to life; 2) To preserve biological diversity; 3) To foster sustainable development by working for the respect of species and ecosystems.

Nature Québec works on the health of ecosystems and conservation issues in general through the publication of memoirs and reports used as reference tools for their public interventions and their awareness-raising projects.

Nature Québec is one of the few nongovernmental organizations in Québec working on regional binational issues such as the renegotiation of the Great Lakes Water Quality Agreement and the Great Lakes – St. Lawrence River Basin Sustainable Water Resources Agreement.

Great Lakes United
(http://www.glu.org/)

Great Lakes United (GLU) was established as a binational coalition for the environmental community and immediately added greatly to the influence and momentum of the environmental movement during the early to mid-1980s. GLU was formed when a group called Save the Rivers formed an alliance with the Michigan United Conservation Clubs (MUCC) over studies and proposals surrounding winter navigation projects that would make it possible for year-round shipping in the Great Lakes and St. Lawrence River (Botts and Muldoon 2005). GLU is a coalition of individuals as well as other environmental organizations and the strength of its agenda was its effective advocacy for policies and legislation based upon proposals submitted by member organizations at GLU annual meetings. GLU also established itself as a leading advocate for the amended GLWQA in 1987.

Alliance for the Great Lakes
(http://www.greatlakes.org/)

Another organization that was created during the environmental movement that occurred in late 1960s and early 1970s was the Lake Michigan Federation now known as the Alliance for the Great Lakes.

The Alliance works to conserve and restore the Great Lakes through policy, education and local efforts aimed at preserving the Great Lakes region as a national treasure. The Alliance is well-known for its grassroots organizing and citizen activism. It works with the region's residents, teachers, scientists, economists, legal specialists, government representatives, communities and individuals.

Council of Great Lakes Industries
(http://www.cgli.org/)

The Council of Great Lakes Industries is a non-profit organization representing the shared interests of U.S. and Canadian industrial organizations from the manufacturing, utilities, transportation, communications, financial services and trade sectors that have investments in the Great Lakes Basin. The Council works to ensure that industry is a substantive partner in the Great Lakes region's public policy development process.

The Council has been an active and effective voice for industry in the Great Lakes and has been a visible presence in the deliberations of the many strategies and agreements that have been developed and formally adopted by state and provincial government over the past 15 years.

The Council also develops policy position statements providing industry's viewpoints on a variety of topics and also leads issue-specific initiative in areas related to things like water footprinting and sustainable development.

National Wildlife Federation – Great Lakes Regional Office
(http://www.nwf.org/Regional-Centers/Great-Lakes)

The National Wildlife Federation (NWF) is America's largest conservation organization working with more than 4 million members, partners and supporters in communities across the United States on issues such as protection and restoration of wildlife habitat, global warming and conservation. The NWF has supported a Great Lakes Regional Center Since 1982. NWF's Great Lakes Regional Center has been a leader in protecting the Great Lakes for wildlife and humans that depend on this invaluable freshwater resource and has been extremely active in helping to shape regional policy related to many of the issues, agreements and initiatives discussed in this chapter. The Great Lakes Regional Center has gained recent notoriety for its leadership in coordinating activities of the Healing Our Waters Campaign to protect and restore the Great Lakes.

Healing Our Waters – Great Lakes Coalition
(http://www.healthylakes.org/)

The Healing Our Waters – Great Lakes Coalition administers the Healthy Lakes, Healthy Lives public awareness campaign. It was launched in 2004 with the support of the Wege Foundation a Michigan based foundation that has supported numerous causes, including the environment and community development.

In May 2004, the Wege Foundation helped to convene a meeting of more than 70 leading environmentalists, ecologists, scientists and academicians. During the two-day summit, these leaders in the conservation arena worked together to outline a plan for restoring and protecting the Great Lakes.

That conference and the ensuing report that came out of it – "Healing Our Waters: An Agenda for Great Lakes Restoration", provided the inspiration to form a large coalition, whose mission was to push for a comprehensive plan to restore the Great Lakes and the U.S. federal funding needed to implement it (Wege Foundation 2004).

The Great Lakes Healing Our Waters Coalition is a broad-based predominantly U.S. network of national, regional and state organizations dedicated to Great Lakes restoration.

The Coalition's membership totals more than 120 organizations. It includes a technical advisory committee comprised of scientists, business leaders, economists and other experts. The Coalition is co-coordinated by the National Parks Conservation Association (NPCA) of Washington, DC and the National Wildlife Federation (NWF) which facilitates HOW activities through NWF's Great Lakes Natural Resources Center in Michigan.

The work of the HOW has been very instrumental in organizing and unifying groups and organizations with similar but not identical missions and mandates behind a common Great Lakes restoration agenda. HOW's ability to mobilize its large coalition of environmental and conservation groups contributed greatly to the momentum behind the Great Lakes Regional Collaboration Restoration Strategy and its effective advocacy was instrumental in helping to secure an unprecedented U.S. federal commitment to support Great Lakes restoration through the Great Lakes Restoration Initiative (GLRI) which was signed by President Obama in February 2009 with an initial appropriation of US$475 million for restoration projects the Great Lakes.

The Evolution of Regional Agreements, Strategies and Initiatives in the 21st Century

Earlier in this chapter several of the salient treaties, agreements, compacts and initiatives of the 20th century were described. With the ushering in of the 21st century numerous additional agreements, strategies and initiatives have been entered into having great significance and important ramifications for the future of governance in the Great Lakes – St. Lawrence River Basin. Space does not permit discussing every agreement or strategic plan that guides actions for the many organizations mentioned in this chapter. However, a few of these agreements are noteworthy because of their scope, their visibility in the region and their contributions to participatory governance. The most significant of these agreements are summarized below.

Great Lakes – St. Lawrence River Basin Water Resources Compact (U.S.) and the Great Lakes – St. Lawrence River Basin Sustainable Water Resources Agreement (Canada)

On 13 December 2005, the Great Lakes Governors and Premiers signed the Great Lakes – St. Lawrence River Sustainable Water Resources Agreement and the Governors committed to seek passage of the Great Lakes – St. Lawrence River Basin Water Resources Compact to protect the region's ecosystem by ensuring the sustainable use and management of the waters of the Great Lakes – St. Lawrence River. The Agreement and the Compact reaffirm the commitment to the water management principles set forth in the Great Lakes Charter of 1985 and the Charter Annex of 2001, earlier non-binding water resources management agreements (Great Lakes – St. Lawrence River Water Resources Regional Body 2005).

On 3 October 2008, President George W. Bush signed a joint resolution of Congress providing consent to the Great Lakes – St. Lawrence River Basin Water Resources Compact. The President's action marked the final step in the Compact's approval process thus enabling the Compact to become law.

During 2007 and 2008, each of the eight Great Lakes State legislatures ratified the Compact through the passage of their own state-specific legislation. U.S. federal legislative approval was completed by the U.S. Senate on 1 August 2008, and by the U.S. House of Representatives on 23 September 2008. In Canada, provincial legislation has been enacted in both Ontario and Québec to provide the authority to implement the Agreement.

These landmark documents, which provide the specificity for the states and provinces to implement the provisions of the Compact and the Agreement, are the culmination of many years of work by a water management working group that was established after the signing of the Great Lakes Charter Annex in 2001.

While the work of this committee was state and province driven, it also included participation and opportunities for input from many other organizations (both governmental and nongovernmental) involved with water management and water availability issues with an interest in or mandate to protect and conserve Great Lakes water resources.

The speed with which the U.S. Congress and the President acted to ratify the Great Lakes Basin Water Resources Compact in striking especially when you consider that it took more than five years for Congressional consent to be achieved for the Great Lakes Basin Compact that formed the GLC. This speedy consent occurred for two primary reasons. First, were the importance of the issue to the region and the high visibility of the work of the Water Management Working Group which was formed in 2001 to develop the plan to implement the provisions of the 1985 Charter and the 2001 Charter Annex.

Throughout the process political, public and media attention were very high and this contributed to a sense of urgency and the need for accomplishment to protect the water resources from adverse impacts associated with new and increased withdrawals, diversions and consumptive use of water.

In addition, there was a high-level of buy in for the Compact and Agreement resulting from the broad stakeholder and public participation that was in place throughout the process. The Water Management Working group received input throughout the process from an advisory committee, resource groups and observers that provided technical input and advice and through ongoing dialogue and consultation with the region's tribes and first nations. When the state legislation was finally enacted, members of Congress were educated and already aware of the importance of the issue and the passage of the compact moved swiftly through the political process to completion.

Canada Ontario Agreement (COA) Respecting the Great Lakes Basin Ecosystem

The Canada Ontario Agreement Respecting the Great Lakes Basin Ecosystem (COA) outlines how the governments of Canada and Ontario will cooperate and coordinate their efforts to restore, protect and conserve the Great Lakes Basin ecosystem. It builds on the actions taken through previous COAs, and focuses on priorities for future action. The sixth COA was signed in 2007 and the seventh COA is currently being negotiated. It was set to expire in March 2010 but was extended for one year to March 31, 2011 to accommodate the GLWQA review and renegotiation process. The COA was extended again to June 24, 2012 in the hope that there will be a new GLWQA in place by then.

The COA establishes an action plan and clear roles and responsibilities between federal and provincial ministries, and helps Canada to meet its commitments under the Great Lakes Water Quality Agreement. This one-year extension was approved primarily to allow Canada and Ontario to continue their important work to protect and restore the Great Lakes while the governments of Canada and the U.S. negotiate amendments to the Agreement.

COA priorities include protecting and restoring habitat, preventing pollution, cleaning up environmental hotspots (Areas of Concern) and working with partners to advance goals and objectives for lakewide sustainability.

The strength of COA is that it provides the foundation for Canada and Ontario for future implementation of restoration priorities once the new Water Quality Agreement is renegotiated.

The Great Lakes Regional Collaboration of National Significance

In May 2004 President George W. Bush issued an Executive Order, which recognized the Great Lakes as a "national treasure" and created a U.S. federal Great Lakes Interagency Task Force (a coalition of 11 U.S. federal agencies) to improve federal coordination on the Great Lakes. The Order also directed the U.S. Environmental Protection Agency Administrator to convene a Regional Collaboration of National Significance for the Great Lakes. This collaborative process was needed to develop, by consensus, the national restoration and protection action plan for the Great Lakes called for under the Executive Order.

After extensive discussions, the Interagency Task Force, Council of Great Lakes Governors, Great Lakes Cities – St. Lawrence Cities Initiative, Native American Tribes, and Great Lakes Congressional Task Force signed a Declaration and agreed to a framework document that signified the convening of the Collaboration in December 2004.

During the year from December 2004 to December 2005, a coalition of more than 1,500 federal, state and local officials and interested nongovernmental stakeholders came together to develop a comprehensive restoration blueprint listing key goals for restoring and protecting the Great Lakes. The Great Lakes Regional Collaboration Strategy to Restore and Protect the Great Lakes was released in December 2005 and provides detailed recommendations for achieving critical restoration goals. The strategy called for federal action to: 1) immediately stop the introduction of aquatic invasive species; 2) increase habitat conservation and species management, including restoration of 550,000 acres of wetlands; 3) protect nearshore waters and coastal areas by cleaning up pollution from sewers; 4) clean up highly contaminated Areas of Concern; 5) improve water quality by reducing polluted runoff; 6) eliminate the discharge of, and reduce exposure to, persistent toxic substances; 7) increase monitoring and coordination; and 8) ensure long-term, sustainable growth and economic development (U.S. Federal Interagency Task Force 2005).

The Great Lakes Regional Collaboration Strategy to Restore and Protect the Great Lakes was viewed by most (at least on the U.S. side) as a success in that it accomplished two important things. One, it established a unified restoration agenda that had been previously lacking. Several years earlier, the U.S. Congress (through the House and Senate Great Lakes Task Force) had challenged the region's leading institutions to come together with a more unified Great Lakes message and a prioritized list of Great Lakes restoration needs. The GLRC helped to achieve this.

Second, the GLRC helped forge new partnerships, build momentum and galvanize support for Great Lakes issues at a critical time. The large number of stakeholders that participated through the year-long process became a visible force, both within their local communities but also one that had a more regular presence on Capitol Hill in Washington through the coordinative efforts of groups such as the Healing Our Waters Coalition described earlier.

The GLRC was criticized by some because of a lack of follow through and a perceived weak response from the U.S. federal government mostly due to a lack of financial commitment to implement the strategy recommendations. At the time, many took notice of the omission in referring to the Great Lakes as an international or binational treasure rather than a recognized "national treasure" and saw this as a further erosion of the binational cooperation and well-established binational approach promoted by the Great Lakes Water Agreement.

Prescription for Great Lakes Ecosystem Protection and Restoration

At the same time that the GLRC restoration strategy was released a scientific white paper titled Prescription for the Great Lakes Ecosystem Protection and Restoration was published by a group of scientists with support from the Wege and Joyce Foundations. The findings of this paper indicate that the Great Lakes are in jeopardy and that immediate emergency actions are needed to restore the ecosystem particularly in nearshore/tributary zones where a suite of stresses that have occurred over time may be leading to an ecosystem breakdown and even potentially irreversible ecosystem collapse. It concludes that the self-healing mechanisms of the lakes have been impaired; and that the key to restoring those mechanisms is to restore the health of the nearshore coastal communities and key tributaries (Bails et al. 2005).

This paper was endorsed by hundreds of scientists from both countries and sets forth the causes of, impacts of and remedies for the damage to Great Lakes ecosystem health. It caused a stir within the policy and scientific community and provided additional food for thought as the action plan was being developed for the Great Lakes Restoration Initiative (GLRI). In fact, this white paper included recommendations that were used to help the federal agencies set the plan's goals, objectives, and principal actions.

Great Lakes Restoration Initiative Action Plan

As previously mentioned, federal funding to implement the GLRC strategy was not provided at a level necessary to achieve success. This changed in 2009 when President Obama proposed, and the U.S. Congress approved, funding for the Great Lakes Restoration Initiative (GLRI), an unprecedented, multi-year program to restore the Great Lakes. The GLRI will help implement a restoration plan, relying heavily on the GLRC strategy of 2005. The cost to implement the GLRI is US$20 billion over five years for cleaning up the Great Lakes. If fully funded over the timeframe, the GLRI, coupled with other critical Great Lakes programs, will help achieve the goals set forth in the GLRC strategy and accelerate on-the-ground action to clean up and restore the Great Lakes. In FY 2010 alone (1 October 2009 through 30 September 2010), the federal government has provided US $475 million for Great Lakes restoration.

The GLRI funding will strategically complement existing federal programs and strengthen collaboration with state, local and nongovernmental partners. It is directed at solving urgent problems facing the Great Lakes in the following ways: 1) shutting the door on aquatic invasive species; 2) cleaning up contaminated sediments in Areas of Concern; 3) improving nearshore health by controlling nonpoint pollution, and cleaning up the region's beaches; 4) restoring degraded wetlands, and 5) conserving and enhancing valuable fish and wildlife resources (U.S. Federal Interagency Task Force 2010).

The GLRI is being led by the U.S. Environmental Protection Agency (U.S. EPA) and coordinated through the Great Lakes Interagency Task Force (continued from the earlier GLRC process) which is charged with improving coordination of federal programs. The GLRI leverages contributions and expertise from state, local and nongovernmental partners. It is guided by a five-year action plan that sets outcome-oriented performance goals, criteria for measuring progress, and clear standards of accountability. The GLRI will also engage an independent scientific review panel to ensure the best available science is used to guide Great Lakes restoration efforts.

Public participation and stakeholder involvement is also a priority under the GLRI. In July and August, 2009 EPA convened a series of meetings across the Great Lakes Basin to review the proposed priorities for the GLRI and secure feedback on the draft action plan outline.

Separate meetings were held in each state for the public and state agencies, and regional conference calls were held for Great Lakes tribes and cities. The meetings were attended by more than 1,000 participants, engaging the Great Lakes states, cities, tribes and other stakeholders. Input from these stakeholders contributed to the final action plan which was released in February 2010.

To ensure accountability, the EPA has developed an internet-based accountability system to track and report on progress in achieving objectives in the action plan. Entities receiving GLRI funds will be required to enter basic data about their projects such as location, funding, projected outcomes, and relationship to the GLRI Action Plan. In addition, U.S. EPA will establish an independent scientific review panel to ensure the best available science is used to guide Great Lakes restoration efforts.

It would not be an overstatement to say that the GLRI has created extreme levels of excitement, enthusiasm and anticipation of achieving visible and measurable progress toward Great Lakes restoration and improvement of ecosystem health. While the program was still quite new in 2010, challenges in budgeting at the state and federal level, concerns about a sluggish U.S. economy and the difficulty faced by EPA to administer funds quickly and efficiently to partner agencies and project applicants has created some doubt about the future of the GLRI. That being said, the GLRI is, at least for the moment, being viewed as a monumental achievement and testimony to what can be accomplished with a unified message, good collaboration and constant coordination among and between agencies both inside and outside of government.

Observations for the Future of Participatory Governance

This chapter has summarized the main treaties, agreements, conventions, compacts and other arrangements that have shaped Great Lakes governance and the institutional framework. It is appropriate to end with some observations regarding strengths of Great Lakes governance, perceived needs in the area of governance and institutional arrangements and challenges that will likely need to be met as we enter further into the 21st century.

The Great Lakes governance and the institutional framework has developed and evolved over time and this chapter has attempted to provide a brief summary and history of changes that have occurred since the signing of the Boundary Waters Treaty more than 100 years ago.

Although the governance structure and institutions in the region have shown some innovation and some ability to adapt to change, there are numerous factors working against change that should be noted.

Although there is a common understanding about the need to manage resources on a watershed basis rather than on a jurisdictional basis, there are still many challenges to be faced when dealing with the complexities of hydrologic versus political boundaries. When dealing with Great Lakes – St. Lawrence River management, as previously pointed out, there are two sovereign countries, eight states and two provinces (with their own degree of sovereignty) and numerous international and regional governmental and quasi-governmental agencies with some advisory, management or regulatory authority over some aspect of the air, land or water resources of the basin. Countries and sovereign governments have to fight the urge to manage based on their own political interests rather than for the good of the resource and the region.

Adding to this point is the difficulty to think and act regionally. The bumper sticker slogan "Think Globally–Act Locally" might more appropriately read "Think Globally–Act Regionally" when referring to the Great Lakes – St.

Lawrence River. Often and especially in a now globally competitive world, the tendency is to protect one's self-interest in order to establish economic, financial or political superiority.

It is the opinion of this author that the region's ability to think and act regionally has not increased over time but rather has waxed and waned. This is probably due to a variety of factors including the perception of the issues at hand, the political climate and the make up of the political leadership both nationally and jurisdictionally, the state of the economy within each country and the extent to which the political leaders (e.g., the Great Lakes – St. Lawrence governors and premiers) are in relationship with one-another. Also, there are very real constraints when changes in political regimes occur. Different ideologies of the political parties and the need to educate and inform new leaders and parties about the issues can cause setbacks in binational relationships and regional cooperation. In addition, regional cooperation seems to be at its highest when public involvement and citizen activism surrounding the issue is high, visible and effective. For instance, regional cooperation was at a high level in the mid-1980s around the time that groups like Great Lakes United, the Lake Michigan Federation, Great Lakes Tomorrow and others were mounting successful grass roots campaigns in support of an amended Great Lakes Water Quality Agreement which occurred in November 1987.

There is a tendency toward governmental and institutional inertia which creates the dynamic of being unable to adapt in a rapidly changing world. The inherent nature and make up of most governments and institutions causes them to resist change. Even when change does occur, it is usually incremental and prolonged. Governments and institutions have trouble responding to regional trends and stresses with creativity and changes to the status quo but rather revert to prolonged research and study of issues that are often already well-understood (Donahue 1987). No wonder that the battle cry of stakeholders during the development of the GLRC and GLRI action plans was "we don't need any more studies, we need on the ground action!"

While governments and institutions tend to resist change they also tend to engage in crisis-response management. The historical study of the region shows this to be true, with governments and institutions tending to react to problems and crises rather than trying to anticipate and avoid them. A crisis-management approach by governments also creates the danger of returning to issue specific approaches to solving problems. The Boundary Waters Treaty was ahead of its time in recognizing the inefficiencies in addressing issues on a narrow problem specific basis, there is still a tendency for governments and institutions to try to solve problems individually rather than on an ecosystem basis.

Donahue (1987) describes an early report to the National Water Commission on the need to preserve the Great Lakes stating that the scope and seriousness of Great Lakes problems appear to be beyond the powers of present programs and policies. Likewise, in the Prescription for the Great Lakes Ecosystem

Protection and Restoration report described earlier, the authors note "when faced with a particularly damaging human perturbation in the Great Lakes, our corrective response has generally been to focus on a particular cause of stress and not on the integrated sources of stress that allowed it to occur" (Bails et al. 2005). In other words, we react to the individual problem rather than trying to seek the larger cause and overall solution to the problem. Examples of this crisis-response management approach continue today on issues such as spill prevention and response, beach health, aquatic invasive species introduction and spread, excessive nutrient loadings causing algal blooms and many others.

The institutional framework also promotes issue specific problem solving. One of the outcomes of the process leading to the GLRC strategy in 2005 was the recognition of the problem of redundancy in governmental programs and funding in some areas and huge gaps in programs and funding in other areas. At the time, there was great resistance to begin reviewing different programs to examine opportunities for combining authorities and funding as well as integrating programs to reduce redundancy and mission overlap to provide a more comprehensive approach to Great Lakes protection and restoration. This is an ongoing need that is critical to address both in the U.S. and Canada.

While governments are resistant to change they are also often fascinated by newness. Political leaders often find it more palatable and exciting to create new institutions than to revamp and refine existing ones (Donahue 1987). The easier or more politically expedient solution, especially when a new issue or problem has emerged or when things are not working as desired with an existing institution is to create a new one.

This chapter review summarized the development and evolvement of existing institutions but did not speak of the many ones that have come and gone over the years. Groups and agencies like the Great Lakes Basin Commission, Great Lakes Tomorrow, Center for the Great Lakes and the Great Lakes/St. Lawrence Maritime Forum are examples of organizations that were established and well-known for a period of time but no longer exist due to a loss or change in mission, lack of political support, lack of funding or because other organizations were formed to take their place.

Funding of Great Lakes institutions and programs has always been a challenge to governments in both the U.S. and Canada. On the U.S. side, the establishment of President Obama's Great Lakes Restoration Initiative (GLRI) and Action Plan in 2009 has created unprecedented excitement, enthusiasm and even hope for the future of the Great Lakes through the establishment of a formal restoration plan but equally important with significant dollars available for implementation.

In FY 2010, US$ 475 million dollars in new funding was made available that strategically targets programs and projects that address the most significant problems in the Great Lakes ecosystem. Congress was clear in FY 2010 that the GLRI is intended to supplement, but not supplant, base funding for federal programs that support ongoing management and conservation of the lakes. Funding for the GLRI's first year approximately doubled the amount of federal dollars to the Great Lakes through the annual appropriations process to nearly US$ 1 billion. The Canadian federal contribution to Great Lakes protection and restoration efforts is unknown but is significantly less.

While the GLRI is on a good trajectory for success, there is a real danger that funding inequities between the U.S. and Canada may negatively impact binational cooperation as well as binational programmatic achievement.

Concerns have arisen that the GLRI funding may not address the highest priority needs because of the pressure to support implementation projects that will show results quickly (i.e., are "shovel ready"). Also because much of the funding is being distributed competitively to a variety of different agencies and groups through a request for proposals (RFP) process, there is concern about a developing dynamic where agencies, organizations and groups will modify or even abandon their historic missions in pursuit of available funding. On the Canadian side, concerns are being raised about a lack of federal commitment to the Great Lakes. The Canadians have the plan (under COA) to accomplish Great Lakes restoration priorities, but the commensurate financial resources are currently not available.

This chapter will conclude with some observations regarding the importance of the renewed commitment and process to renegotiate the next Water Quality Agreement which was officially begun when representatives of the governments of the U.S. and Canada met on the Peace Bridge, which spans the Niagara River at the east end of Lake Erie, on 13 June 2009 during the official celebrations commemorating the 100th anniversary of the Boundary Waters Treaty of 1909. There on the bridge, the Honorable Lawrence Cannon, Minister of Foreign Affairs and U.S. Secretary of State Hillary Rodham Clinton committed the two countries to amend the Great Lakes Water Quality Agreement. The announcement was made after both countries had received considerable input, after a lengthy Water Quality Agreement review process, from state and provincial governments, municipalities, nongovernmental groups, tribes and first nations and other stakeholders. This review of the Agreement, which occurred during 2006 and 2007, was procedurally triggered by the release of the IJC's 12th Biennial Report in September 2004. Unlike earlier reviews, which were deferred or delayed for a variety of reasons, there was widespread support within the Great Lakes community for a substantive review of the Agreement (Agreement Review Committee 2007). The two countries recognized this growing interest and in response conducted the broad-based and inclusive review of the Agreement's operation and effectiveness. The formal renegotiation process began in 2010. With it, comes a potential new chapter to be written and hopefully some new opportunities for renewed relationships and a renewed commitment to binational Great Lake governance.

During the last review, a Binational Agreement Review Committee (ARC) was formed to lead and coordinate the 2007 review and a Review Report was prepared, and submitted to the BEC, synthesizing the key findings, results and recommendations that were gleaned from a series of review working groups and a workshop on the Agreement's governance and institutions.

In its September 2007 report to the BEC, the binational Agreement Review Committee said this:

> "The work of restoring and maintaining the Great Lakes Basin Ecosystem and preventing further degradation must continue. Citizens; federal, state, provincial and local agencies; Tribes and Aboriginal Peoples; elected officials; and stakeholder groups, including non-governmental organizations, industry groups and the agricultural community, play a vital role in the protection and restoration of the Great Lakes Basin Ecosystem and should again be guided by a Great Lakes Water Quality Agreement that is visionary. The Governments of the U.S. and Canada have an opportunity to revitalize the Agreement to once again set a clear path for making the Great Lakes greater" (Agreement Review Committee 2007).

Let's hope that the governments take this admonition seriously and that this opportunity for a renegotiated Agreement will lead to greater and more effective participation in the governance and management of the greatest freshwater resources on earth.

REFERENCES

Agreement Review Committee. 2007. Review of the Canada–U.S. Great Lakes Water Quality Agreement. Report to the Binational Executive Committee, vol 1. Canada.

Bails, J. and A. Beeton, J. Bulkley, M. DePhilip, J. Gannon, M. Murray, H. Regier, and D. Scavia. 2005. Prescription for Great Lakes Ecosystem Protection and Restoration: Avoiding the Tipping Point of Irreversible Changes. USA.

Botts, L. and P. Muldoon. 2005. Evolution of the Great Lakes Water Quality Agreement. Michigan State University Press, USA.

Broun, C.N. and M.L Buenger, M.H. McCabe, and R.L. Masters. 2006. The Evolving Use and the Changing Role of Interstate Compacts: A Practitioner's Guide. ABA Press, USA.

Colborn, T.E. and A. Davidson, S.N. Green, R.A. Hodge, C.I. Jackson, and R.A. Liroff. 1990. Great Lakes Great Legacy? The Conservation Foundation and the Institute for Research on Public Policy, U.S. and Canada.

Donahue, M.J. 1987. Institutional Arrangements for Great Lakes Management: Past Practices and Future Alternatives. Michigan Sea Grant College Program, USA.

Gauthier, R.L. and C. Manninen. 1999. Living with the Lakes: Understanding and Adapting to Great Lakes Water Level Changes. The U.S. Army Corps of Engineers (Detroit District) and the Great Lakes Commission, USA.

Government of Canada and U.S. Environmental Protection Agency. 1995. The Great Lakes: An Environmental Atlas and Resource Book. 3rd edn. U.S. and Canada.

Great Lakes Commission. 1968. The Great Lakes Basin Compact with State and Federal Legislative History. USA.

Great Lakes Commission. 1992. Advisor, vol 5 no 1. Special Edition: 1991 Annual Report. USA.

Great Lakes Commission. 1995. Advisor, vol 8 no 1. Special Edition: 1994 Annual Report. USA.

Great Lakes Fishery Commission. 2006. Annual Report. USA.

Great Lakes - St. Lawrence River Water Resources Regional Body. 2005. Great Lakes - St. Lawrence River Basin Sustainable Water Resources Agreement. USA.

Hildebrand, L.P. and V. Pebbles and D.A. Fraser. 2002. Cooperative ecosystem management across the Canada–U.S. border; approaches and experiences of transboundary programs in the Gulf of Maine, Great Lakes and Georgia Basin/Puget Sound. Ocean & Coastal Management 45: 421-457.

International Joint Commission. 2005. A Guide to the Great Lakes Water Quality Agreement: Background for the 2006 Governmental Review. Canada.

U.S. Federal Interagency Task Force. 2005. Great Lakes Regional Collaboration Strategy to Restore and Protect the Great Lakes. USA.

U.S. Federal Interagency Task Force. 2010. Great Lakes Restoration Initiative Action Plan. USA.

Wege Foundation. 2004. Healing Our Waters: An Agenda for Great Lakes Restoration. USA.

CHAPTER 3

Governance in the Great Lakes – A Regime in Transition

Paul Muldoon

Environmental Lawyer
E-mail: Paul.muldoon@sympatico.ca

INTRODUCTION

Since the conclusion of the Great Lakes Water Quality Agreement (GLWQA) between the United States and Canada in 1972,[1] the Great Lakes regime has been the subject of many studies and reports on its governance structures and processes (Jackson and Sloan 2008, Krantzerg et al. 2007, Dempsey 2004, Nowland and Bakker 2010). In this chapter, the governance structures and processes for the Great Lakes is referred to as the "Great Lakes regime." The notion of "regime" refers to the broader formal and informal structures and processes that describe the governance system for a region (see Scharm Stokke 1997).

Over its nearly 40-year history, the Great Lakes governance framework has been in constant, albeit often subtle and gradual, transition and evolution. Governance regimes are organic and need to change to respond to current circumstances and that is certainly true for the Great Lakes. The ecological, social and economic stresses facing the region are both similar to the past, but in other ways, profoundly different. The governance regime must now evolve to

[1]Agreement Between Canada and the United States of America on Great Lakes Water Quality, April 15, 1972, 23 U.S.T. 301; amended Nov. 22, 1978, 30 U.S.T. 1384, Protocol amending the 1978 Agreement between Canada and the United States of America on Great Lakes Water Quality as amended on October 16, 1983, November 18, 1987, T.I.A.S. No. 11551.

address existing and new ecological stresses; it must respond to the changing political priorities of the various governments at all levels within the regime; and it must be cognizant of the disparate perspectives of those that live within the Great Lakes. The evolving regime must respond to these challenges while attempting to preserve and promote the key values and principles embedded both in the GLWQA and its legal foundation, the Boundary Waters Treaty (BWT).[2] One vehicle for the continuing evolution of the regime is the formal review process for the GLWQA,[3] the latest one that commenced in 2005 and, as of now, remains ongoing.

When reviewing its long history, what are the key themes or the lessons learned in the maturation and evolution of the Great Lakes governance regime under the GLWQA? This chapter will provide a brief review of the Great Lakes governance regime under the GLWQA. The latter part of the chapter will seek to identify some of those key themes or lessons learned that could at least provide a launching pad into a more detailed and comprehensive discussion in the context of the review of the GLWQA with respect to a revamped or updated governance regime.

AN OVERVIEW TO THE GREAT LAKES ECOSYSTEM

A Brief Description

The Great Lakes comprise five large lakes – Superior, Huron, Michigan, Erie and Ontario, along with the St. Lawrence River, and holds about one-fifth of the world's freshwater surface water [Environment Canada and United States Environmental Protection Agency, Botts and Krushelnicki (1995) Great Lakes Information Network] and indeed, some 95% of the surface freshwater in the U.S. (Hall 2009). The Great Lakes and their connecting channels include over 22,700 kilometers of water and 17,000 kilometers of shoreline. The Great Lakes were formed by the melting of glaciers that receded over what we know call the Great Lakes some 10,000 to 15,000 years ago (Botts and Krushenlnicki 1995). The 300,000 square-mile Great Lakes drainage basin includes about one-third covered by water with the basin containing hundreds of tributaries and inland lakes. It is interesting to note that "waters" of the Great Lakes are often considered to be more of a finite resource than something that can be easily renewed. Water in Lake Superior has a residency of over 200 years while the time is much less for the swallow lakes like Lake Erie. Most important

[2]Treaty Relating to Boundary Waters and Questions Arising with Canada, United States and United Kingdom, January 11, 1909, 36 U.S. Stat. 2448.

[3]Under Article X(4) of the GLWQA, the Parties must conduct a comprehensive and effectiveness of the Agreement following every third biennial report of the IJC.

though is that only 1% of the Great Lakes waters are renewed each year (Farid et al. 1997, Sanderson 1993, Pearse et al. 1985).

There are over 35 million people who live in the basin and more than half depend on the lakes for water supply (Botts and Muldoon 2005). About 60% of North American steel and one-fifth of all manufacturing goods are produced in the Great Lakes (Botts and Muldoon 2005, Botts and Krushelnicki 1995). One fourth of the manufacturing capacity of the US rests within the Great Lakes. It is estimated that the annual value of recreational boating, sport fishing and tourism is over 10 billion (US).

Present Stresses and Future Threats

The ecological state of the Great Lakes presents a mixed picture. It has been argued that significant progress has been made since the earlier 1970s to "save the Great Lakes" and that the GLWQA was one of a number of factors that can take credit for the improvement in the Great Lakes.[4] However, all of the threats that were identified in the early 1970s have not been totally resolved and new ones seem to constantly appear (Bails et al. 2005). A very cursory look at some of the ongoing issues related to just to water gives some insight to the breadth of the challenges for the Great Lakes decision-makers and public.

A few years ago, a number of leading Great Lakes scientists endorsed a summary of findings on the state of the Great Lakes with the document starting with the statement that there is "widespread agreement that the Great Lakes presently are exhibiting symptoms of extreme stress from a combination of sources that include toxic contaminants, invasive species, nutrient loading, shoreline and upland land use changes, and hydrologic modifications" (Bails et al. 2005). One of the key messages of that report state:

> In large areas of the lakes, historical sources of stress have combined with new ones to reach a tipping point, the point at which ecosystem-level changes occur rapidly and unexpectedly, confounding the traditional relationships between sources of stress and the expected ecosystem response. There is compelling evidence that in many parts of the Great Lakes are increasingly experiencing ecosystem breakdown, where intensifying levels of stress from a combination of sources have overwhelmed the natural processes that normally stabilize and buffer the system from permanent change (Bails et al. 2005).

The IJC concluded in its 2006 13th Biennial Report that the future of the Great Lakes remains "uncertain" (International Joint Commission 2006).

[4]In Botts and Muldoon (2005), the argument is made that a number of factors contributed to the improvement of the Great Lakes from 1970, and that GLWQA is certainly one of those factors. In other words, the improvement in the physical condition of the Great Lakes would not have been to the extent it has but for the conclusion and reliance on the Agreement. See pp. 188-191.

A number of findings reinforce that, despite past progress, there are continuing threats to the health of the Great Lakes:

- There are ongoing threats to the wetlands considering that over 90% of pre-settlement wetlands along Lake Huron/Lake Erie corridor have been lost. U.S. Environmental Protection Agency (2005), U.S. Environmental Protection Agency and Environment Canada (2007), Jackson and Sloan (2008) Since the European settlement about 70 to 80 % of original wetlands in southern Ontario have been lost (Botts and Krushelnicki 1995).
- Since the middle of the last century, over 180 aquatic and at least 157 terrestrial non-native species have been identified in the Great Lakes, including zebra mussels, sea lamprey, purple loosestrife, some with dramatic impacts, and others with consequences yet unknown (Environment Canada 2006).
- Using information from the US Toxic Release Inventory and the Canadian National Pollutant Release Inventory, it is estimated that more 627 million kilograms of pollutants released into the Great Lakes and St. Lawrence River Basin by Canadian and U.S. facilities in 2002 (Environmental Defence and Canadian Environmental Law Association 2006).
- It is estimated that there are over 90 million cubic metres of contaminated sediments in the Great Lakes (International Joint Commission 10th Annual Reports).
- There are various reports which outline that in Areas of Concern in the Great Lakes, there are reproductive problems, reduced stress response, disrupted thyroid function, suppressed immune function, among other impairments in wildlife (snapping turtles, herring gulls) and reproductive failures in various species of fish (including lake trout, sturgeon, lake herring) (Jackson and Sloan 2008, Environment Canada, 2003, Bails et al. 2005).
- "General water quality degradation, associated algal blooms, Type E botulism in fish and waterfowl, and contamination of drinking water" (Bails et al. 2005)

These threats suggest that there is a continuing, and perhaps more urgent, need for the Great Lakes oversight and management. The next section briefly describes the present governance arrangements.

THE GREAT LAKES REGIME – AN OVERVIEW TO THE INSTITUTIONAL AND GOVERNANCE FRAMEWORK

The Great Lakes regime has its legal foundation in the BWT of 1909, which in turn, provides the basis for the GLWQA. This section provides an overview of the Great Lakes legal framework and the institutional setting for the regime. This background is useful since the next section of the chapter suggests a number of principles that can be derived from the evolving legal and institutional framework. For example, this chapter argues in a subsequent section that the principle of "binationalism" is an operating principle of the Great Lakes regime, although the term is not used anywhere in the formal legal and policy context.

The Boundary Waters Treaty and the GLWQA

The Governance Structure at a Glance

There is nothing simple about the governance regime for the Great Lakes. Its complexity can be attributed to a number of factors. First, the sheer number of jurisdictions involved. The most obvious jurisdictions include the two federal governments and the agreements they have made between each other with respect to the Great Lakes that now span well over a century. Then, there are the eight U.S. states and the two provinces that both legally and practically have enormous authority to address Great Lakes issues. In order to understand how to effect action on any one issue, such as shoreline protection measures, it is often not only necessary to understand the international and bilateral obligations, but to have an appreciation of the federal laws and policies in each country and the laws and policies of the states and provinces.

Second, the GLWQA is an agreement between the two federal governments, although the Agreement itself recognizes, and in some ways, formally includes the states and provinces within the regime. However, there are more governmental actors in play than just federal, state and provincial governments. First Nations and tribes have, and will increasing have important, if not fundamental, implications for future Great Lakes management regimes. While the key legal agreements discussed below do not formally recognize the role or influence of First Nations and tribes, their political influence and emerging legal authority in both countries cannot be underestimated as the Great Lakes governance regime continues to mature. Similarly, the role and influence of municipal governments have been recognized in recent years (Great Lakes Cities Initiative). Municipalities dealing with Great Lakes issues are on the "front line" since they often have to deal with their constituencies on issues like drinking water, beaches, sewage, shoreline and nearshore protection, among so many others.

The third reason for the complexity of the system is that the "regime" not only comprised governments as actors, but a broad and diverse set of stakeholders, such as environmental non-governmental groups, industry and industrial associations, community organizations, organized labour groups, academics and scientists, journalists, charitable foundations, to name but a few. It is the collection of governmental and non-governmental efforts directed toward a common purpose that made the regime effective, if not unique, and has been referred to as the Great Lakes community.[5]

The Great Lakes regime is complex not only because of the fragmentation of the jurisdictions and the multitude of actors, all of whom have some role in its governance, but also because it is evolving. Although the core governance structures and processes have essentially been in place for some decades, there have been subtle, but important, changes to the governance regime over the past two decades as noted below.

The Boundary Waters Treaty and Its Key Principles

The starting point to describe the governance structure for the Great Lakes is the BWT since it provides the legal and institutional framework for the Great Lakes. It has been noted that the BWT has "provided an international model for transboundary environmental protection and management" (Hall 2009). The BWT was agreed to between Canada and the U.S. in 1909 owing to concerns over the apportionment of water for producing hydropower. Other issues of concern at the time that provided the motivation for a more formal approach to addressing issues along the "common frontier" ranged from sewage effluent, navigation, use and diversion of boundary and transboundary waters. These issues that seem to constantly arise between the two countries convinced the governments to consolidate prior agreements and treaties into an overarching treaty that would govern all boundary and transboundary water for the purposes of resolving existing issues and avoiding conflicts in the future.

Article IV of the BWT establishes the International Joint Commission (IJC) which is a six member body with three members appointed by the Prime Minister of Canada and three from the President of the U.S. Under the Treaty, the IJC is given a number of functions. It is given a quasi-judicial function that empowers the Commission whether to approve construction or other measures that will affect the levels and flows of boundary and transboundary waters. Article VIII sets out guiding principles for consideration of approvals, including the principle that both countries have "equal and similar rights to the use of the water." An "order of precedence" is outlined for various uses when there is a conflict over competing uses, including the limit control of waters levels in Lakes Superior, Lake Ontario and Lake Erie via locks and control structures. One of the key implied goals with respect to the Articles relating to levels and flows was to set "rules of game" in advance to avoid conflicts between the two countries as boundary and transboundary water resources were developed over time.

[5]The Need for Inclusion: The Great Lakes as a Community.

Up until the 1960s, most of the Commission's work dealt with authority under the approval power. However, once many of the major diversions, such as hydro and seaway works, had been completed that included boundary and transboundary waters, the IJC's work focused more on its "reference" power. Under this power, the IJC has the authority to undertake studies or investigations of any issue along the "common frontier" upon the request of both governments. If the governments give the Commission a reference, it is to undertake studies and make recommendations on that specific problem or issue.

The reference power given to the IJC is particular germane to the Great Lakes because the GLWQA is a "standing reference" under the BWT. As noted below, the GLWQA also gives additional powers to the IJC under the Agreement that identifies and clarifies the Commission's role and duties.

Another function of the IJC is its arbitral power to resolve disputes under Article X, although this function has yet to be used.

The IJC has a number of important operating principles which have been explained in detail elsewhere. One principle is that, despite the enormous difference in population and economic power between Canada and the US, the two countries maintain parity and equality between each other within the regime. Further, the IJC is expected to work free from nationalistic considerations and "to seek the best solution to common problems based objectively on results of joint fact-finding studies" (Botts and Muldoon 2005). This principle is supplemented by another principle which is the reliance on common fact-finding as an effective way to either investigate matters or resolve disputes.[6]

A Short History of the GLWQA

The next building bloc in explaining Great Lakes regime is the GLWQA. The Agreement, first concluded in 1972, was a response to a list of serious problems in the Great Lakes that prompted action.[7] The problems were well-documented by a number of references to the IJC dating back to 1912[8], including a reference in 1964 dealing with deteriorating conditions in Lake Ontario and Lake Erie.[9] There was also significant public attention directed to the pollution of the Great Lakes both in Canada and the US in the 1960s and early 1970s. Moreover, environmental issues generally become more recognized as a legitimate public policy issue with the passing of various new statutes, including the U.S. National Environmental Policy Act in 1970 and the formation of new environmental agencies and departments at the federal and state/provincial levels.

[6]The "fact-finding" approach as a mechanism of dispute resolution and conflict avoidance is discussed in: Munton, "Paradoxes and Prospects."

[7]For a full discussion on the history of the Great Lakes, see: Leonard B. Dworsky 1988.

[8]For a review of the early work of the Commission with such issues, see: R.B. Bilder 1972.

[9]International Joint Commission, Docket 54, 55, and 83.

The GLWQA was first signed in June, 1972 under great fanfare, and for good reason. The blatant pollution problems on both sides of the border had reached a point where political attention was expected and needed. The 1972 Agreement focused on a number of key institutional and oversight provisions that were carried on subsequent iterations of the Agreement discussed below. For the most part, though the Agreement dealt with phosphorous loading reductions in order to address the most visible problem in the Great Lakes, the eutrophication of a number of the Great Lakes.

The success of the 1972 GLWQA is seldom questioned particularly in reducing phosphorous loadings. Within a relatively short time, the Agreement took credit for reversing the declining ecosystem. The success of the 1972 Agreement set the stage for the re-negotiation of the Agreement that resulted in the 1978 GLWQA.

Key Elements of the GLWQA

The 1978 Agreement incorporated a number of key innovations. No attempt is made to undertake a detailed a review of the GLWQA. Instead, Table 3.1 outlines its key provisions while the next subsection outlines the institutional perspective the Agreement.

The Institutional Setting

The Role and Operation of the International Joint Commission

Within the Great Lakes, there are literally dozens of various bilateral institutions or inter-jurisdictional institutions that have some role in the management of the Great Lakes.[10] In addition, there are federal, state/provincial, municipal agencies that also share in this task, along with the various agencies within First Nations and tribes.

The institution that has the broadest mandate is the IJC. Over the years, there has been much discussion and debate on the role of the IJC in Great Lakes issues, and for good reason. Since 1909, the Commission has had a vital role to play in the identification of environmental issues, investigation and fact-finding that has led to addressing some of the problems and administrating programs.

---

[10]For example, some other institutions include the Great Lakes Fisheries Commission, the Great Lakes Commission, the Council of Great Lakes Governors, to name only a few. For a more detailed review, see Regier 1986.

Table 3.1 KEY PROVISIONS UNDER THE 1978 GREAT LAKES WATER QUALITY AGREEMENT (Adopted from: Botts and Muldoon 2005)

Overall Purpose – The purpose of the Parties is to restore and maintain the chemical, physical, and biological integrity of the waters of the Great Lakes Basin Ecosystem.

Article I - Definitions

"Great Lakes Basin Ecosystem" means the interacting components of air, land, water and living organisms, including humans, within the drainage basin of the St. Lawrence River at or upstream from the point at which this river becomes the international boundary between Canada and the United States.

Article II – Purposes – ...to restore and maintain the chemical, physical, and biological integrity of the Great Lakes, and three policy commitments, including the goal to prohibit the discharge of toxic substances in toxic amounts and to virtually eliminate persistent toxic substances.

Article III – General Water Quality Objectives: These objectives provide narrative objectives, such as ensuring the waters are free from substances in concentrations that are toxic for humans, animals, or aquatic. Other objectives deal with putrescent sludge material, floating debris, materials that may cause a nuisance.

Article IV – Specific Water Objectives: These provisions commit to meeting specific numeric objectives for levels of reduction of pollutants as set out in the Annexes to the Agreement.

Article V – Standards and Other Regulatory Requirements: This article commits the governments to have their domestic standards consistent with the Agreement.

Article VI – Programs and Other Measures: These provisions commit the governments to the development of programs directed toward the achievement of the water quality objectives. These include programs for pollution controls from municipal and industrial sources and eutrophication from agricultural, forestry and other land-use activities; and pollution from shipping, dredging, and onshore and offshore facilities, among others.

Article VII – Powers, Responsibilities and Functions of the IJC: These provisions outline power and responsibilities of the Commission, including the data and information collection on the state of the Great Lakes, advice tendering and reporting to the Parties biennially.

Article VIII – Joint Institutions: This provision mandates the establishment of a Water Quality Board and a Science Advisory Board, as well as a regional office located in the Great Lakes basin.

Article IX – Submission and Exchange of Information: This article outlines requirements between the governments and the IJC concerning data and other information relating to water quality.

Article X – Consultation and Review: This article requires consultation among the IJC and the governments. It commits to a comprehensive review of the agreement after five years.

Article XI – Implementation: This article commits the government to appropriating adequate funds to implement the agreement, and to enacting any necessary legislation to implement the agreement.

Annexes

While their mandate covers both boundary and transboundary waters, the IJC is given special and specific duties under the GLWQA as a standing reference under the BWT. As such, in the exercise of its roles and responsibilities under the Agreement, the IJC has the same powers it is given under the BWT. Article VII of the GLWQA outlines the key roles and responsibilities of the Commission. In summary, they include:

- To collate, analyze and disseminate data and information concerning Great Lakes water quality and the operation and effectiveness of programs and measures established under the Agreement;
- To verify independently data or information given to it by the governments;
- To tender advice and recommendations to the Parties and provincial and state governments on problems relating to water quality, including recommendations concerning the General and Specific Objectives, legislation, standards and other regulatory requirements, programs and other measures relating to water quality;
- To provide assistance relating to the coordination of joint activities under the Agreement and to research;
- To investigate matters relating to the Great Lakes when requested by the Parties.

The key, and perhaps the most recognized duty of the IJC is to report to the federal governments every two years on the progress in achieving the goals under Article X of the Agreement. The report must include an assessment of the effectiveness of programs and other measures undertaken by the agreement and advice and recommendations. The IJC may, at any time, publish any report in the discharge of its functions and can independently verify the data and other information submitted by the federal, state or provincial governments. To assist in the carrying out of its duties, the Agreement establishes a Regional Office, the Science Advisory Board, and the Water Quality Board.

The Science Advisory Board: An advisory board made up of academic, governmental and other experts in an array of fields. The mandate of the board is to review existing and emerging scientific issues relating to the Great Lakes.

The Water Quality Board: This board is comprised of the federal, state and provincial governmental representatives from each of the jurisdictions within the Great Lakes. Its overall task is to provide advice to the IJC.

The Council of Great Lakes Research Managers: This advisory body is not founded under the Agreement but initiated through the IJC and provides an inventory and evaluation of research programs in the Great Lakes. It comprised managers of research programs pertaining to the Great Lakes.

In addition, the Agreement establishes an IJC Regional Office, located now in Windsor, Ontario, that provides support for the Commission to fulfil its mandate.

Through its Biennial Reports, and the reports of the Commission's advisory bodies, there is a very rich documented history on the issues facing the Great Lakes and the effectiveness of efforts by governments to realize the goals of the GLWQA. As noted below, this task has very much evolved over the years, and according to some commentators, the reporting role of the IJC has lost is authority for a number of factors.[11] First, since the early 1990s, there has been a subtle, but real, diminishing reliance by governments on the role of the IJC. The diminishing role can be measured in various ways, but notably by the lack of resources given to the institution, the development of alternative inter-governmental arrangements (discussed in the next section), decreasingly delegating "issues of the day" to the Commission and the fact that the governments give less deference to words of Commission than they have in past decades. While the IJC Biennial Reports were once feared and revered in the past, the governments receive them today with little fanfare, and some would say, with too little serious attention. One of the consequences of the decreasing reliance on the Commission by governments is that the Commission has less capacity to fulfil its role under the Agreement in terms of resources, political authority and public attention.

Second, the Commission itself has become more "politicized" over the years, and in particular, the appointment of Commissioners. Further, it continues to search for an appropriate or "renewed" niche for itself within the bilateral relationship. Since the early 1990s, the governments of the U.S. and Canada, has relied less on the Commission, although that point is by no means a consensus one. In searching for its role, it appears that the IJC, at times, is careful in not being to be too critical of government action (or lack thereof) in fear that it will further alienate and distance itself from the governments it seeks to serve. On the other hand, its relevance and mandate lies directly with its ability to further the principle of accountability by making thorough, timely and, at times, critical assessments of government action or inaction. In the past decade or so, it continues to search for its role in this regard.

The Inter-governmental Regime

While the influence of the IJC may be in transition over the past years, the role of other inter-governmental arrangements have increased, and most notably, the Binational Executive Committee (BEC). The BEC was formed in the early 1990s and meets twice a year to provide coordination of the work of the government agencies under the GLWQA. BEC includes key federal

[11]See: Botts and Muldoon 2005, Chapter 6.

agencies from Canada and the U.S. along with state and provisional agencies and First Nations and tribal governments. Some of core roles of the BEC are to coordinate governmental activities under the Agreement; set priorities and strategic directions for binational programs, respond to new and emerging issues, evaluate progress under the Agreement.

Another institution that has emerged in the 1990s is the "State of the Lakes Ecosystem Conference" (SOLEC). Every two years or so, the governments sponsor this major conference and focus on some current ecological issues facing the Great Lakes. Experts from the Great Lakes and beyond bring leading edge research to this conference and the conference proceedings contribute to the body of knowledge on the topic. Since its inception in the early 1990s, SOLEC has accumulated an impressive amount of information concerning the Great Lakes.[12]

Other Players in the Regime

As noted above, there are a number of important players in the Great Lakes regime apart from the governmental actors. There are a number of other bodies that are made up mostly of governments that have direct interests in the Great Lakes, such as Great Lakes Commission and Great Lakes and St. Lawrence Cities Initiative. Similarly, other important interests include non-governmental organizations that represent specific interests, such the Council of Great Lakes Industries and public interest organizations, such as Great Lakes United and the Alliance for the Great Lakes. But there are also other important contributors to the regime, such as the charitable foundations, scientific organizations and even media interests that focus on the Great Lakes. All of these organizations are in addition to the industrial, scientific, public interest and community organizations within each state and province.

One of the innovations in the Great Lakes regime was the role non-government groups played in the negotiation of the 1987 Protocol where members of the public interest groups were part of the negotiation process (Manno 1993). The role of the various stakeholders on various advisory bodies with governments and the IJC, the role of the public at the Biennial meetings, and the attention and energy the stakeholders have invested in Great Lakes issues suggest that public participation and involvement is a deep cultural feature of the Great Lakes regime.

Great Lakes as a Laboratory for Policy Innovation

The Great Lakes regime has had a long reputation for policy innovations that have had implications well beyond the Great Lakes. This chapter will identify

[12]For more information, see: www.epa.gov/solec

two such innovations: the ecosystem approach to Great Lakes management and the goal of virtual elimination of persistent toxic substances.

The Ecosystem Approach to Great Lakes Management*:* The 1972 Agreement focuses primarily on water quality and more particularly, the reduction of phosphorous loadings. Following the conclusion of the agreement, a new reference was given to the IJC. The IJC initiated a new reference entitled the Pollution from Land-Use Reference Group (PLUARG).[13] The extensive study will not be summarized, but the thrust of the study is to recognize the interrelationship between the sources of pollution from land, water and air. The study is said to lay the foundation for the development of the ecosystem concept for management and its incorporation into the 1978 Agreement (Botts and Muldoon 2005).

Article I of the Agreement refers to the "Great Lakes Basin Ecosystem" which in turn defines it as "the interacting components of air, land, water and living organism, including humans, within the drainage basin of the St. Lawrence River at or upstream from the point at which this river becomes the international boundary between Canada and the United States."

The ecosystem approach to environmental management, a concept that would be almost obvious now, was seminal at the time and has had a tremendous influence in approaching environmental issues (Botts and Muldoon 2005). Although the title of the Agreement still refers to "water quality," the breadth of the Agreement recognized the influence of the entire drainage system, literally twice the size of the waters of the Great Lakes along with the interaction of the air and living organisms.

"Virtual Elimination" of Persistent Toxic Substances*:* Another innovation of the 1978 Agreement pertains to one of its goals to "virtually eliminate" the discharge of persistent toxic substances." More specifically, Article II of the Agreement states that "the discharge of toxic substances in toxic amounts [shall] be prohibited and the discharge of any or all persistent toxic substances [shall] be virtually eliminated." Annex 12 of the Agreement goes on to state that the "the philosophy adopted for control of inputs of persistent toxic substances shall be zero discharge."

During the early 1970s, scientists were becoming aware of the dangers of persistent toxic substances, that is, substances, like PCBs, that can persist in the environment for a long period of time and affect fish, wildlife and humans. Science was in its infancy in terms of the full understanding of the impacts of persisting toxics, but the drafters of the Agreement undertook the bold initiative by including the term in the Agreement. The actual origins of the term remains elusive, however, the best explanation is that the term probably originates from the wording in the US Clean Water Act.

---

[13]For a description of this reference and its implications, see: Becker 1993.

It is probably safe to say that the drafters at the time had no idea how controversial, the amount of attention, and the impact of this term eventually would have over the years. First, from the early 1980s to the late 1990s, the term "virtual elimination" and "zero discharge" became a rallying cry for non-governmental groups to promote government action to implement the concept. Both the IJC and non-governmental groups released a variety of reports highlighting the environmental and human health impacts of persistent toxic substances and progress by governments in dealing with persistent toxics (National Wildlife Federation and Canadian Institute for Environmental Law and Policy 1988). It is fair to say that heightened attention to both the issue of toxic chemicals and the term "virtual elimination" gradually turned the issue into a key public policy issue in both countries. Hence, another consequence of the term was the fact that there were a number of governmental efforts that transpired and probably would not have happened but for the public's new found awareness of the issue. In the U.S., the Great Lakes Initiative became a major effort that sought to tighten water quality standards in each state. In Ontario, the 1986 Municipal-Industrial Strategy for Abatement," a program to develop new water quality standards, had "virtual elimination" as its goal. At the federal level, new provisions in the 1999 Canadian Environmental Protection Act specifically recognize the need to virtually eliminate some substances and sought to "categorize" or identify persistent, bioaccumulative and inherently toxic substances in Canada.

The third consequence was that the term had implications outside of the Great Lakes regime. The term was commonly used and referred to in the negotiations of the 2004 treaty pertaining to persistent organic pollutants (POPs), now called the Stockholm Convention. on Persistent Organic Pollutants. The influence of the term should not be surprising since many Great Lakes activists work for organizations that focused on the negotiations and the impetus and relevance of the Convention can be traced to Great Lakes scientists and their ground breaking research (Botts and Muldoon 2008).

THEMES OR LESSONS EMERGING FROM THE GLWQA

Introduction: Past Successes and New Challenges

In the Evolution of the Great Lakes Water Quality Agreement, the authors exhaustively canvassed the progress or the "evolution" of the Agreement from its inception to 2005. The authors then identified the factors that contributed to the success of the Great Lakes regime and those factors that present a challenge to the regime. Since 2005, there would not appear to be any substantial events or circumstances to suggest that these "past successes" and "new challenges" have changed.

The authors identified the following "successes" of the regime:

- binationalism
- promotion of community
- equality and parity in structure and obligations
- common objectives
- joint fact-finding and research
- accountability and openness in exchange information
- flexibility and adaptability to changing circumstances (Botts and Muldoon 2005)

The challenges and opportunities for the regime included:

- diminished political commitment and resources to the Great Lakes
- reduced political power of the Great Lakes community
- new threats to the Great Lakes
- decreased funding for science
- changes in the operation for the IJC and in the relationship between the IJC and the governments;
- lack of communication with other international institutions (Botts and Muldoon).

This section of the chapter is going to build upon these findings and identify three "themes" or lessons arising from almost 40 year experience with the GLWQA. The themes blend many of the successes and challenges described above and thus at times can be seen as both a success and a challenge. These are: binationalism; accountability and oversight; and the notion of scale.

The Principle of Binationalism

Perhaps one of the most important, and least recognized, themes of the Great Lakes regime is the notion of binationalism. The term, "binationalism," is neither used in the BWT nor the GLWQA, however, it can be safely stated that binationalism is the consequence of the mechanisms established by those agreements and the traditions or practice in implementing them. Essentially, binationalism emanates from the regime's philosophy that issues or problems referred to the IJC must be resolved jointly, with all involved acting as international civil servants as opposing defining national interest, while using common fact-finding approaches. The theme has been described in this way:

> The theory of binationalism in this context is that it creates a forum in which the IJC commissioners and government officials on both sides can serve in effect as international civil servants. This approach provides opportunity for agency policymakers to discuss matters without home-agency constraints, and fashion

> solutions that can be presented to the home agencies with an understanding of how other jurisdictions will approach the problem. Binationalism is conducive to consensus, since separate interests are set aside for the broader mutual purposes (Botts and Muldoon 2005).

One of the key features of binationalism is that it tends to "de-politicize" an issue because the focus is on trying to come to agreement on the nature, scope and sources of the issue by experts in the field. A Canadian government report noted that the "fact-finding process which de-politicizes each problem and unites both technical staffs in the search for the basic facts of the situation" (Standing Senate Committee on Foreign Affairs, "Canada – United States Relations: The Institutional Framework for the Relationship" 1972, Canadian Senate, Ottawa, Canada). Once the experts have come to a set of findings, the focus is then on problem-solving rather than "name-calling" or fault-finding.

Binationalism, in this context, is not limited to a feature or characteristic of the IJC and its advisory bodies. Indeed, many of the key actors in the Great Lakes regime also, to varying degrees, take a binational approach in that their positions transcend nationalistic concerns and work toward common goals within that organization. Perhaps the most notable is Great Lakes United, a coalition of some 150 community, environmental, labour and other groups from both sides of the border. The collective goal of the organization is the protection and preservation of the Great Lakes and urging all of the governments to further this goal. It is for this reason some of the Canadian based groups feel quite comfortable speaking of the issues in the U.S. and vice versa.

There is a considerable body of literature that suggests binationalism is one of the more important reasons for the success and enduring features of the Great Lakes regime. However, it is a feature that is perhaps least recognized and gradually being dismantled in favour of a "bilateral approach." In a bilateral approach, issues and problems are by each side attempting to further to the greatest extent possible their national interest. This is not to say that a bilateral approach is either inappropriate or necessarily ineffective. It is just different.

Literally, a large chapter on the Great Lakes regime can be written about history, effectiveness and lessons learned from a binational approach. However, the governments now seem content to further a bilateral approach with its results being carefully monitored. It was recently noted that:

> Binationalism is not the same as bilateralism. With binationalism, the common interest supersedes separate interests. Bilateralism involves negotiation between two parties with national interest as the controlling factor. Experience under the Great Lakes regime has demonstrated the strength of the binational approach for dealing with such a large ecosystem across an international border but it has also confirmed challenges (Botts and Muldoon 2008).

A number of reasons have been cited for the decline of the binational approach in the Great Lakes regime. It has been recently noted that "…the most well-recognized is the decline of the stature, and more important, government reliance on, the IJC" (Botts and Muldoon 2008). It was also noted that sides of the border: "...the funding for Great Lakes science on both has shrunk and the IJC's ability to carve its role in promoting cutting-edge research has diminished. Even the non-governmental groups have struggled to maintain a binational voice (Botts and Muldoon 2008).

A good illustration in the difference between the "binational" versus the "bilateral' approach relates to the formation of the BEC. As described above, the BEC is a body of senior bureaucrats from Canada and the U.S. formed in the early 1990s and was intended to be the body to coordinate the Great Lakes programs from each country within the regime. It has been a long standing suspicion that the formation of the BEC was a means to gradually move influence and relevance from the IJC. The implications of the approach is that the principles and mode of operation embedded in the IJC that has led to a binational approach are absent within the BEC framework. Rather than relying on the IJC's board and agencies to investigate issues, fact-find, and give advice to policy-makers, the BEC provides a new, virtually non-accountable, mechanism outside of the Agreement (Botts and Muldoon 2008).

The subtle changes have been described as follows:

> Technical experts and program managers no longer have the opportunity to step out of their parochial roles in working jointly to reach consensus. Instead, they always act as agency representatives, which potentially undermines their ability to reach consensus and likely influences the nature of that consensus.
>
> Another concern about the new arrangement is whether the IJC and the public are receiving enough information and whether sufficient data are being collected and analyzed to allow independent scrutiny of the effectiveness of programs.... Still another concern is whether the IJC retains the capacity to evaluate the information that come from the parties. Without assistance from its own advisory boards, the IJC is hampered in its ability to very independently and to analyze whatever data are submitted (Valiante et al. 1997).

Few would suggest that BEC and SOLEC do not serve both a legitimate role and a useful contribution to the Great Lakes regime. However, that is not why concerns have been raised about these institutions. Concerns have been raised because the roles the institutions play, at least in part, have been the roles traditionally reserved or occupied by the IJC through a binational approach. The binational approach has yielded significant results over the years, and it will be interesting to assess the effectiveness of the move to bilateralisms with the U.S.-Canada relationship. What is perhaps needed is an in-depth discussion as to the respective roles for both approaches under the Agreement.

The Notion of Scale: RAPs, LAMPs, Ecosystem

One of the most significant challenges and successes of the Great Lakes regime is that it has evolved with a number of incredible records of success in many realms despite the mere complexities of the jurisdictional fragmentation alone. The regime includes two federal governments, eight states, two provinces, a large number of influential municipalities and a number of First Nations and tribes. Often, one may ask how there can be any agreement on any issue in light of this jurisdictional fragmentation, let alone all of the other competing tensions within the regime.

One of the interesting themes throughout the evolution of the regime has been the attempts to tackle Great Lakes issues with a view to what "scale" is appropriate to address the vital issues at hand. For example, the 1972 Agreement clearly focused at the large, Great Lakes-wide scale in order to address phosphorus loading to the Great Lakes. The 1978 Agreement took a similar approach with respect to the persistent toxic substances.

The 1987 Agreement, however, attempted to take a slightly different approach with respect to scale in Annex II with the development of Remedial Action Plans (RAPs) and Lake Wide Management Plans (LAMPs). RAPs would focus on remedial measures for 43 Areas of Concern (sometimes called "toxic hotspots"), that is, areas where the specific problems in a region requiring special and focused attention. Under Annex II, a process was established with the intention of developing measures to address those issues that gave the region the designation of Area of Concern. LAMPs are to focus at a lake-wide level rather than at a specific locality. One of the unique approaches employed in Annex II was the requirement for public involvement in developing remedial plans, since there was an assumption that most remedial plans would require a set of "trade-offs" and the public should have a voice in understanding what trade-offs were appropriate.

Another dimension of this issue is when the Great Lakes governments and the IJC recognized that the problem of persistent toxic chemicals could not be resolved solely by governmental action within the Great Lakes since a large contribution to the Great Lakes was from beyond its borders. Hence, the governments and IJC supported the effort to develop a legally binding global agreement to deal with the most dangerous toxic substances. Through a number of biennial reports, for example, the IJC recommended a broader, international approach to problem of toxics since the source of a large percentage of Great Lakes contaminant was far outside the boundaries of the Great Lakes basin ecosystem.

While a global agreement did eventually conclude on the topic, the Stockholm Convention on Persistent Organic Pollutants (2004), it is fair say there were many Great Lakes fingerprints on the agreement and the process leading to the agreement. For example, Great Lakes science and scientists provided significant impetus and a knowledge base for the Convention. For years, the impacts of toxic substances on fish, wildlife and humans had been studied and debated in the Great Lakes. Even though many of the same issues were raised at the international level, the Great Lakes legacy in dealing with such issues provided a strong informational basis for the broader international community and probably accelerated the learning curve for all involved. Moreover, many Great Lakes activists and policy-makers contributed to, and in fact, played a significant role in the development of the Stockholm Convention.

Scale, as a concept, is fundamentally important because it focuses on how best to deal with a problem. Due to the large geographical area that the Great Lakes ecosystem covers and the diversity of jurisdictions and issues, there has to be some thinking as to what scale is necessary to address the problem – basin-wide like with phosphorous; local as in RAPs or lake-wide as in LAMPs. Who is most affected by the problem? Who has to be involved? Who has the authority to deal with the issue? Who can bring resources to address the issue? Where can a political constituency be built to add support for action?

In short, scale is important to ensure there is appropriate political and community support for the efforts. In other words, few politicians or communities work on a "basin-wide" scale; instead, they work on the scale closest to their own constituency, although they may understand how their community is so intimately connected to the whole system. For example, beach closures are a local problem for communities. Often, however, the problem in one community is mirrored in others. The challenge is that the solution may have to be regional, if not larger, in nature. A local community, therefore, may need to spark debate on the need for work on the issue, and then join forces with other communities to seek out a common remedy.

More recently, a new approach with respect to "scale" is being discussed in the context of the renegotiation of the Agreement. In Ontario, the province passed the Clean Water Act[14] that requires the development of "source water protection plans" for 36 watersheds in the province. Although the source water protection plans are focused on drinking water protection, the model may provide another, and perhaps more ecologically appropriate, approach to the issue of scale. A watershed approach is simply another attempt to calibrate what scale is appropriate for the issue at hand.

The Need for Inclusion: The Great Lakes as a Community

Related very closely to the notion of scale is that of community. Early section of this chapter, under the title "the Great Lakes Regime," the formal governance structure within the Great Lakes regime, is outlined. In most regimes, scholars and students focus on these formal arrangements that often emanate from the legal and policy framework However, one of the endearing features and unique aspects of the Great Lakes is the sense of community that has evolved over the 40 or so years of the regime. The sense of community has been defined as follows:

[14]Statutes of Ontario, 2006, Chapter 22.

> The kind of basin-wide community that developed among environmental organizations also can be found among scientists and scientific institutions, the funding institutions, and industry. Within each community of interest, the Great Lakes are the organizing theme and focus of efforts that often transcend national borders and cultural differences. By the late 1980s, Canadians participated in lobby days in Washington, and Americans could be found engaging Canadian policymakers in Toronto and Ottawa on Great Lakes issues (Botts and Muldoon 2005).

The importance of community is multi-fold. First, it provides a mutually reinforcing exchange of information among the different sectors: industry becomes more cognizant of the concerns of public interest groups, who in turn begin to appreciate the limits and constraints of business decisions; all sectors become more aware of the science behind the issues and where there is less or more uncertainty with respect to those issues. The development of a community does not necessarily mean that there is more agreement among the parties, but it may mean there is greater understanding of the respective positions and the basis for more informed, balanced arguments before decision makers. It has also been noted that:

> Few commentators would question that the Great Lakes scientific community evolved the understanding of complex large lake science to a point where the Great Lakes region was viewed as a laboratory for ground-breaking research. Not only did the community evolve the science, the community also evolved the responses to challenges identified by fashioning innovative new approaches… (Botts and Muldoon 2008).

Second, the notion of community is a mechanism for accountability. One sense of accountability is that all the different stakeholders will hold government accountable first of all, to make appropriate and timely decisions, and second, to make decisions in full view of the stakeholders. The notion of community also holds stakeholders within the community more accountability in the sense that one stakeholder will make it known if the other has unfair access to decision-makers or presents misleading information.

Further, there is a broader benefit of the development of a community within a regime as commentators describe:

> …the formal regime and the community have developed a synergy that is the most important factor contributing to the regime's effectiveness in improving Great Lakes water quality. That is, a strong and diverse nongovernmental community developed as a result of the formal structures, and the formal structures have been energized and legitimized because of the continuing active involvement of that community (Valiante et al. 1997).

The notion of community within the regime is an unrecognized asset. This asset should be carefully reviewed as the Agreement is being renegotiated to ensure that the notion of community is not only recognized but enhanced in light of the positive attributes it has shown over past decades.

CONCLUSION

The Great Lakes regime continues to be in transition. It must respond to both the existing and emerging ecological stresses as well as the ever changing political and social priorities. The regime has demonstrated enormous innovation and success over the years, in part, because it has evolved to respond to changing circumstances. There is no doubt, however, as the U.S. and Canadian governments continue to renegotiation a new GLWQA, the regime is at cross roads and it will be of great interest to assess whether the changes to the regime will allow it to regain its effectiveness to respond to the needs and stresses of the upcoming decades.

REFERENCES

Bails, J., A. Beeton, J. Bulkley, M. De Philip, J. Gannon, M. Murray, H. Regier and D. Scevia. 2005. Prescription for great lakes ecosystem protection and restoration: avoiding the tipping point of irreversible changes, prepared for the National Wildlife Federation, available at http://online.nwf.org/site/DocServer/prescriptionforgreatlakes_1_.pdf?docID=2621

Becker M.L. 1993. The International Joint Commission and Public Participation: Past Experiences, Present Challenges, Future Tasks Natural Resources Journal 33: 235.

Bilder, R.B. 1972. Controlling great lakes pollution: a study in United States - Canada environmental cooperation. Michigan Law Rev. 70: 469-556.

Billups, S., T. Eder, J. Jachsm, P. Muldoon and M. Murray. 1998. Treading water: A review of government progress under the Great Lakes water quality agreement (part I) - a report to the International Joint Commission. Toledo J. of Great Lakes' Law, Science and Pol. 91: 245-297.

Botts, L. and B. Krushelnicki. 1995. The Great Lakes: An Environmental Atlas and Resource Book. 3rd ed. Great Lakes National Program Office, U.S. Environmental Protection Agency, Chicago, USA.

Botts, L. and P. Muldoon. 2005. Evolution of the Great Lakes Water Quality Agreement. Michigan State University Press, East Lansing, Michigan. USA. pp. 2-191.

Botts, L. and P. Muldoon. 2008. Using the Boundary Waters Treaty for the 21st century: revitalizing the Great Lakes governance regime. Wayne Law Rev. 54(4): 1553-1570.

Colborn, T., A. Davidson, S. Green, R.A. (Tony) Hodge, C. Ian Jackson and R. Liroff. 1990. Great Lakes great legacy? The Conservation Foundation and the Institute for Research on Public Policy, Washington and Ottawa.

Colborn, T. and D. Dumanoski and P. Myers. 1996. Our Stolen Future. Dutton, NewYork, USA.

Dempsey, D. 2004. On the Brink - The Great Lakes in the 21st century. Michigan State University Press. East Lansing, Michigan, USA.

Dworsky, L.B. The Great Lakes: 1955-1985. *In:* L.K. Caldwell [ed] 1988. Perspectives on Ecosystem Management - A Reader. SUNY, Albany, N.Y. USA. pp. 59-113.

Environment Canada and United States Environmental Protection Agency (Great Lakes National Program Office). 1995. An Environmental Atlas and Resource Book. 3rd ed. Accessed at: http://www.epa.gov/glnpo/atlas/index.html

Environment Canada, Great Lakes. 2003. Fact sheet: fish and wildlife health effects in the Canadian Great Lakes areas of concern.

Environment Canada. 2006. How are the lakes doing? Accessed at: http://www.ec.gc.ca/grandslacs-greatlakes/default.asp?lang=En&n=9889A192-1#a5

Environmental Protection Agency. Great Lakes Monitoring. Great Lakes Factsheet No. 1: Physical Features and Populations. Accessed at http://www.epa.gov/glnpo/atlas/gl-fact1.html

Environmental Defence and Canadian Environmental Law Association [PollutionWatch] 2006. Partners in Pollution: An Assessment of the Continuing Canadian and United States Contribution to Great Lakes Pollution available at http://www.cela.ca/files-/uploads/GLreptFeb06.pdf]

Farid, C. and J. Jackson and K. Clark. 1997. The Fate of the Great Lakes Sustaining or Draining the Sweetwater Seas? Accessed at: http://www.cela.ca/sites/cela.ca/files/fate-of-the-lakes.pdf

International Joint Commission. 2006. 13th Biennial Report on Great Lakes Water Quality p. 1.

Great Lakes Cities Initiative. http://www.glslcities.org/aboutus.cfm

Great Lakes Information Network. Facts and Figures. Accessed at: http://www.great-lakes.net/lakes/ref/lakefact.html

Hall, N. 2009. Introduction: Canada – United States Transboundary Environmental Protection. 26 Windsor Review of Legal and Social Issues 1.

Jackson J. and K.K. Sloan. 2008. A Way Forward: Strengthening Decision-Making and Accountability under the Great Lakes Water Quality Agreement Great Lakes United.

Krantzberg, G. and J. Manno and C. de Boer 2007. Great Lakes St. Lawrence Governance: Report on the Expert Workshop.

Manno, J. 1993. Advocacy and Diplomacy in Negotiating the Great Lakes Water Quality Agreement Buffalo Law Journal 1, No. 1.

Munton, D. Paradoxes and Prospects. p. 88. *In:* R. Spencer and J. Kirton and K.R. Nossal [eds.] 1981. The International Joint Commission Seventy Years On. Centre for International Studies, Toronto, Canada.

Nowland, L. and K. Bakker 2010. Practising Shared Water Governance in Canada: A Primer UBC Program on Water Governance.

Pearse, P.H. and et al. 1985. Currents of Change: Final Report of the Inquiry of Federal Water Policy Ottawa, Canada. p. 27.

Prescription for a Healthy Great Lakes. 1988. National Wildlife Federation and the Institute for Environmental Law and Policy.

Regier, H. 1986. Progress with Remediation, Rehabilitation and the Ecosystem Approach Alternatives 13, No. 3: 54-64.

Sanderson, M.1993. Climate Change and Water in the Great Lakes Basin. Canadian Water Resources Journal 18: 4.

Scharm Stokke, Olav. Regimes as Governance Systems. pp. 27-63. *In:* Global Governance – Drawing Insights from the Environmental Experience. 1997. MIT Press, Cambridge, Mass. USA.

Schaefer, M. 2004. Multiple Actors in Canada – U.S. Relations: The Role of Non-Governmental Organizations in Canada – U.S. Relations 30 Canada – U.S. Law Journal 69.

Standing Senate Committee on Foreign Affairs. 1972. Canada – United States Relations: The Institutional Framework for the Relationship Canadian Senate, Ottawa, Canada. p. 42

Stockholm Convention on Persistent Organic Pollutants (POPs), available at http://chm.pops.int

Valiante, M., L. Botts and P. Muldoon. Ecosystem governance: lessons from the Great Lakes. pp. 197-225 *In:* Oran R. Young [ed.] 1997. Global Governance – Drawing Insights from the Environmental Experience, MIT Press, Cambridge, Mass., USA.

U.S. Environmental Protection Agency and Environment Canada, 2005. State of the Great Lakes.

U.S. Environmental Protection Agency. Great Lakes factsheet no. 1: physical Great Lakes monitoring, features and populations. Accessed at http://www.epa.gov/glnpo/atlas/gl-fact1.html

U.S. Environmental Protection Agency and Environment Canada. 2007. State of the Great Lakes highlights.

CHAPTER 4

Innovations in Decision Making

Chris McLaughlin

Walter G. Booth School of Engineering Practice, McMaster University,
Hamilton, Ontario, Canada L8S 4K1
E-mail: chris.mclaughlin@mcmaster.ca

> A philosophic reconstruction which should relieve men of having to choose between an impoverished and truncated experience on one hand and an artificial and impotent reason on the other would relieve human effort from the heaviest intellectual burden it has to carry.
>
> John Dewey, 1920
> *Reconstruction in Philosophy*

Deficiencies in how we conceptualize and manage collaborative decision processes are "intellectual burdens" on Great Lakes governance and the reform of policy to cope and learn effectively in the face of uncertainty. Relieving these burdens will certainly require a continued focus on the content of science and policy. But my focus is not on the *what* of our understanding, but on the *how*. Substantive research into the ecological knowledge needed for Great Lakes restoration fills journal pages and conference agendas. Yet despite that increased understanding, "there is widespread agreement that the Great Lakes presently are exhibiting symptoms of extreme stress from a combination of sources that include toxic contaminants, invasive species, nutrient loading, shoreline and upland land use changes, and hydrologic modifications," and that a "failure to understand the ecosystem-level disruptions caused by the combination of multiple stresses have led to the false assumption that the Great Lakes ecosystem is healthy and resilient" (Bails et al. 2005). In such circumstances, the content of research is "impoverished and truncated" unless matched to science-policy processes that are more closely attuned to the properties of complex Great Lakes social-ecological systems and capable of making use of that research content to the greatest effect.

The processes of management and decision making involved in Great Lakes restoration are at once political, bureaucratic, scientific, and social. However, these processes now occur on a scale of complexity that has outgrown the capacity of traditional management of segregated resource uses and narrow agency mandates (Scholz and Stiftel 2005). As such, new institutional models are needed that are able to accommodate emerging conflicts and complexity more effectively. My main contention is that such models must become more adaptive to better reconcile the complexity of competing priorities and their underlying uncertainties.

An adaptive governance model is one designed to resolve collective action problems that occur between different types of resource users in ways that lead to sustainable uses of natural systems (Scholz and Stiftel 2005). Such models make use of techniques to facilitate a broad participation in problem definition and decision making, and deal with uncertainties about both the natural systems being managed and the human organizations themselves. Dewey (1920) said that reason is not "something laid from above upon experience," but rather it is "employed through inventions in a thousand ways to expand and enrich experience." To that end, the focus of this chapter is on two such inventions – adaptive management and decision analysis – proposed to have potential to "expand and enrich" Great Lakes governance. My purpose is to clarify the principles and fundamental character of these two techniques as a means to developing an institutional framework for participatory and adaptive governance in the Great Lakes that more directly links science to policy and management activity to program outcomes.

GOVERNANCE

Critiques of current shortcomings in Great Lakes governance have come from academia (e.g., Manno and Krantzberg 2008, Krantzberg 2009, MacDonagh-Dumler 2009), civil society (e.g., Botts and Muldoon 2005, Benevides et al. 2007, Jackson and Sloan 2008), a combination of both (e.g., Johns et al. 2005, Barr et al. 2010), and from government authorities (OAG 2001, GAO 2003, ARC 2007). Commonalities among proposed remedies with relevance for this discussion include stronger accountability, greater flexibility, and public engagement that is sustained and more meaningful. Each of these demands that we more effectively address and incorporate the diversity of views and information now required of evolving pluralist governance models. More specifically, Manno and Krantzberg (2008) stipulate that "governance for sustainability" includes active public participation, integrated management, multi-jurisdictional collaboration, and a shared sense of responsibility. They suggest that this mix of ingredients was recognized at the outset of the regime formed four decades ago around the binational Great Lakes Water Quality Agreement, and yet a decline in the effectiveness of governance in the Great Lakes has resulted nonetheless. Sources of institutional stress leading to that decline include ideological change, scale mismatches of problem and solution, and inadequacies in funding, monitoring, and program evaluation. Further, recent analyses have demonstrated significant deficits in Agreement-related implementation (i.e., numerous unmet policy objectives), with related underlying problems including non-existent program evaluation, program incompatibility, lack of coordination among programs, authorities that establish and then abandon public participatory initiatives, and disincentives from the inappropriate choice of policy mechanisms or inadequate level of support for appropriate mechanisms (McLaughlin and Krantzberg 2011, 2012).

The linkages between science, policy, and management have become increasingly complex in recent decades (or at least our perception of those linkages has sharpened). A major reason for that complexity is the emergence of "boundary organizations" and other increasingly nuanced arrangements that orchestrate activities in the worlds of science and policy and that mediate complex mixtures of discourse and subjectivity (Miller 2001). In the Great Lakes, these arrangements include a long-standing and high-functioning non-governmental sector, and innovations in policy ranging from an ecosystem approach institutionalizing the people-in-nature view, to dozens of local Remedial Action Plans (RAPs) in the most degraded locations that involve diverse stakeholders. But even the Agreement's most laudable features are failing. For example, a lack of careful definition in the application of the ecosystem approach under the Agreement has seen the concept misused (Manno and Krantzberg 2008). According to the International Joint Commission's (IJC 2003b) most recent program assessment, "key challenges" remaining for the RAP program are securing the resources to implement the plans, identifying accountability and responsibility, defining restoration targets where they do not exist, setting priorities, and monitoring recovery – essentially the entire program.

Great Lakes governance must embrace opportunities to invigorate the processes of management and develop new frames of reference for treating uncertainty and engaging with people. Such processes must complement traditional decision making (Wondolleck and Lewis 2000), but at the same time enhance those traditions through reciprocally responsible action (Westley 1995, Regier et al. 1999). A key to such a renewal of governance is learning. Later in this chapter I discuss a specific type of learning related to uncertainty that is accomplished by combining adaptive management and decision analysis, but for now the focus is more broadly on policy learning as an outcome of systems thinking. Policy learning provokes "a relatively enduring alteration of thought or behavioural intentions" that are related to (and perhaps revise) "the precepts of a policy core belief system" (Sabatier 1993). Figure 4.1 is a model of this idea of policy learning as it could occur in a complex problem environment. The process illustrates where fundamental changes in governance could both foster and reflect a significant new conceptualization of social-ecological complexity and new approaches to cope and learn effectively in the face of uncertainty. Figure 4.1 illustrates that thinking about the problem situation as one system has the potential to more broadly consider the elements and issues that define it, and that this would also stimulate a more sophisticated portrayal of the behaviour of those elements, particularly where relationships are evolving. Westley (1995) suggests that this is the essence of management itself, to synthesize a diversity of old knowledge and form new perspectives, and that

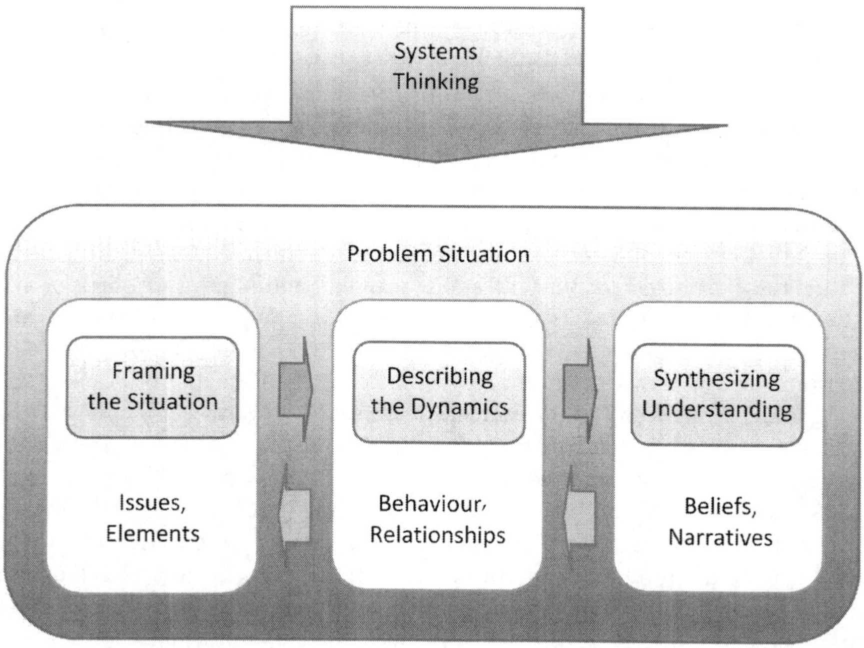

Fig. 4.1 Three phases in addressing a problem situation with systems thinking (created from text in Waltner-Toews et al. 2008). The application of a systems thinking perspective to a problem situation involves three phases. To frame the situation is to describe or map elements and issues that define the situation. A description of the dynamics of a situation portrays the behaviour of those elements, particularly where relationships evolve. And synthesizing understanding from the first two phases can allow for both resolution and new perspectives and questions. The direction of the darker-shaded arrows in the problem situation indicates the flow of phases one through three. The feedback created by a new understanding and possible re-examination is indicated by the lighter-shaded arrows in the problem situation.

done well, this is also the essence of leadership. The cycle continues in that the reformation of perspectives, behaviours, and relationships leads us to reconsider the elements and issues that we use to possibly redefine the problem situation.

DECISION MAKING

A primary reason for our traditional misconceptions of social-ecological processes and behaviour is our tendency to characterize them simplistically as linear and straightforward (Holling 1995). But where complex processes such as *deciding* may give the impression of a singular act, decisions usually result from an ongoing determination involving priorities and judgments in a stream of new and usually unpredictable information. This is true for both individual and collective decision making processes (Beratan 2007). This section briefly considers individuals in a sketch of decision making, and makes a few observations about people and their ability to influence collective decisions.

Individual Decisions

Cognitive processes determine how we absorb information, recognize problems, and imagine solutions. Beratan (2007) explains that the human mind is thought to have a dual process of reasoning where a rapid nonconscious or preconscious mode is accessed effortlessly to contextualize the present at any given moment. Such access is automatic and comprises the vast majority of what our brains are doing continuously. Processing in this nonconscious mode is based on an individual's background beliefs and learning history. By default, we base decisions in the present on the assumption that the outcome will be similar to past results of similar decisions in similar situations. There is also a second, parallel mode of conscious thought that is effortful, but comprises as little as 2% of brain activity (Beratan 2007). We rely on this mode to deal with novelty or to strategize beyond what is routine, but accessing this mode requires motivation and cognitive capacity. Our nonconscious mental mode processes quickly and efficiently, but for that we sacrifice flexibility and creativity. Beratan (2007) explains that the human mind integrates new information gained from conscious processing into our experiential knowledge, and thus such learning then becomes available for fast future use by the nonconscious mode.

Beyond the architecture of the mind, people employ a "bounded rationality" to make decisions that are reasonable given the limited and imperfect quality of the information usually available (Simon 1972). We are not rational optimizers, but rather we make "satisficing" decisions where we attempt to *satisfy* our needs by accounting *sufficiently* for whatever information we can acquire (Simon

1957). The idea of satisficing is captured by the 80:20 rule, which states that endless detail is most often not necessary and that you can generally achieve 80% of a desired output with 20% of the effort. This concept is applicable for management, whereby the top few highest ranked priorities are selected and managers invest the initial 20% of effort needed to solve a problem completely. The hope is that the majority of desired correction (80%) does indeed come from the first 20% of effort, thereby allowing the manager to deploy what would have been a largely unproductive effort to more productive output elsewhere (B. Feltmate, personal communication). A difficulty for adaptive governance is in the application of this concept more broadly to management decisions. What might make sense for one individual in a social system is often detrimental within a broader context where that decision undermines the condition of the wider system (Meadows 2008). The challenge therefore will be to use the techniques of adaptive governance to resolve the most contentious issues most effectively.

Collective Decisions

The history of finding balance in natural resource management can be described largely as one of *command and control* (Holling 1995). This approach typically targets a single environmental variable (such as the concentration of phosphorus, for example) and management activities are directed at maintaining it at an "optimal" level. The success of command and control methods requires significant assumptions about the characteristics of ecological systems, that they are "well-bounded, clearly defined, relatively simple, and generally linear with respect to cause and effect" (Holling and Meffe 1996). These methods almost always hit a point of diminishing returns (or worse) in a world that is instead found increasingly to be complex, nonlinear, and unpredictable – and management is left therefore with best judgments, not certainties (Holling 2001). Whether acknowledged explicitly or not, a reliance on judgment in the face of uncertainty is present in all management decisions and actions, and is the fundamental reality that underpins the concepts and methodologies described in this chapter.

There is an increasing appreciation and concern for the role of uncertainty in the management of dynamic environments in the Great Lakes (e.g., Bails et al. 2005). Holling (1973) first replaced the traditional equilibrium notion of resilience as stable ecological systems, with a view of resilience as the capacity for ecological function to persist in the face of change. Holling proposed that "resilience determines the persistence of relationships within a system and is a measure of the ability of these systems to absorb changes of state variables, driving variables, and parameters, and still persist." The complexities and resilience of such environments has attracted considerable attention as a perspective for understanding the dynamics of social-ecological systems (e.g., Gunderson and Holling 2002, Folke 2006, Norberg and Cumming 2008), particularly with respect to the design of governance institutions (e.g., Lebel et al. 2006).

Decision making is an intensely political process, however, and an appreciation of complexity inherent to a problem situation can be severely constrained by competing interests (Miller 1999). In this social context, the descriptions of individual conscious and nonconscious decision processes have parallels to collective processes: institutional decision making is habitual, and increased flexibility therefore comes only with significant effort (Miller 1999, also Holling and Meffe 1996). In addition, an increasing heterogeneity of stakeholders in environmental management (although not all involved in decision making) has resulted in an increase of the "intellectual and interpersonal complexity" facing those responsible for decision making (Miller 1999). Further still, complex environmental decisions that involve social-ecological interactions also combine the characteristics of complexity with a continuum in time and space where feedbacks continually alter conditions as we attempt to account for them. The behaviour of complex systems includes non-linear dynamics and self-organizing properties that create inherent uncertainties (Gunderson 2000). Decisions taken in the face of uncertainty, however, can demonstrate the flexibility with which people respond to continual (if only slight) changes in task conditions that affect their preferences, judgments, and choices among alternative actions (Payne et al. 1993). The choice of governance model or policy tool can either suppress or leverage that flexibility.

PRACTICES FOR A BETTER UNDERSTANDING

Despite the Agreement's policy innovations and measured success at improving water quality in the Great Lakes, significant shortcomings in implementation risk its "irrelevancy" (Botts and Muldoon 2005, also Krantzberg 2009). At the same time, symptoms of extreme stress in the Great Lakes from a combination of regional and global sources, if not addressed with "great urgency," pose "potentially irreversible" harm (Bails et al. 2005). This assessment of environmental and policy realities is not unique to the Great Lakes but follows a global trend of fraying ecologies, insufficient outcomes, and uncertain futures (MA 2005, also Gunderson et al. 1995). Such assessments have identified traditional command and control approaches to management as undermining resilience in social-ecological systems such as the Great Lakes. This suggests that those traditions are inadequate for making significant progress towards meeting the purpose of the Agreement to restore and maintain the integrity of the waters of the Great Lakes Basin ecosystem. Most significantly for the Great Lakes, jurisdictional and administrative myopia create a social fragmentation that limits our ability to fully realize the potential of an ecosystem approach to addressing the complexities of ecological problems and their underlying human provenance (Francis and Regier 1995).

Societal responses to environmental decline in preceding decades, as expressed through governmental policy and management, have passed through three overlapping epochs. These are characterized by command and control regulation and compliance, efficiency-based reforms that introduced preventative approaches, and most recently a more comprehensive inclusion of social elements into management (Mazmanian and Kraft 2009). The history of the Great Lakes regime has mirrored this trajectory, from initial end-of-pipe pollution management through to a model of governance "haltingly evolved" to relate more responsibly to broader ecosystemic properties (Regier et al. 1999). Significant next steps in this evolution must be in the direction of adaptive governance (Manno and Krantzberg 2008, Folke et al. 2005). The third epoch represents an "Emerging Era" that is self-organizing and open (Regier et al. 1999), characteristics that traditional management frameworks will be unable to comprehend and reconcile (Lubell et al. 2009, Moran 2010). Devolved power, integration, and collaboration are characteristic of the Emerging Era – qualities requiring a model of governance with more "social responsiveness to ecosystem dynamics" (Folke et al. 2005) and a new role for science that probes uncertainty and facilitates social learning (Regier et al. 1999).

People learn in order to acquire a better understanding of some thing or event. According to Argyris (1993), people organize (in the form of natural resource management agencies, for example) in order to act and to accomplish intended goals (e.g., policy outcomes) that represent or contribute to preferred states of nature (that individuals strive to "satisfice" by acting). These preferences can be inferred through observation to drive and guide those individual or collective actions (and are not the underlying beliefs people espouse). Figure 4.2 illustrates Argyris's (1993) two learning cycles defined by the degree of reflection and fundamental change provoked by learning. Single-loop learning occurs when there is a match between intentions and outcomes, or a mismatch is corrected by changing actions. Double-loop learning occurs when mismatches between intentions and outcomes are corrected by examining and altering the governing variables (i.e., underlying preferences and beliefs) and then the actions.[1] Argyris (1993) is careful to emphasize that learning may not be said to occur through the discovery of a new problem or the invention

[1]It may be that societal transformation in the Emerging Era only results from triple-loop learning (Hargrove 2002, also Pahl-Wostl 2009), but because organizational change is overwhelmingly informed by a single-loop (Argyris 1993, also Parson and Clark 1995, Miller 1999), a double-loop reframing of Great Lakes issues, especially where those issues constitute barriers to restoration would be considered a paradigmatic change in the character of our collective mode of understanding.

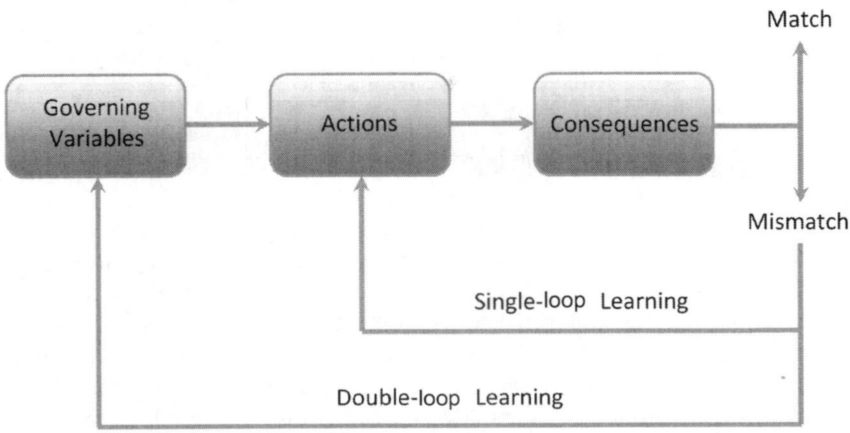

Fig. 4.2 Single-loop and double-loop learning (adapted from Argyris 1993; reproduced with permission of the author). People organize (in the form of natural resource management agencies, for example) in order to undertake collective action to accomplish goals. Governing variables are preferred states that individuals strive to "satisfice" by acting. Single-loop learning occurs when there is a match between intentions and outcomes, or a mismatch is corrected by changing actions. Double-loop learning occurs when mismatches between intentions and outcomes are corrected by examining and altering the governing variables and then the actions. Learning occurs only when a match or mismatch is produced.

of a solution to a problem. Learning occurs when the invented solution is actually put into practice – an important distinction implying that discovering problems and inventing solutions are necessary but not sufficient conditions for learning.

It is significant, therefore, that language added to the Agreement in 1978 instructs the Great Lakes institutional regime, and the federal governments as its signatories in particular, to create such learning. The revisions introduced the often repeated purpose of the federal governments of Canada and the United States: "to restore and maintain the chemical, physical, and biological integrity of the waters of the Great Lakes Basin Ecosystem." To achieve this, however, requires the often overlooked remainder of the purpose, that the governments are "to make a maximum effort to develop programs, *practices* and technology necessary for a *better understanding* of the Great Lakes Basin Ecosystem..." (emphasis added; see McLaughlin and Krantzberg 2006). I draw this out to emphasize the relationship between learning and practice in the purpose of the Agreement. The phrase "better understanding" acknowledges explicitly the principal insight of Argyris (1993), that a better understanding of the Great Lakes system (i.e., learning) will result from a continual and maximal effort to refine our practice in the service of new knowledge. Finally, learning in this context is not necessarily limited to technical exercises; the concept is widely described as "a set of frameworks, ideas, tools, information, styles, language, stories, and documents" shared by members of a community, be that community defined by technical competencies or more broadly (Wenger et al. 2002).

ADAPTIVE MANAGEMENT

Renowned American forester-philosopher Aldo Leopold (1930) followed in the pragmatic tradition of John Dewey in foretelling adaptive management with the idea of learning through experience in natural resource management: "We conservationists... have many ideas as to what needs to be done, and these ideas quite naturally conflict. We are in danger of pounding the table about them, instead of going out on the land and giving them a trial." This insight is only one of the reasons why Leopold was ahead of his time, and in many ways, is still ahead of our time.

Adaptive management (AM) is principally about learning – but a particular type of learning that Crawford et al. (2005) describe as *the reduction of key uncertainties through management*. A key uncertainty is a "knowledge gap that prevents selection of the suitable management option from alternatives because the impact of various management actions on indicators cannot be predicted with adequate certainty" (AMET 2003). In this light, the goal of adaptive management is ultimately to fill that knowledge gap.

What AM Is and Not

Every key uncertainty in management is a potential learning opportunity (Grumbine 1994, Murray and Marmorek 2003b). The practice of adaptive management can provide reliable information to managers and policy makers about "what works and what does not work" (Stolnack et al. 2005) because policies are intentionally designed as hypotheses and management is implemented as experiments to test the understanding on which the policies were developed (Holling[2] 1978, Walters 1986, Lee 1993). Crawford et al. (2005, 2010) provide a tidy history of AM as a series of evolving operational frameworks (Lancia et al. 1996, Schreiber et al. 2004, and Gregory et al. 2006 are also similarly suggested). For example, Walters (1986) included a structured description that formalized the links between science and policy. Walters described explicitly how the cyclic character of scientific inquiry (i.e., hypotheses, predictions, tests, and probabilities) was consistent with the cyclic character of policy decision making where "designing adaptive management strategies appears to involve four basic issues:

[2]C.S. Holling was editor but not sole author of this seminal text. His co-authors are almost always overlooked: A. Bazykin, P. Bunnell, W.C. Clark, G.C. Gallopin, J. Gross, R. Hilborn, D.D. Jones, R.M. Peterman, J.B. Rabinovich, J.H. Steele, and C.J. Walters.

1. Bounding of management problems in terms of explicit and hidden objectives, practical constraints on action, and the breadth of factors considered in policy analysis;
2. Representation of existing understanding of managed systems in terms of more explicit models of dynamic behaviour, that spell out assumptions and predictions clearly enough *so that errors can be detected and used as a basis for further learning* (emphasis added);
3. Representation of uncertainty and its propagation through time in relation to management actions, using statistical measures and imaginative identification of alternative hypotheses (models) that are consistent with experience but might point toward opportunities for improved productivity;
4. Design of balanced policies that provide for continuing resource production while simultaneously probing for *better understanding* and untested opportunity" (emphasis added).

Adaptive management generates learning because management programs are designed to identify and reduce key uncertainties about resource dynamics by iteratively using feedback information (to inform policy redevelopment) from the system being managed. Figure 4.3 illustrates this analogy between the policy cycle and the scientific method – a fundamental challenge to (and the basis on which to overturn) the traditional perception of a policy as constituting an answer, rather than being itself a question.

Active AM imposes rigorous procedures and expectations on managers and participants. Nudds et al. (2003) sketch the basic tenets of this active approach: models of the system under study are developed to explicitly incorporate sources of uncertainty; the development of a management plan that postulates each of those uncertainties as testable hypotheses maximizes available results and potential learning opportunities; and monitoring of implemented management options is used to evaluate the expected performance of those choices. The results of this process are used to adapt and learn – to essentially complete the feedback loop whereby the uncertainties associated with future choices are reduced with the use of new knowledge. It is this final point – completion of the feedback loop with reliable knowledge that informs management – that is a critical and distinguishing feature of active adaptive management. AM is not an attempt to necessarily identify the "best" policy option, but to "embrace uncertainty" by using models and analysis to identify a set of candidate policy options that are all defensible and to eliminate options that are likely inadequate to meet management goals (Walters 1986). Ideally these candidate policy options are each tested by applying them to the managed system as a set of experimental treatments, either sequentially over time or across a set of experimental locations. The U.S. Department of the Interior (DOI; Williams et al. 2009) has refined a clear articulation of adaptive management in a standard protocol that follows these steps:

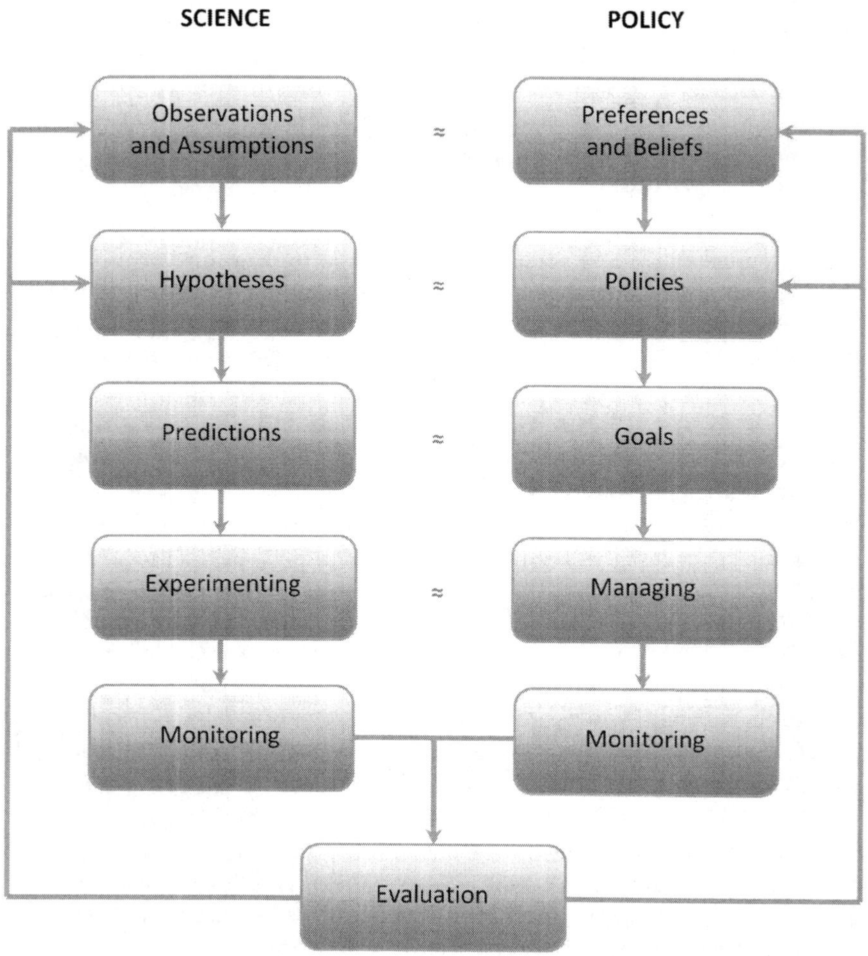

Fig. 4.3 Policy as hypothesis, management as experiment (adapted from Nudds 1999, Crawford et al. 2005; reproduced with permission of the authors). The columns contrast steps of the conventional scientific method with stages of the policy cycle. Policies are formulated to achieve certain goals. Once adopted, those goals direct management actions to be taken, and monitoring informs policymakers on the effectiveness of those actions at meeting those goals. In this approach, resource management as practiced has the potential to become an experimental test of policies. The monitored outcomes of management activities enter a feedback loop that provides reliable knowledge for the potential reformulation of those policies in a manner that parallels how experimental results in science provide for a reconsideration of assumptions and hypothetical explanations of the phenomena under study. Applied to policy cycles in complex systems, this approach to management presumes that policy goals are questions, not answers.

1. involve stakeholders;
2. establish objectives;
3. identify potential management actions;
4. identify models (i.e., hypotheses) about cause-effect;
5. identify and design monitoring plans;
6. select management actions;
7. monitor system responses;
8. assess predicted changes against observed changes; and
9. iterate through the adaptive management cycle.

Crawford et al. (2010) suggest that "it would be difficult to find a more succinct and structured presentation of adaptive management principles and processes for use by resource management agencies."

Alternatively, *passive* AM typically employs historical data from the managed or comparable areas, and a "best guess" hypothesis and preferred actions are developed. Outcomes are monitored and the data are updated with the new information. Gregory et al. (2006) note that this approach can make good sense when uncertainty is low, or when regulatory constraints (on ecological variability) are high – but in practice, they explain, passive AM is often simply basic trial and error learning that is slow and superficial due to an absence of explicit hypotheses and adequate monitoring, and is without clear implications for management practices. There is a real potential for error, they warn, "because complex interactions and cumulative effects may confound results and analyses." Whether active or passive, adaptive management is done when the simultaneous goals of achieving management objectives and acquiring reliable knowledge are accomplished (Lancia et al. 1993, 1996) and new knowledge is incorporated purposefully into a decision process that allows for the reduction of uncertainty and therefore learning.

As a tool to learn deliberately from the outcomes of management action, adaptive management is intuitively attractive. To employ a scientific approach to managing ecosystems in the face of unwieldy uncertainty is impressive and "easy to understand – almost" (Lee 1993). *Almost* because adaptive management is widely misrepresented, and often misunderstood fundamentally (Lee 1993, Murray and Marmorek 2003a, Schreiber et al. 2004, Allan et al. 2008). Common misconceptions about adaptive management include that it is (after Murray and Marmorek 2003a):

- trial-and-error (i.e., without a directed research component to management);
- spurious adaptation of policies as you go (simply in reaction to unfolding events);
- sophisticated modelling skills and tools alone;

- consensus from all stakeholders (even when informed by information feedbacks),
- something only scientists do; or
- a panacea that can solve all problems.

Despite the continued evolution of step-wise procedures and the growing enthusiasm for the process of *learning while doing*, adaptive management "has been much more influential as an idea than as a way of doing conservation" (Lee 1999, also Allan and Stankey 2009). In the Great Lakes that enthusiasm for adaptive management is present in plans to develop a nearshore framework to address some of the most acute ecological degradation that occurs at the human-environment interface. Indeed, the "cornerstone" of this framework is to be the "adoption of an adaptive-management strategy" (IJC 2009b).

When to Use AM and Not

Argent (2009) states that successful adaptive management requires "clear objectives, data and knowledge, the right participants, science skill, willing partners, and money and time." But because uncertainty and unpredictability in social-ecological systems can also produce political, social, and economic surprises, we must be prepared to rationalize the use of any management tool in a broader context, including the use of adaptive management. Gregory et al. (2006) suggest that because adaptive management is as widely misunderstood as it is widely promoted, its track record is weak and many adaptive management initiatives exhibit few or none of the characteristics generally considered essential (see also Lee 1999, Walters 2007). There are legitimate examples of adaptive management, according to Gregory et al. (2006), but "other cases share little more than the same name." Or as Crawford et al. (2010) suggest: "When in doubt regarding different ideas called 'Adaptive Management', we strongly recommend that the reader keep their attention focused on the essential role of hypotheses (possible cause-effect explanations, models) and predictions in the AM learning cycle; if these concepts are absent or vague in the discussion, then – in our opinion – the authors are not discussing Adaptive Management."

Gregory et al. (2006) suggest that the misapplication of AM techniques – the "dark side" of the attraction to AM – means that projects are often not clear on what constitutes learning or how much of it is required. Further problems include that the initiation of an AM program will allow for postponement of difficult decisions that hide behind resource constraints and scientific uncertainty, and that AM will result in the design of large, long, costly experiments that come at the expense of other important environmental, social, or economic objectives. The British Columbia Forest Service successfully used adaptive management to evaluate forest-harvest techniques with limited geographical scope beyond the treatment locations (Sit and Taylor 1998).

Schmiegelow and Hannon (1993) and Schmiegelow et al. (1997) describe an experimentally-manipulated commercial forest harvest in Alberta over large spatial scales. Similarly, AM programs in the Florida Everglades (Walters et al. 1992) and the Columbia River Basin (Lee 1993) have been conducted over large geographical regions. In fact, Williams and Johnson (1995), Williams et al. (1996), and Johnson et al. (1997) document a North America-wide AM program for the renewable harvest of mallard ducks.

Formal and thoroughly-documented examples of adaptive management in the Great Lakes are rare and usually related to fisheries (Jones 1999, M. Jones, personal communication). Walters and Green (1997) reported that of 25 major AM planning exercises that he had participated in, only seven resulted in experiments of an appropriate scale, and only two could be considered well-planned in terms of statistical design. The others, he noted, either "vanished with no visible product" or became "trapped in an apparently endless process of model development and refinement." How are the managers, analysts, and participants of future adaptive management initiatives to avoid being likewise misled or unproductive? Gregory et al. (2006) suggest that the most significant obstacle to successful adaptive management is the context for identifying and defining the problems to which AM might be applied, and that this obstacle may have more to do with the indiscriminate choice of contexts within which AM is applied than with AM itself. To help avoid this misapplication, Gregory et al. provide criteria to help practitioners decide on the appropriateness of using AM in a given management situation (Table 4.1). A full discussion of their criteria is beyond the scope of this chapter, but the criteria are provided here because they are an excellent overall sketch of the social-ecological context for adaptive management. Finally, in order to provide a brief look at AM in action, Box 4.1 provides a short overview of the use of AM in a science-policy problem situation involving the management of the invasive sea lamprey in the Great Lakes (based on Hansen and Jones 2008).

DECISION ANALYSIS

Decision analysis (DA) is principally about uncertainty. Lindley (2006) explains the principal feature of DA: in the event of a problem of which you are uncertain, "you have to do something, you have to act. Thinking about the act involves not only uncertainty... but also the possible consequences of your action." And so what we require is a measure of how desirable the outcome could be, a method "to explore the manner in which desirability and uncertainty may be combined to produce a solution to your problem. This method is called decision analysis because it permits you to analyze the manner in which you ought sensibly to decide." Put another way, decision analysis (DA) is a structured way to think about decisions, providing a procedural structure within which a decision

Table 4.1 Summary of proposed criteria for deciding whether and how to use adaptive management (AM) (adapted from Gregory et al. 2006; reproduced with permission of the Ecological Society of America).

Topic-area consideration	Criteria questions
Spatial and temporal scale	
Duration	Is the project timeline to obtain verified results compatible with management decision-making requirements?
Spatial extent and complexity	If spatial extent or complexity is large, are there opportunities to apply AM on a subset of the problem and scale up?
External effects	Have potential issues related to background trends and cumulative effects of management actions been addressed in the AM design?
Dimensions of uncertainty	
Parameter uncertainty	Has the AM design been pared down to focus on only those uncertainties most likely to influence management decisions?
Structural uncertainty	Are there profound structural uncertainties? If so, how will surprise outcomes be managed?
Stochastic uncertainty	How do low-probability random natural and other causal events affect the AM design and expected outcomes?
Confidence in assessments	If the confidence in the proposed AM design is low, can expert judgment or other techniques help?
Costs, benefits, and risks	
Specifying benefits and costs	Can all the costs and benefits (and risks) be documented and communicated in a manner understandable to all stakeholders?
Magnitude of effects	Will the information collected through AM have sufficient predictive ability to make a difference to managers?
Multiple objectives	Does the design and assessment of AM plans explicitly address the multiple goals of stakeholders (rather than only scientists)?
Perceived risks of failure	Can stopping rules and clear thresholds identify and/or minimize the perceived risks of failures, to species and to institutions?
Stakeholder and institutional support	
Leadership	Is there explicit policy guidance and leadership support for AM? Will stakeholders see AM as an effective way to deal with uncertainty?
Flexibility in decision making	Is there sufficient management flexibility (and continuity) to incorporate new information in revised experimental designs?
Avoidance of taboo trade-offs	Does the proposed AM design involve any trade-offs that might be considered taboo by some stakeholders?
Institutional capacity	Are sufficient analytical skills available (staff or contractors) to design, evaluate, and monitor AM plans?

maker can develop beliefs and feelings into those subjective judgments that are critical for a good solution (Clemen and Reilly 2001).

A decision may also involve several and varied uncertainties that may interact "in tangled ways" to determine ultimate consequences (Hammond et al. 1999). To be useful, uncertainty must therefore be simplified to capture essential information about its effect on possible choices by describing the key uncertainties, the possible outcomes of these uncertainties, the chances of each outcome occurring, and the consequences of each outcome. DA is intended to

BOX 4.1 Learning to manage sea lamprey.

The Great Lakes have proved vulnerable to the introduction of aquatic invasive species (AIS), or non-native plants and animals, many of which (such as the sea lamprey) have significant ecological and economic consequences. The IJC (2009a) estimates that there have been about 180 such introductions into the Great Lakes. Francis and Regier (1995) note that each of the Great Lakes has seen a widespread disappearance of most native trout and salmon species, with predation by the sea lamprey a primary cause of the extirpations (along with excessive predation by humans). The Great Lakes Fishery Commission (GLFC) was established in 1955 primarily to direct control of the sea lamprey, an eel-shaped organism that parasitizes other fishes. The GLFC is the major decision-making authority in sea lamprey management, and more than US 15 million is spent annually on controlling sea lamprey populations using periodic treatment of spawning streams with a lampricide. Generally, high uncertainty surrounds our limited knowledge of AIS biology, behaviour, and ecosystem impacts, however, complicating the relationship between AIS research initiatives and the development of AIS management protocols (IJC 2009a). The two sections below highlight aspects of recent Great Lakes research relating to AIS protocols that employed adaptive management and decision analysis. The purpose is to illustrate rare examples where management objectives inspired research design and the findings explicitly informed a rigorous reconsideration of the objectives.

Sea lamprey and adaptive management

Hansen and Jones (2008) developed and evaluated an alternative method (rapid assessment or RA) of assessing the extent and location of lampricide application. Sea lamprey life history does not require that every stream be treated each year in order for treatment to be maximally effective. However, natural variation in population characteristics makes it impossible to predict with certainty when each stream will require treatment. Their objective was to compare the performance of the RA method against the current assessment method (quantitative assessment sampling or QAS). They assumed that the RA method would be less accurate but less costly (due to fewer requirements of the sampling and ranking methods), and that assessment cost savings could be added directly to the treatment budget in order to treat additional streams. They hypothesized that allocating fewer resources in assessment would therefore mean greater suppression of sea lamprey across the Great Lakes. They provide a summary of the significant differences between QAS and RA, and descriptions of their larval sampling and stream ranking procedures.

Hansen and Jones determined that using the RA method to select streams for treatment would result in at least as much or more sea lamprey mortality than the QAS method. Their study is a rare example of adaptive management implemented properly because the two differing management activities (the RA and QAS assessment to inform lamprey treatment options) are posed explicitly as competing hypotheses with testable predictions. They do note that their

Box 4.1 contd. ...

Box 4.1 contd. ...

"experiment is not a traditional example of adaptive management, because assessment options rather than control options are being compared. However, because we compared assessment methods that have a minimal effect on the system being observed, we were able to apply both assessment methods to the same set of streams in each year of the study and directly comparethe results." Determining the current status of a population is a critical component of many resource management programs, they further note, and suggest that "this research is broadly applicable not only to sea lamprey management, but also to managed systems in general."

Moreover, Hansen and Jones report that beyond the insights from their research with direct linkages to future management, the GLFC (which had provided funding for the research) decided to adopt the RA protocol for ranking streams beginning in 2008, and that "in the spirit of adaptive management, managers and biologists will continue to monitor populations of sea lampreys and native fishes to assess the performance of RA and adjust the protocol as necessary to best meet management goals." Indeed, the GLFC saved approximately US $300,000 by employing the RA method rather than the QAS method that had formally been used, and similar savings have been realized during the years following and were then used to conduct additional lampricide treatments (Michael Siefkes, GLFC, Ann Arbor, Michigan, USA, pers. comm.). Hansen and Jones conclude by noting that "the use of adaptive management to test new methods of assessment and resource allocation is a means through which the optimal balance of resource demands can be determined and should be applied to other systems."

Sea lamprey and decision analysis

Haeseker et al. (2007) designed a decision analysis (DA) to assist the GLFC (which partially funded the research) in facing the challenge of developing feasible long-term control strategy for effective control of sea lamprey in the St. Marys River. The project went on the assumption that a DA "could provide a process for formally considering the uncertainties that made development of an optimal strategy so difficult" to reconcile with many competing management priorities.

Haeseker et al. followed the eight-step DA process outlined by Peterman and Anderson (1999) and also endorsed by Crawford et al. (2005). They convened a meeting of scientists, management agency personnel, and other stakeholders to describe the DA process and identify candidate sea lamprey control actions, and describe subsequent discussions of the attributes and costs (i.e., the uncertainties) of each possible action. Defining the management objectives proved more difficult, they report, although with a carefully selected group of experts (i.e., any stakeholder substantively engaged in the issue) they are able to narrow the list of uncertainties to a small number that dominate in the decision process. Haeseker et al. then used a variety of methods to assign probabilities to uncertain future states of nature and then modelled forecasts of possible outcomes based on the range of management options. The decision problem was visualized using a decision tree depicting "the management options and the critical uncertainties

Box 4.1 contd. ...

Box 4.1 contd. ...

that may influence the expected performance of different management options." Haeseker et al. used a variety of performance measures to rank the outcomes of 10 management options (which in a DA are described in terms that relate directly back to the management objectives, such as economic value or mortality rate). To determine whether the rank order of the 10 management options (as preferences, all factors give weight in the decision) changed with different assumptions, Haeseker et al. performed a sensitivity analysis that showed that the conclusions of the analysis were independent of the choice of performance indicator, meaning that the conclusions were robust.

As in the adaptive management example, the GLFC used the research findings to guide future management decisions. Haeseker et al. explain that there were three consequential outcomes to their analysis: "First, our explicit consideration of uncertainty altered the choice of management option made by the GLFC... . Second, the analysis allowed the GLFC to justify a compromise management option that reflects a balance of short- and long-term interests. Third, the analysis cemented support for ongoing assessment in the St. Marys River to inform future decisions."

deal with surprises in this regard by explicitly and quantitatively considering the implications of uncertainties for decisions (Peterman and Peters 1998). It is this unambiguous and quantitative approach to decisions that can reduce the arbitrary nature of those decisions by explicitly incorporating uncertainty into the evaluation of alternatives to maximize the probability of choosing the most beneficial option (Raiffa 1968). More specifically, decision analysis treats both our belief in a particular outcome in terms of a probability, and our satisfaction with that outcome. Lindley (2006) emphasizes that belief and satisfaction are quite separate elements of the same event, and that decision analysis is "a method of making the uncertainties and the qualities of the outcomes combine, leading to a sensible, coherent way of deciding how to act."

DA uses probabilities to quantify the consequences of a set of possible actions given the uncertainties in our current knowledge, with the benefit that "uncertainty is handled in a consistent and systematic way rather than being ignored" Hilborn and Ludwig (1993). It has been suggested (Peterman and Anderson 1999; emphasis added) that managers of natural resources have historically addressed uncertainty with the following methods:

- *best point estimates* for parameters and state variables that describe system dynamics (although uncertainty is not considered once the parameters are estimated);
- maintaining the *status quo* (when managers are reluctant to revise their policies precisely because outcomes of actions are very uncertain);
- *aggressive policies* for management interventions when the pressure for large, short-term economic yields is high and potential negative consequences cannot be demonstrated convincingly;

- application of somewhat *arbitrary safety margins* or other precautionary actions (when risk is assumed to be high); and
- explicitly and quantitatively consider the implications of uncertainty for decisions using a technique known as *decision analysis* to unambiguously reduce the arbitrariness of decision-making.

It is recommended that readers consult Crawford et al. (2005) for a concise overview of the historical development of DA – "making hard decisions in the face of uncertainty" – and Miles (2007) for a thorough history. Additionally, a review of the development of DA operational cycles is beyond my purpose, although this was done by Crawford et al. (2005). They include the cycle of Clemen and Reilly (2001), a standard DA reference in the natural resource management literature, but favour a variation developed in Peterman and Peters (1998) and Peterman and Anderson (1999).

Embracing uncertainty, or at least attempting to handle it, means that we must measure our beliefs through probability, both to see how they combine (i.e., how the beliefs of more than one person can be made to cohere with one another) and how those beliefs change with new information (Clemen and Reilly 2001, Lindley 2006). It is critical to appreciate that DA deals with "beliefs in action" by analyzing "how you might decide between different courses of action, without saying what the decisions should be, only how they should be organized" (Lindley 2006). As with AM, formal and thoroughly documented examples of decision analysis in the Great Lakes are rare and usually related to fisheries (see Crawford et al. 2005, Jones and Bence 2009, and Jones et al. 2008; but Schleen et al. 2003 on sea lamprey). Finally, in order to provide a brief look at DA in action, Box 4.1 provides a short overview of the use of DA in a science-policy problem situation involving the management of the invasive sea lamprey in the Great Lakes (based on Haeseker et al. 2007).

DAAM

Despite the strengths of adaptive management to generate learning, any particular process of investigation through management, adaptive or not, must usually consider more than one candidate policy option. Each of these options will inherently embody uncertainty from the complexity and variability of the problem situation, and from both the sampling and modelling error that occurs in trying to quantify and describe the system (Peterman and Peters 1998). Resource management has historically failed to address uncertainty, or has done so with arbitrary means in the absence of good understanding or appreciation (Hilborn and Ludwig 1993, Peterman and Anderson 1999).

Adaptive management is a set of techniques to improve understanding, but alone it is not a decision process or model (Argent 2009). Adaptive management can *contribute* reliable knowledge to decision making through the generation of learning, but the modification of behaviour (such as some element of management) subsequent to learning must result from a decision process. Decision analysis is such a process. AM and DA are more than complementary: "they need each other" because together (i.e., DAAM) they are "all about making hard decisions in the face of uncertainty *while* reducing key uncertainties through management" (Crawford et al. 2005; their emphasis). Although the scientific aspect of AM can leave managers and other stakeholders believing that important non-scientific questions or issues are subservient to some higher calling to learn, the DA process in fact helps AM to capture those issues and beliefs in a very transparent and accountable way. Similarly, DA on its own can be a useful tool for managers to organize their possible actions into a framework that explicitly incorporates their key uncertainties about the likely consequences of various decisions to act. But if an uncertainty is significant enough to be included in a formal DA, then it is important enough to use AM as a tool to reduce that uncertainty.

Figure 4.4 illustrates how the principles of DA and AM combine to create four shared principles of DAAM that reflect the shared history of DA and AM from a "common lineage of thought" on the role of uncertainty in resource management: non-exclusive participation, explicit management objectives, probabilities of competing hypotheses, and valued reliable information (Crawford et al. 2005). Jones and Nudds (2003) capture the essence of DAAM in that "AM without DA is inefficient and DA without AM is unfinished business" – inefficient because information gained through an AM program without DA would not necessarily lead to improved policy revisions in future, and unfinished business because for reduced uncertainty to be achieved a DA process requires the reliable information gained through AM. The cyclic, rigorous power of DAAM derives from the use of DA for organizing and choosing actions and AM for assessing their effectiveness. This ultimate meaning of the DAAM approach was articulated by Holling (1978) in stating that "it should be clear that the real problem of evaluation is not one of technique, but of meaning... [that] the ultimate goal is not to produce a set of numerical rankings, but to understand the strengths and weaknesses of alternative policies' performances. For it is on the basis of such understanding that meaningful, adaptive steps can be taken toward policy modification, improvement, and eventual implementation."

Furthering a better understanding as it has been described in this chapter will be the hallmark of an adaptive governance model. This defining characteristic is captured in the term *value of learning* (VOL), as it recognizes learning explicitly as a core management objective with benefits to a decision process that can be defined and evaluated (McDaniels and Gregory 2004). The VOL recognizes that many aspects of a decision can benefit from learning, including a better characterization of the objectives, the suggestion of new options (including

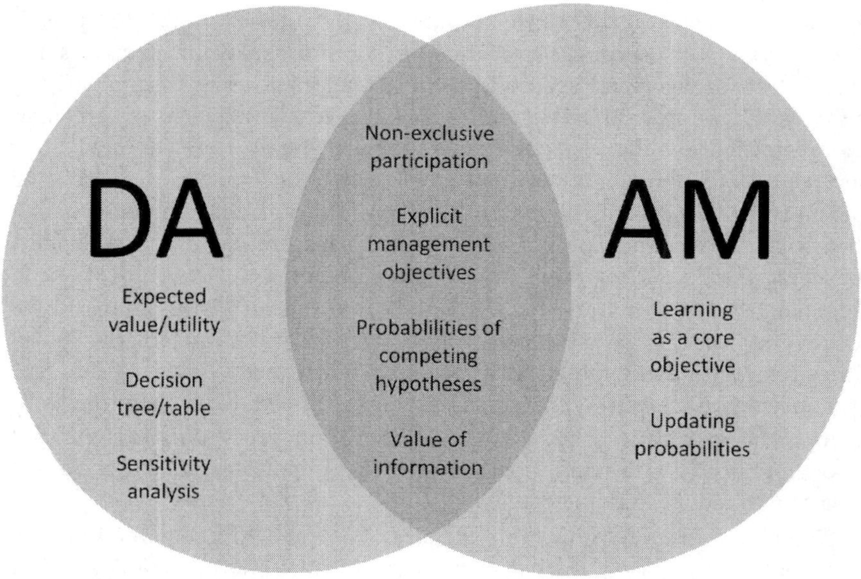

Fig. 4.4 Essential principles of Decision Analysis, Adaptive Management, and those of DAAM as combined and illustrated by Crawford et al. (2005; reproduced with permission of the authors). According to Jones and Nudds (2003), DA and AM are more than just complementary, as that would imply that they are not necessarily co-dependent but exist merely adjacent to each other. Indeed, AM without DA is inefficient, and DA without AM is unfinished business. The four shared principles represent the full breadth of DAAM, involving fully people, uncertainty, choices, policy, management, and learning.

ways to better implement existing options), creation of a performance measure for learning, and improved understanding about consequences and tradeoffs between learning and other objectives. In addition, learning can occur for related decisions outside any particular decision context, and hence, the VOL can be extremely high where many related decisions that have recurring elements, common features, and high stakes are affected. The VOL is perhaps the most elegant description of how AM can be integrated explicitly with DA in resource management decision-making (Crawford et al. 2005).

Research Context for DAAM in the Great Lakes

Actual applications of active processes of DAAM in the Great Lakes—management activities *and* decision processes to make hard decisions in the face of uncertainty *while* reducing key uncertainties through management—have been exceptionally rare. Elsewhere, major components of both DA and AM were present in an integrated approach to an Australian multispecies fishery (Sainsbury 1987, 1988, 1991, Sainsbury et al. 1993, 1997). These researchers demonstrated how DA could usefully compare experimental and non-experimental strategies with respect to economic performance of management options. In an unpublished Masters thesis at the University of British Columbia, Ohlson (1999) explicitly combined DA and AM to apply to a land use conflict with chronic uncertainty. Benefits of DAAM in the circumstances that Ohlson investigated were its ability to reveal the effects of key uncertainties on the ranking of alternative management actions, and to increase confidence among stakeholders because of the significant demands that DAAM places on transparency and accountability. David Marmorek and colleagues, principally at ESSA Technologies Ltd. in Vancouver (e.g., Deriso et al. 2001, Peters et al. 2001, Marmorek and Peters 2002), have developed and implemented arguably the most sophisticated DAAM process for recovery of depleted and endangered populations of Columbia River chinook salmon. Overall, their research demonstrated that robust recovery strategies could be identified before uncertainties were fully resolved, which may not occur until the stocks have disappeared.

The purposeful development of DAAM in the Great Lakes region has been driven by Mike Jones of Michigan State University and Tom Nudds of the University of Guelph (e.g., Jones and Nudds 2003), and by Steve Crawford at the University of Guelph in conjunction with the Chippewas of Nawash First Nation on the Bruce Peninsula in southwestern Ontario and the Ontario Commercial Fisheries' Association (e.g., Crawford et al. 2005). Linkov et al. (2006) were also developing DAAM-related concepts, but with only marginal and non-specific reference to the Great Lakes. Adaptive management is an organizing principle in the Canada-Ontario Agreement Respecting the Great Lakes Basin Ecosystem (Canada and Ontario 2007), the IJC's International Upper Great Lakes Study (IJC, n.d.), and the IJC's developing Nearshore Framework priority (IJC 2009b). There are other examples of AM programs in the region (e.g., MacDonald et al. 1999), but none are coupled explicitly with DA. Perhaps the one area of Great Lakes management with an effective history of active AM is sea lamprey control, where many sea lamprey control programs have contained elements that are arguably consistent with adaptive management principles of using science to inform policy by formally evaluating performance of management and adjusting policy as learning is achieved (M. Jones, personal communication). Decision analysis has also been applied in a Great Lakes fisheries context, including the management of the salmon fishery in Lake Michigan and of the walleye fishery in Lake Erie (S. Crawford, personal communication). Again, the reader is referred to Box 4.1 (mindful that these two examples are not explicitly DAAM). Currently, there are three active research programs focused on the development and application of DAAM for the Great Lakes fisheries management, including the Lake Huron fisheries management, Lake Erie walleye and yellow perch fisheries management, and the Great Lakes Fishery Commission's ecological review of exotic salmon stocking in the Great Lakes Basin (S. Crawford, personal communication).

Policy Context for DAAM in the Great Lakes

It follows from the purpose of the Agreement – "to restore and maintain the chemical, physical, and biological integrity of the waters of the Great Lakes Basin Ecosystem" – to ask: restore and maintain at what identifiable endpoint? This question represents a fundamental tension in ecological restoration because defining what measureable degree of restoration is both attainable and acceptable is highly contestable. Adaptive governance is critical to resolving this type of question – especially when we must cope with uncertainty in our knowledge and with competing preferences and beliefs as the primary context for decision making.

The IJC's Science Advisory Board's (SAB) contribution to the IJC's 1999–2001 Priorities Report (IJC 2001) recommended that the IJC "comprehensively identify and review emerging issues." A principal objective of the SAB's emerging issues workshop in 2003 was to identify "specific initiatives that represent the most promising future opportunities for sustaining progress" under the Agreement (IJC 2003a). The workshop concluded that the institutions of the Great Lakes are fundamentally unable to stop new and re-emerging stressors without a renovation of their governance structure, and that the Agreement should be examined for opportunities to improve integration between science and policy. Science and policy have a tendency to be "overcompartmentalized, with policy frequently lagging behind current scientific understanding" it was also concluded, and therefore greater institutional capacity is required to reduce scientific uncertainty (IJC 2003a). However, there is a lack of flexible institutional structures that would enable us to synthesize and learn from future research, and these structures will be required so as to form coherent policy based on a more systemic understanding of Great Lakes systems for decision making (McLaughlin and Krantzberg 2006).

Could a reinvigoration of Remedial Action Plans (RAPs) provide the type of institutional structure needed to pursue DAAM initiatives? RAPs have proven difficult to develop and implement over the long-term because of difficulties in sustaining progress and in maintaining continuous and vigorous oversight. However, RAP governance models could be well-suited to incorporate DAAM processes because RAPs were designed to be mechanisms to "harmonize and achieve" the interdependent goals of environmental, social, and economic development (Hartig and Zarull 1992, also Sproule-Jones 2002, Krantzberg 2003, 2006). RAP processes may also be suitable for DAAM because beyond institutional structure, mere participation in RAPs exhibited uncertainties: cost uncertainty (e.g., who is going to pay for this, and will it be more than projected?), cause and effect uncertainty (e.g., do we understand what is going on, and do scientists and lay stakeholders share the same concerns on issues such as sediment remediation or habitat restoration?), trust (of each other) and other elements of social cohesion (such as defining and sharing responsibilities), valuations of nature (e.g., how to agree on ranking one resource use over another, and whether a restoration endpoint is sufficient?), and uncertainty that all the effort will be worth it in the end.

Is there evidence that an adaptive governance approach could be possible in the Great Lakes? In an authoritative chronicle of the Agreement's history, Botts and Muldoon (2005) suggest that several features of the Great Lakes regime have contributed to its success. Three of those features I highlight below as possible contributions to creating the conditions for adaptive governance with a more rigorous and inclusive approach to collective decision making in the Great Lakes.

Joint Fact-finding and Research

The Agreement obliges both federal governments to contribute to research and monitoring needs identified through the IJC in addition to their individual research programs. The Agreement also recognizes the necessity for ongoing research and joint monitoring to measure progress and identify emerging problems. Joint fact-finding has evolved with the regime, contributed to the development and acceptance of new concepts (such as the weight of evidence approach), stimulated debate about how to make decisions in the face of incomplete knowledge, and enhanced the credibility of the IJC (Botts and Muldoon 2005). Joint fact-finding benefits our collective construction of knowledge and understanding (Andrews 2002). However, as responsibilities have evolved with revisions of the Agreement, a decline in the capacity of the IJC to collect and analyze data makes the future of joint fact-finding uncertain. In particular, the change in the role of the IJC's Water Quality Board (WQB) from evaluating government programs and progress toward meeting Agreement objectives to acting simply as policy advisor to the IJC created a gap in gathering and analyzing information (Botts and Muldoon 2005).

Accountability and Openness in Information Exchange

There are several mechanisms in the Agreement intended to foster accountability and openness, which in turn promotes involvement and a sense of community among numerous and diverse Great Lakes interests. Two such mechanisms are the requirement for regular progress reports by the advisory boards to the IJC and by the IJC to the governments (to be followed by the responses of the governments to the IJC), and the requirement for periodic review of the Agreement itself (Botts and Muldoon 2005). All aspects of those processes are (relatively) open to public scrutiny (although see Krantzberg 2011), and the Great Lakes Regional Office is directed to provide a public-information service on the programs of the Agreement. The governments have allowed observers to attend Binational Executive Committee (BEC) meetings and the State of the Lakes Ecosystem Conference (SOLEC), although they have not actively sought or promoted public participation at these events. Additionally, the public participation in its biennial meetings that has been promoted historically by the IJC, including participation from industry, civil society, professional organizations, has diminished sharply since the 1990s (Botts and Muldoon 2005). And the superficial engagement of non-governmental stakeholders in the 2010-2012 governmental renegotiation of the Agreement has been the subject of significant criticism based on concerns that it will undermine confidence in the final revised Agreement (e.g., GLU 2010, Krantzberg 2011).

The joint nature of many research programs under the Agreement has encouraged scientists to share research outcomes not just with each other, but with program managers, policymakers, legislators, and activists in the larger community. This practice of social learning has engendered at times a camaraderie among government-agency officials, IJC staff, scientists, staff and members of environmental organizations, and political leaders and their staff (Botts and Muldoon 2005). Furthermore, because the Great Lakes policy regime is founded on the 1909 Boundary Waters Treaty that established the IJC, those relationships between stakeholders benefits from historical, relatively stable, and ongoing established processes that would be vital to establishing adaptive governance structures that combine science and society—largely because of the commitment and leadership that is required, and because of the local and regional processes in which an organization such as the IJC is logically and historically positioned to play a central role. "The triumph of using the IJC to avoid disputes about use of the waters under the treaty," state Botts and Muldoon (2005), "set the stage for the challenge of restoring ecological integrity under the Great Lakes Agreement."

Flexibility and Adaptability to Changing Circumstances

Most international agreements are adopted to solve a specific problem or resolve a specific dispute and have little ability to accommodate significant related issues. The flexibility of the Agreement results from the periodic reviews that are required to assess progress and to allow consideration of changes to any element of its programs (Botts and Muldoon 2005). Those potential changes provide opportunities to apply new knowledge and adapt the objectives of the Agreement to changing circumstances, although arguably these opportunities have been underutilized in the past. It is this ability to apply new knowledge that suggests that the Agreement may be amenable to DAAM processes and feedbacks in the manner illustrated in Box 4.1.

Elements of DAAM Accountability

The federal governments have various obligations to report, explain, and/or justify actions they take to fulfil the purpose of the Agreement. Some of that responsibility is shared (under varying circumstances and jurisdictional authority) with stakeholders such as provincial and state agencies and a growing list of others with a desire to participate in management and restoration activities. These additional stakeholders represent a cross section of society, such as civil society groups, First Nations and Tribes, industry and trade organizations, municipal governments, professional organizations and academics, and individual citizens. However, it is critical to appreciate that despite the challenges involved in facilitating a wider participation in decision making, the principles of DAAM actively require this participation. The shared principles of DAAM as illustrated in Figure 4.4 – non-exclusive participation, explicit management objectives, probabilities of competing hypotheses, and the value of information – are the foundation for accountability in DAAM processes, as they direct who is involved, and require substantial trust, openness, and transparency. The principles also incorporate two overarching characteristics of management that require special focus under DAAM: the *technical design* elements of DAAM necessary for proper implementation, and the institutional insights to improve the *governance structure*, particularly transparency and accountability, necessary to operationalize DAAM. Technical design without adequate attention to governance will significantly undermine the robustness of the process, particularly as DAAM is an inherently social process.

Norton (1995) states that "environmental management faces a crisis" in that domestic legislation and international agreements all urge that natural systems be managed to protect system health and integrity. However, the design of such laws and institutional arrangements have not yet developed a consensus in operational management directives, and therefore we have not typically seen the kind of significant changes in management and decision-making necessary for DAAM initiatives. The challenge of creating such change is hindered in the Great Lakes by multiple uncertainties and multiple institutional actors and priorities. I emphasize that the practicality of applications of DAAM (i.e., careful case selection in future), the importance of stressing process and not prescribing recipes, and the importance of cultivating relationships and commitments across stakeholder groups will underpin whether attempts to incorporate DAAM into existing or future Great Lakes management programs succeed. Academics and practitioners indicate that common themes of DAAM success include the importance of leadership, partnerships, "closing the loop" to management, and organizational commitment and resources. These themes are all critical issues of accountability, too, because where DAAM is concerned with the appropriate balance between management objectives and learning objectives, compromise is usually required to maximize learning and minimize risk (Murray and Marmorek 2003a).

Allan et al. (2008) detail three strategies for building adaptive organizations and processes that engage those with the primary responsibility for programs with support staff and outside stakeholders, and I abbreviate these three strategies as follows. Strategy 1 encourages practitioners and their leaders to define adaptive management, focus and be purposeful. As a pre-requisite, proper adaptive management programs "must be recognized as a radical departure from established ways of managing natural resources. Adaptive management is not 'business as usual', nor should it be seen as an excuse to muddle through management problems." In this respect, organizations and their leaders must create the conditions for managers to be adaptive and reflective by defining the context in which appropriate management actions can occur. DAAM will require champions who have (or who have been given) the time, resources, capacities, and responsibility to influence how policy is devised and how it is implemented. When such leadership is in place purposeful activities can be developed and implemented (Allan et al. 2008).

Strategy 2 advocates that practitioners and their leaders encourage and support evaluation. Evaluation considered as a stage in learning and not as mere auditing is central to the adaptive management process. For example, research and coordination can often mean conducting inventories, which does not meet the requirement for learning. Evaluation also requires that the necessary institutional space must be actively created to allow genuine reflection on processes and outcomes. Creating such space requires a reassessment of institutional incentive systems, and there must be an acceptance of the limits of knowledge and the possibility of errors and mistakes (Allan et al. 2008).

Strategy 3 is a directive to collaborate and integrate. Participatory approaches require a recognition and acceptance of the multiple ways of knowing and understanding (Allan et al. 2008). However, social learning processes cannot be simply lifted from the page and applied (Keen et al. 2005). The legitimacy of knowledge and learning that underpins management decisions must be clarified, especially when integration of scientific and other forms of knowledge generation are proposed as it is explicitly in DAAM. Ultimately, DAAM strategies turn accountability from a mere reporting requirement response to an incentive to lead.

UNCERTAINTY AND LEARNING IN THE GREAT LAKES

The focus of this chapter has been on how we might better integrate priorities and uncertainties about the future of the Great Lakes, and in the process improve the accountability, responsiveness, and efficacy of restoration policy and decision making. Research into decision making has traditionally focused on how individuals choose among alternatives to make the "right" decision. Three traditional assumptions about the creature known as *homo economicus*

(Lee 1971) are that the rational decision maker is thoroughly informed, is infinitely sensitive and discriminating, and possesses the rational capacity for ordering alternatives so as to maximize something. However, we now know that decisions are almost always taken in the context of incomplete knowledge, a reality imposed by inherent uncertainties in complex systems of people and nature. Whether people individually make rational choices is somewhat beside the point, at least in the context of requiring a forum for the open exchange and reconciliation of a broad range of differing and often conflictual perspectives. A corollary of that diversity of opinion is the expectation of disagreement about what we should maximize (and why).

Decision making cannot be thought of as an event, decoupled from our continuous coping in the face of the complexity and uncertainty that characterizes human-environment interactions. So how do we properly attend to significant threats that undermine social-ecological resilience in the Great Lakes and elsewhere? And how can we do justice to the volume of effort among science and society to restore the Great Lakes and create a sustainable future? At the root of this dilemma is a fundamental question that we are obligated to answer: how does our experience stand in relation to our problems? I say that we are obligated because a *burden* is not only a load that must be borne – a serious task for example, with the potential for serious consequences – but it also implies a duty or responsibility.

To appreciate the significance of DAAM for accountability and collective success at solving very real problems hindering restoration of the Great Lakes, consider that *all* management decisions are taken in the face of uncertainty (Ludwig et al. 1993), and that *all* key uncertainties in management are learning opportunities (Murray and Marmorek 2003a). A primary concern is whether the institutional capacity exists (or can be created) to embrace a fundamental and systemically different approach to management that explicitly acknowledges that we always lack sufficient knowledge to act with a full understanding of consequences. DAAM demands that we elevate the role of double-loop learning activities such as monitoring and evaluation beyond the cosmetic and superficial attention they too often receive. The institutionalization of DAAM is dependent on building capacity, support, and enthusiasm for a culture of learning, and systems that enable institutional memory of what is learned. The ability and willingness to institute such a capacity is perhaps the ultimate accountability in considering the question of how to begin to incorporate DAAM into existing or new Great Lakes management paradigms. But the present is a good time for those capable of dealing with uncertainty, suggest Westley et al. (2006) in a hopeful tone, as a willingness to embrace it offers the possibility of transformation, and some relief from the sense of "being stuck" that burdens so much of what collaborative decision processes try to accomplish.

ACKNOWLEDGMENTS

I thank Mike Jones and Tom Nudds for helpful discussions about the development and application of DAAM, and Steve Crawford for a thorough review of an earlier version of this chapter. Research for portions of this chapter was supported by the International Joint Commission, with particular appreciation to John Gannon.

REFERENCES

Allan, C. and A. Curtis, G. Stankey, and B. Shindler. 2008. Adaptive management and watersheds: a social science perspective. Journal of the American Water Resources Association 44: 166-174.

Allan, C. and G. Stankey. [eds.]. 2009. Adaptive Environmental Management: A Practioner's Guide. Springer Science and Business Media B.V., Dordrecht, The Netherlands.

AMET [Adaptive Management Experiment Team]. 2003. Glossary. Accessed 29 Jun 2006 at http://www.ameteam.ca/glossary.html

Andrews, C.J. 2002. Humble Analysis: The Practice of Joint Fact-Finding. Praeger Publishers, Westport, USA.

ARC [Agreement Review Committee]. 2007. Report to the Great Lakes Binational Executive Committee: Volumes 1, 2, and 3. http://binational.net/glwqa_2007_e.html

Argent, R.M. Components of adaptive management. pp. 11-32. *In*: C. Allan and G. Stankey [eds.]. 2009. Adaptive Environmental Management: A Practioner's Guide. Springer Science and Business Media B.V., Dordrecht, The Netherlands.

Argyris, C. 1993. On Organizational Learning. Blackwell Business, Cambridge, USA.

Bails, J. and A. Beeton, J. Bulkley, M. DePhilip, J. Gannon, M. Murray, H. Regier, and D. Scavia. 2005. Prescription for Great Lakes Ecosystem Protection and Restoration: Avoiding the Tipping Point of Irreversible Changes. Prepared for the Wege Foundation and the Joyce Foundation, Grand Rapids and Chicago, USA.

Barr, J. and T.J. Hall, H.J. Harris, G. Krantzberg, and S. Sowa.2010. Review report of the SOLEC independent expert panel. Presented to Environment Canada and the U.S. Environmental Protection Agency. Accessed 30 Nov 2010 at http://msep.mcmaster.ca/epp/publications/SOLEC_Report.pdf

Benevides, H. and C. Davis, F. de Leon, R. Gilbert, J. Jackson, J. Nalbone, J. Potter, D. Stack, and P. Wilder. 2007. Promises to Keep, Challenges to Meet. Perspectives from Citizens. Prepared for Alliance for the Great Lakes, Biodiversity Project, Canadian Environmental Law Association, Great Lakes United. Accessed 15 Oct 2010 at http://s.cela.ca/files/uploads/553GLWQA_promises.pdf

Beratan, K.K. 2007. A cognition-based view of decision processes in complex social–ecological systems. Ecology and Society 12(1): art27.

Botts, L. and P. Muldoon. 2005. Evolution of the Great Lakes Water Quality Agreement. Michigan State University Press, East Lansing, USA.

Canada and Ontario. 2007. Canada-Ontario Agreement Respecting the Great Lakes Basin Ecosystem.

Clemen, R.T. and T. Reilly. 2001. Making Hard Decisions with Decision Tools. Duxbury/ Thomson Learning, Pacific Grove, USA.

Crawford, S. and S. Matchett, and K. Reid. 2005. Decision Analysis/Adaptive Management (DAAM) for Great Lakes Fisheries: A General Review and Proposal. Draft discussion paper presented to the International Association for Great Lakes Research Conference, Ann Arbor, MI. Accessed 15 Mar 2010 at http://www.uoguelph.ca/~scrawfor

Crawford, S. and C.A. Wehkamp, and N. Smith. 2010. Translation of Indigenous/Western Science Perspectives on Adaptive Management for Environmental Assessments. Canadian Environmental Assessment Agency, Research and Development Monograph Series.Acccessed 8 Jan 2011 at http://www.ceaa.gc.ca/7F3C6AF0-docs/TIWSPAMEA-eng.pdf

Cundill, G. and C. Fabricius. 2009. Monitoring in adaptive co-management: toward a learning based approach. Journal of Environmental Management 90: 3205-3211.

Deriso, R.B. and D.R. Marmorek, and I.J. Parnell. 2001. Retrospective patterns of differential mortality and common year-effects experienced by spring and summer chinook salmon (*Oncorhynchus tshawytscha*) of the Columbia River. Canadian Journal of Fisheries and Aquatic Sciences 58: 2419-2430.

Dewey, J. 1920. Reconstruction in Philosophy. Henry Holt and Company, New York, USA.

Folke, C. 2006. Resilience: the emergence of a perspective for social-ecological systems analyses. Global Environmental Change 16: 253-267.

Folke, C. and T. Hahn, P. Olsson, and J. Norberg. 2005. Adaptive governance of social-ecological systems. Annual Review of Environment and Resources 30: 441-473.

Francis, G. and H.A. Regier. Barriers and bridges to the restoration of the Great Lakes basin ecosystem. pp. 239-291. *In:* L.H. Gunderson, C.S. Holling, and S.S. Light [eds.]. 1995. Barriers and Bridges to the Renewal of Ecosystems and Institutions. Columbia University Press, New York, USA.

GAO [United States General Accounting Office]. 2003. Great Lakes: An Overall Strategy and Indicators for Measuring Progress are Needed to Better Achieve Restoration Goals. Report GAO-03-515 to Congressional Requesters.

GLU [Great Lakes United]. 2010. Governments remain silent on lack of public input. Accessed 19 Dec 2010 at http://www.glu.org/en/press/governments-remain-silent-lack-public-input

Gregory, R. and D. Ohlson, and J. Arvai. 2006. Deconstructing adaptive management: criteria for applications to environmental management. Ecological Applications 16: 2411-2425.

Grumbine, R.E. 1994. What is ecosystem management? Conservation Biology 8: 27-38.

Gunderson, L.H. 2000. Ecological resilience – in theory and application. Annual Review of Ecology and Systematics 31: 425-439.

Gunderson, L.H. and C.S. Holling. [eds.]. 2002. Panarchy: Understanding Transformations in Human and Natural Systems. Island Press, Washington, USA.

Gunderson, L.H. and C.S. Holling, and S.S. Light. [eds.]. 1995. Barriers and Bridges to the Renewal of Ecosystems and Institutions. Columbia University Press, New York, USA.

Haeseker, S.L. and M.L. Jones, R.M. Peterman, J.R. Bence, W. Dai, and G.C. Christie. 2007. Explicit consideration of uncertainty in Great Lakes fisheries management: decision analysis of sea lamprey (*Petromyzon marinus*) control in the St. Marys River. Canadian Journal of Fisheries and Aquatic Sciences 64: 1456-1468.

Hammond, J.S. and R.L. Keeney, and H. Raiffa. 1999. Smart Choices: A Practical Guide to Making Better Decisions. Harvard Business School Press. Boston, USA.

Hansen, G.J.A. and M.L. Jones. 2008. A rapid assessment approach to prioritizing streams for control of Great Lakes sea lampreys (*Petromyzon marinus*): a case study in adaptive management. Canadian Journal of Fisheries and Aquatic Sciences 65: 2471-2484.

Hargrove, R. 2002. Masterful Coaching. Revised Edition. Jossey-Bass/Pfeiffer, San Francisco, USA.

Hartig, J.H. and M.A. Zarull. Keystones for success. pp. 263-279. *In:* J.H. Hartig and M.A. Zarull [eds.]. 1992. Under RAPs: Toward Grassroots Ecological Democracy in the Great Lakes Basin. University of Michigan Press. Ann Arbor, USA.

Hilborn, R. and D. Ludwig. 1993. The limits of applied ecological research. Ecological Applications 3: 550-552.

Holling, C.S. 1973. Resilience and stability of ecological systems. Annual Review of Ecology and Systematics 4: 1-23.

Holling, C.S. [ed.]. 1978. Adaptive Environmental Assessment and Management. Wiley, Chichester, USA.

Holling, C.S. What barriers, what bridges? pp. 3-34. *In:* L.H. Gunderson, C.S. Holling, and S.S. Light [eds.]. 1995. Barriers and Bridges to the Renewal of Ecosystems and Institutions. Columbia University Press. New York, USA.

Holling, C.S. 2001. Understanding the complexity of economic, ecological, and social systems. Ecosystems 4: 390-405.

Holling, C.S. and G. Meffe. 1996. Command and control and the pathology of natural resource management. Conservation Biology 10: 328-337.

IJC [International Joint Commission]. 2001. Priorities 1999–2001: priorities and progress under the Great Lakes Water Quality Agreement. Science Advisory Board, Chapter 2. Accessed 30 Nov 2010 at http://www.ijc.org/rel/boards/sab/pr9901/index.html

IJC [International Joint Commission]. 2003a. Priorities 2001–2003: priorities and progress under the Great Lakes Water Quality Agreement. Science Advisory Board, Chapter 5. Accessed 30 Nov 2010 at http://www.ijc.org/php/publications/pdf/ID1568.pdf

IJC [International Joint Commission]. 2003b. Status of Restoration Activities in Great Lakes Areas of Concern: A Special Report. Accessed 30 Nov 2010 at http://www.ijc.org/php/publications/html/aoc_rep/english/report/pdfs/chapter1-e.pdf

IJC [International Joint Commission]. 2009a. Work Group Report on Binational Aquatic Invasive Species Rapid Response Policy Framework, Special Publication 2009-04. Great Lakes Water Quality Agreement Priorities 2007-09 Series. Accessed 22 Feb 2010 at http://meeting.ijc.org/reports

IJC [International Joint Commission]. 2009b. Work Group Report on the Nearshore Framework, Special Publication 2009-01. Great Lakes Water Quality Agreement Priorities 2007-09 Series. Accessed 22 Feb 2010 at http://meeting.ijc.org/reports

IJC [International Joint Commission]. n.d. International Upper Great Lakes Study. Accessed 8 Jan 2011 at http://www.iugls.org

Jackson, J. and K.K. Sloan. 2008. A Way Forward: Strengthening Decision-making and Accountability under the Great Lakes Water Quality Agreement. Accessed 31 Oct 2008 at http://www.glu.org/sites/default/files/Governance%20GLU%20Jan08.pdf

Johns, C. 2005. Transboundary Water Governance: Future Challenges. Report on a Workshop. Royal Society of Canada and McMaster University, Ottawa and Hamilton, Canada.

Johnson, F.A. and C.T. Moore, W.L. Kendall, J.A. Dubovsky, D.F. Caithamer, J.R. Kelley, and B.K. Williams. 1997. Uncertainty and the management of mallard harvests. Wildlife Management 61: 202-216.

Jones, M.L. Adaptive management and Great Lakes fisheries. *In:* G.B. MacDonald, J.Z. Fraser and P. Gray [eds.]. 1999. Adaptive Management Forum: Linking Management and Science to Achieve Ecological Sustainability. Ontario Ministry of Natural Resources, Science Development and Transfer Series No. 001, Toronto, Canada.

Jones, M. and T. Nudds. 2003. Draft Principles of Decision Analysis and Adaptive Management. Prepared for the SaugeenOjibway/Canada/Ontario Plenary. Ontario.

Jones, M.L. and J.R. Bence. Uncertainty and fishery management in the North American Great Lakes: Lessons from applications of decision analysis. *In:* C.C. Krueger and C.E. Zimmerman [eds.]. 2009. Pacific Salmon: Ecology and Management of Western Alaska's Populations. American Fisheries Society, Bethesda, USA.

Jones, M.L. and J.R. Bence, E.B. Szalai, and W. Dai. Assessing stocking policies for Lake Michigan salmonine fisheries using decision analysis. pp. 81-88. *In:* D.F. Clapp and W. Horns [eds.]. 2008. The State of Lake Michigan in 2005. Great Lakes Fishery Commission Special Publication 08-02.

Keen, M. and V.A. Brown, and R. Dyball. 2005. Social Learning in Environmental Management: Building a Sustainable Future. Earthscan Publications, Oxford, UK.

Krantzberg, G. 2003. Keeping remedial action plans on target: lessons learned from Collingwood Harbour. Journal of Great Lakes Research 29: 641-651.

Krantzberg, G. 2006. Sustaining the gains made in ecological restoration: case study Collingwood Harbour, Ontario. Environment, Development and Sustainability 8: 413-424.

Krantzberg, G. 2009. Renegotiating the Great Lakes Water Quality Agreement: the process for a sustainable outcome. Sustainability 1: 254-267.

Krantzberg, G. 2011. Renegotiation of the 1987 Great Lakes Water Quality Agreement: from confusion to promise. Sciforum Electronic Conferences Series, Proceedings of the 1st World Sustainability Forum, November: 1-30.

Lancia, R.A. and T.D. Nudds, and M.L. Morrison. 1993. Adaptive resource management: policy as hypothesis, management by experiment. Opening comments: slaying slippery shibboleths. Transactions of the North American Wildlife Natural Resources Conference (Washington D.C., 19-24 Mar 1993) 58: 505-508.

Lancia, R.A. and C.E. Braun, M.W. Collopy, R.D. Dueser, J.G. Kie, C.J. Martinka, J.D Nicol, T. D. Nudds, W.R. Porath, and N.G. Tilghman. 1996. ARM! for the future: adaptive resource management in the wildlife profession. Wildlife Society Bulletin 24: 436-442.

Lebel, L. and J.M. Anderies, B. Campbell, C. Folke, S. Hatfield-Dodds, T.P. Hughes, and J. Wilson. 2006. Governance and the capacity to manage resilience in regional social-ecological systems. Ecology and Society 11(1): art19.

Lee, W. 1971. Decision Theory and Human Behaviour. Wiley, New York, USA.

Lee, K.N. 1993. Compass and Gyroscope: Integrating Science and Politics for the Environment. Island Press, Washington, USA.

Lee, K.N. 1999. Appraising adaptive management. Conservation Ecology 3(2): art3.

Leopold, A. 1930. The American game policy in a nutshell. Transactions of the 17th American Game Conference, 281-283.

Lindley, D.V. 2006. Understanding Uncertainty. Wiley Interscience, Hoboken, USA.

Linkov, I. and F.K. Satterstrom, G. Kiker, C. Batchelor, T. Bridges, and E. Ferguson. 2006. From comparative risk assessment to multi-criteria decision analysis and adaptive management: recent developments and applications. Environment International 32: 1072-1093.

Lubell, M. and W.D. Leach, and P.A. Sabatier. Collaborative watershed partnerships in the epoch of sustainability. pp. 255-288. *In:* D.A. Mazmanian and M.E. Kraft [eds.]. 2009. Toward Sustainable Communities: Transition and Transformations in Environmental Policy. Second Edition. MIT Press, Cambridge, USA.

Ludwig, D. and R. Hilborn, and C. Walters. 1993. Uncertainty, resource exploitation, and conservation: lessons from history. Science 260: 17-36.

Ludwig, D. and M. Mangel, and B. Haddad. 2001. Ecology, conservation, and public policy. Annual Review of Ecology and Systematics 32: 481-517.

MA [Millennium Ecosystem Assessment]. 2005. Ecosystems and Human Well-being: Synthesis. Island Press, Washington, USA.

MacDonagh-Dumler, J. 2009. Policy innovation for Great Lakes ecosystem management. Journal of Great Lakes Research 35: 477-481.

MacDonald, G.B. and J.Z. Fraser, and P. Gray [eds.]. 1999. Adaptive Management Forum: Linking Management and Science to Achieve Ecological Sustainability. Ontario Ministry of Natural Resources, Science Development and Transfer Series No. 001, Toronto, Canada.

Manno, J. and G. Krantzberg. Rediscovering and revitalizing the Great Lakes governance. pp. 159-170. *In:* K. Bosselmann, R. Engel and P. Taylor [eds.]. 2008. Governance for Sustainability Issues, Challenges, Successes. IUCN Environmental Policy and Law Paper No. 70.

Marmorek, D. and C. Peters. 2002. Finding a PATH toward scientific collaboration: insights from the Columbia River basin. Conservation Ecology 5(2): art8.

Mazmanian, D.A. and M.E. Kraft. The three epochs of the environmental movement. pp. 3-32. *In:* D.A. Mazmanian and M.E. Kraft [eds.]. 2009. Toward Sustainable Communities: Transition and Transformations in Environmental Policy. Second Edition. MIT Press, Cambridge, USA.

McDaniels, T.L. and R. Gregory. 2004. Learning as an objective within a structured risk management decision process. Environmental Science & Technology 38: 1921-1926.

McLaughlin C. and G. Krantzberg. 2006. Toward a 'better understanding' of the Great Lakes Basin Ecosystem. Journal of Great Lakes Research 32: 197-199.

McLaughlin C. and G. Krantzberg. 2011. An appraisal of policy implementation deficits in the Great Lakes. Journal of Great Lakes Research 37: 390-396.

McLaughlin C. and G. Krantzberg. 2012. An appraisal of management pathologies in the Great Lakes. Science of the Total Environment 416: 40-47.

Meadows, D.H. 2008. Thinking in Systems: A Primer. Chelsea Green Publishing, White River Junction, USA.

Miles Jr., R.F. The emergence of decision analysis. pp. 13-31. *In:* W. Edwards, R.F. Miles Jr., and D. von Winterfelt [eds.]. 2007. Advances in Decision Analysis: From Foundations to Applications. Cambridge University Press. Cambridge, UK.

Miller, A. 1999. Environmental Problem Solving: Psychosocial Barriers to Adaptive Change. Springer, New York, USA.

Miller, C. 2001. Hybrid Management: Boundary Organizations, Science Policy, and Environmental Governance in the Climate Regime. Science, Technology, and Human Values 26: 478-500.

Moran, E.F. 2010. Environmental Social Science: Human-Environment Interactions and Sustainability. Wiley-Blackwell, Malden, USA.

Murray, C. and D.R. Marmorek. Adaptive management and ecological restoration. pp. 417-428. *In:* P. Freiderici [ed.]. 2003a. Ecological Restoration of Southwestern Ponderosa Pine Forests. Island Press, Washington, USA.

Murray, C. and D.R. Marmorek. 2003b. Adaptive management: A science-based approach to managing ecosystems in the face of uncertainty. Prepared for presentation at the Fifth International Conference on Science and Management of Protected Areas: Making Ecosystem Based Management Work, Victoria, British Columbia, May 11-16, 2003.

Norberg, J. and G.S. Cumming [eds.]. 2008. Complexity theory for a sustainable future. Columbia University Press. New York, USA.

Norton, B.G. 1995. Evaluating ecosystem states: two competing paradigms. Ecological Economics 14: 113-127.

Nudds, T.D. Adaptive management and the conservation of biodiversity. pp. 179-193, 288-291. *In:* R.K. Baydack, H. Campa III, and J.B. Haufler [eds.]. 1999. Practical Approaches to the Conservation of Biological Diversity. Island Press, Washington, USA.

Nudds, T.D. and Y. Jiao, S. Crawford, K. Reid, K. McCann, and W. Yang. 2003. Decision Analysis and Adaptive Management (DAAM) Systems for Great Lakes Fisheries: The Lake Erie Wall-eye and Yellow Perch Fisheries: Project Background and Work Plan. Prepared for the Ontario Commercial Fisheries Association and the Chippewas of Nawash First Nation. Guelph, Canada.

OAG [Office of the Auditor General of Canada]. 2001. A legacy worth protecting: charting a sustainable course in the Great Lakes and St. Lawrence River basin; Chapter 1 of the Report of the Commissioner of the Environment and Sustainable Development. Public Works and Government Services Canada.

Ohlson, D.W. 1999. Exploring the Application of Adaptive Management and Decision Analysis to Integrated Watershed Management. M.Sc. Thesis, University of British Columbia. Vancouver, Canada.

Pahl-Wostl, C. 2009. A conceptual framework for analysing adaptive capacity and multi-level learning processes in resource governance regimes. Global Environmental Change 19: 354-365.

Parson, E.A. and W.C. Clark. Sustainable development as social learning: theoretical perspectives and practical challenges for the design of a research program. pp. 428-

460. *In:* L.H. Gunderson, C.S. Holling, and S.S. Light [eds.]. 1995. Barriers and Bridges to the Renewal of Ecosystems and Institutions. Columbia University Press, New York, USA.

Payne, J.W. and J.R. Bettman, and E.J. Johnson. 1993. The Adaptive Decisionmaker. Cambridge University Press, Cambridge, UK.

Peterman, R.M. and C.M. Peters. Decision analysis: taking uncertainties into account in forest resource management. pp. 105-127. *In:* V. Sit and B. Taylor [eds.]. 1998. Statistical Methods for Adaptive Management Studies (Handbook No. 42.). British Columbia Ministry of Forestry.

Peterman, R. and J.L. Anderson. 1999. Decision analysis: a method for taking uncertainties into account in risk-based decision-making. Human and Ecological Risk Assessment 5: 231-244.

Peters, C.N. and D.R. Marmorek, and D.B. Deriso. 2001. Application of decision analysis to evaluate recovery actions for threatened Snake River fall chinook salmon (*Oncorhynchus tshawytscha*). Canadian Journal of Fisheries and Aquatic Sciences 58: 2447-2458.

Raiffa, H. 1968. Decision Analysis: Introductory Lectures on Choices under Uncertainty. Addison-Wesley, Reading, USA.

Regier, H.A. and M.L. Jones, J. Addis, and M. Donahue. Great Lakes–St. Lawrence River basin assessments. pp. 133-165. *In:* K.N. Johnson, F.J. Swanson, M. Herring and S. Greene [eds.]. 1999. Bioregional Assessments: Science at the Crossroads of Management and Policy. Island Press, Washington, USA.

Sabatier, P.A. Policy change over a decade or more. pp. 13-39. *In:* P.A. Sabatier and H.C. Jenkings-Smith [eds.]. 1993. Policy Change and Learning: An Advocacy Coalition Approach. Westview Press, Boulder, USA.

Sainsbury, K.J. 1987. Assessment and management of the demersal fishery on the continental shelf of northwestern Australia. pp. 465-503. *In:* J.J. Polovina and S. Ralston [eds.]. Tropical Snappers and Groupers: Biology and Fisheries Management. Westview Press, Boulder, USA.

Sainsbury, K.J. The ecological basis of multispecies fisheries and management of a demersal fishery. pp. 349-382. *In:* J.A. Gulland [ed.]. 1988. Fish Population Dynamics. Second Edition. Wiley, Chichester, USA.

Sainsbury, K.J. 1991. Application of an experimental approach to management of a tropical fishery with highly uncertain dynamics. ICES Journal of Marine Science 193: 301-320

Sainsbury, K.J. and R.A. Campbell, and A.W. Whitelaw. Effects of trawling on the marine habitat on the north-west shelf of Australia and implications for sustainable fisheries management. pp. 137-145. *In:* D.A. Hancook [ed.]. 1993. Sustainable Fisheries Through Sustaining Fish Habitat. Australian Government Publishing Service.

Sainsbury, K.J. and R.A. Campbell, R. Lindholm, and A.W. Whitelaw. Experimental management of an Austrailian multispecies fishery: examining the possibility of trawl induced habitat modifications. pp. 107-112. *In:* E.L. Pikitch, D.D. Huppert and M.P. Sissenwine [eds.]. 1997. Global Trends: Fisheries Management. American Fisheries Society Symposium, vol. 20, Bethesda, USA.

Schleen, L.P. and G.C. Christie, J.W. Heinrich, R.A. Bergstedt, R.J. Young, T.J. Morse, D.S. Lavis, T.D. Bills, J.E Johnson, and M.P. Ebener. 2003. Development and implementation of an integrated program for control of sea lampreys in the St. Marys River. Journal of Great Lakes Research 29 (supp. 1): 677-693.

Schmiegelow, F.K.A. and S.J. Hannon. 1993. Adaptive management, adaptive science and the effects of forest fragmentation on boreal birds in Northern Alberta. Transactions of the North American Wildlife and Natural Resources Conference 58: 584-598.

Schmiegelow, F.K.A. and C.S. Machtans, and S.J. Hannon. 1997. Are boreal birds resilient to forest fragmentation? An experimental study of short-term community responses. Ecology 78: 1914-1932.

Scholz, J.T. and B. Stiftel. Introduction: the challenges of adaptive governance. pp. 1-11. *In:* J.T. Scholz and B. Stiftel [eds.]. 2005. Adaptive Governance and Water Conflict: New Institutions for Collaborative Planning. Resources for the Future, Washington, USA.

Schreiber, E.S.G. and A.R. Bearlin, S.J. Nicol, and C.R. Taylor. 2004. Adaptive management: a synthesis of current understanding and effective application. Ecological Management and Restoration 5: 177-182.

Simon, H. Theories of bounded rationality. pp. 161-176. *In:* C.B. McGuire and R. Radner [eds.]. 1972. Decision and Organization. North-Holland Publishing Company, Amsterdam, The Netherlands.

Sit, V. and B. Taylor [eds.]. 1998. Statistical Methods for Adaptive Management Studies.British Columbia Ministry of Forests, Research Branch, Land Management Handbook No. 42.

Sproule-Jones, M. 2002. Restoration of the Great Lakes: Promises, Practices, Performances. UBC Press, Vancouver, Canada.

Stolnack, S.A. and M.D. Bryant, and R.C. Wissmar. 2005. A review of protocols for monitoring streams and juvenile fish in forested regions of the Pacific Northwest. General Technical Report PNW-GTR-625. U.S. Department of Agriculture, Forest Service, Pacific Northwest Research Station.

Susskind, L. and P. Field, M. van der Wansem, and J. Peyser. Integrating scientific information, stakeholder interests, and political concerns. pp. 181-203. *In:* K.S. Hanna and D.S. Slocombe [eds.]. 2007. Integrated Resource and Environmental Management: Concepts and Practice. Oxford University Press, Don Mills, Canada.

Walters, C.J. 1986. Adaptive Management of Renewable Resources. MacMillan, New York, USA.

Walters, C.J. and R. Green. 1997. Valuation of experimental management options for ecological systems. Journal of Wildlife Management 61: 987-1006.

Walters, C.J. 2007. Is adaptive management helping to solve fisheries problems? Ambio 36: 304-307.

Walters, C. and L.H. Gunderson, and C.S. Holling. 1992. Experimental policies for water management in the everglades. Ecological Applications 2: 189-202.

Waltner-Toews, D. and J.J. Kay, and N.-M.E. Lister [eds.]. 2008. The Ecosystem Approach: Complexity, Uncertainty, and Managing for Sustainability. Columbia University Press, New York, USA.

Wenger, E. and R. McDermott, and W. Snyder. 2002. Cultivating Communities of Practice: A Guide to Managing Knowledge. Harvard Business School Press, Boston, USA.

Westley, F. Governing design: the management of social systems and ecosystems management. pp. 391-427. *In:* L.H. Gunderson, C.S. Holling, and S.S. Light [eds.]. 1995. Barriers and Bridges to the Renewal of Ecosystems and Institutions. Columbia University Press, New York, USA.

Westley, F. and B. Zimmerman, and M.Q. Patton. 2006. Getting to Maybe: How the World is Changed. Random House Canada, Toronto, Canada.

Williams, B.K. and F.A. Johnson. 1995. Adaptive management and the regulation of waterfowl harvests. Wildlife Society Bulletin 23: 430-436.

Williams, B.K. and F.A. Johnson, and K. Wilkins. 1996. Uncertainty and the adaptive management of waterfowl harvests. Journal of Wildlife Management 60: 223-232.

Williams, B.K. and R.C. Szaro, and D.D. Shapiro. 2009. Adaptive Management: The U.S. Department of the Interior Technical Guide (Revised Edition). Adaptive Management Working Group, U.S. Department of the Interior. Accessed 15 October 2010 at http://www.doi.gov/initiatives/AdaptiveManagement/documents.html

Wondolleck, J.M. and L.S. Yaffee. 2000. Making Collaboration Work: Lessons From Innovation in Natural Resource Management. Island Press, Washington, USA.

CHAPTER 5

Transnational Municipal Networks of American and Canadian Local Governments in the Context of Bilateral Environmental Relations: The Emergence of a European Phenomenon in the Great Lakes Basin

Irek W. Kusmierczyk

Department of Political Science, Vanderbilt University,
VU Station B#351817, Nashville, Tennessee 37235, 1- 519-990-3132
E-mail: i.kusmierczyk@vanderbilt.edu

INTRODUCTION

On 16 March 2006, Richard M. Daley submitted testimony to the United States Senate Environment and Public Works Committee in his capacity as Mayor of Chicago and the Chairman of a recently established transnational municipal network called the Great Lakes and St. Lawrence Cities Initiative. In his testimony, Mayor Daley discussed the Great Lakes Regional Collaboration Strategy to Restore and Protect the Great Lakes, a significant document released in 2005 that set priorities, objectives and strategies for environmental management in the basin and laid the groundwork for the bi-national review of the Great Lakes Water Quality Agreement then underway. The Strategy resulted from months of effort on the part of the Great Lakes Regional Collaboration[1], a network which

[1]www.glrc.us

brought together 1,500 stakeholders from the federal, state, provincial and local governments mostly from the United States but also Canada, as well as representatives of First Nations, non-governmental organizations, and industry, in meetings held in cities across the basin in Rochester, Toledo, Traverse City, and Duluth. In discussing the contributions of local government to the Strategy, Mayor Daley submitted two documents to the Senate Committee. The first was a list of the 48 cities from the United States and 37 cities from Canada who, as current or future members of the Cities Initiative, had deliberated on and formed common positions that were represented in the Strategy. The second document was a letter from David Miller, then Mayor of Toronto and the Canadian Chair of the Cities Initiative, supporting the testimony of the mayor of Chicago and presenting to the Committee a very brief update on the work taking place on the Canadian side of the border.

This seemingly insignificant testimony symbolizes the latest development in the evolution of bilateral relations between Canada and the United States. Historically, the dominant paradigm for bilateral environmental relations has been a state-centric, hierarchical, federal-to-federal model which espouses top-down management of environmental issues through international agreements, protocols, compacts and global conferences that minimize or ignore non-central state actors such as local governments (Bulkeley and Betsill 2003). However, the last two decades have witnessed the democratization of bilateral environmental relations including a proliferation in the quantity and variety of governance instruments being applied to transboundary water governance (Norman and Bakker 2009). By the end of the 1990s, scholars had gradually peeled back, and peered below, the layer of the nation-state to find a thick, intense web of sub-national cross-border interactions in North America between provinces and states (Munton and Kirton 1996, Alper 1997, Rabe 1997, Smith 2002), and identified intricate networks of cross-border exchange, a kind of micro-diplomacy not confined to the public sector but including epistemic communities involving non-governmental organizations and industry (Munton and Kirton 1996, VanNijnatten 2003). Consequently, rather than a simple federal-to-federal model, bilateral environmental relations between Canada and the United States have evolved into what Schwartz (2000) describes as a "complex web of interactions on many levels". Mayor Daley's Senate testimony as a delegate of the Great Lakes and St. Lawrence Cities Initiative, a bi-national sub-national organization which itself took part in an expansive multilevel, multi-stakeholder, and bi-national collaborative policymaking network, perfectly illustrates the new complexity of environmental policymaking around the Great Lakes.

Although we have attained a more sophisticated appreciation of this complexity, our understanding of who participates in the bilateral environmental relations between Canada and the United States remains incomplete because our understanding of the role local government plays within this mesh of relations remains a blank space. In part, this is the result of delayed recognition of local government as a bona fide international actor (Latouche 1988), which stems from the absence of formal-constitutional recognition of local government in the United States and Canada. In particular, the lack of formal authority in external relations and environmental policymaking combined has meant that local governments have traditionally been depicted as inconsequential actors in bilateral environmental policymaking and therefore predominantly excluded from bilateral governance structures, thus consequently ignored by scholars (Valiante 2007).

This chapter will contribute to the thin body of literature that directly acknowledges the increasing capacity of local government in transboundary environmental management defined as the degree to which actors can participate in, and influence, governance or decision-making processes (Norman 2009). The focus will be on the transformation of the role of local government in environmental policymaking in the Great Lakes Basin and specifically on a peculiar institutional manifestation of increased local government policy capacity—the Great Lakes and St. Lawrence Cities Initiative—which is the only example of a *transnational municipal network* operating across the border between Canada and the United States. It is through the Cities Initiative that local governments have attained a seat at the decision-making forums in the basin, and it is the Cities Initiative which acts as the major vehicle for the expansion of local government policy capacity.

AN UNFORESEEN EXPANSION OF LOCAL GOVERNMENT POLICY CAPACITY

A remarkable characteristic of the transformation of environmental management within the Great Lakes Basin over the last decade has been the internationalization of local government. Local governments have been engaging more meaningfully in environmental policymaking across the border with greater frequency than a formal reading of their constitutional authority would predict. A formidable example of individual local governments directly engaging in cross-border environmental management is the case of American Electric Power Corporation (AEP). In 2001, the City Council of Toronto instructed the City Solicitor to seek *Amicus Curiae* (Friend of the Court) status in a lawsuit against AEP launched by the United States federal government, eight states and 12 non-governmental organizations in a United States court[2]. The suit claimed that the company violated the Clean Air Act (CAA) by not including best available pollution reduction technologies when undertaking major modifications to its coal-fired electric plants. As one of the largest utilities in the United States, AEP was already the largest emitter of sulfur-dioxide and nitrogen-oxide in the country, pollutants which compromise both air and water quality in the basin. Toronto was granted Friend of the Court status in 2001 and filed an *amicus curiae* brief in 2005 demonstrating the impact of transboundary air pollution from U.S. coal-fired plants on Toronto's population. The report estimated that half of all air pollution affecting Toronto comes from the United States and particularly from Midwest coal-fired plants like those belonging to AEP. That same year, a report published by the Ontario Ministry of the Environment concluded that transboundary pollution contributes to 2,750 premature deaths from air pollution—about half the province's total—and CDN $5 billion in environmental damage (City of Toronto 2007).

In 2007, a landmark out-of-court settlement was reached in the AEP case which mandated emissions reductions for nitrogen-oxide and sulfur-dioxide of 69 and 79% respectively in coal-fired plants belonging to AEP, CDN $4.6 billion of upgrades, CDN $15 million in environmental penalties and CDN $60 million in clean-up costs (City of Toronto 2007). According to a staff report from Toronto's Medical Officer of Health, "this is the largest pollution reduction ever obtained from the owner or operator of a Clean Air Act stationary source" (City of Toronto 2007).

This cross-border activity was also the first time that the City of Toronto had used the American political and judicial system to achieve its domestic environmental and health policy aims. An official from the Environmental Protection Office (EPO) within the City of Toronto Public Health Department stated, "I am not aware of other municipalities in Ontario, or elsewhere in Canada that have used this approach to deal with a transbounary air pollution problem" (EPO personal correspondence 2008). It is also worth noting that the idea for filing an amicus brief originated from within the City of Toronto, from a report released by the Toronto Medical Officer of Health in 2000 and 2004 which studied the economic and personal impacts of illness resulting from air-pollution (City of Toronto 2000, 2004). This report by a local government was

[2]The states include: Connecticut, Maryland, Massachusetts, New Hampshire, New Jersey, New York, Rhode Island, and Vermont. The non-governmental organizations include: Ohio Citizen Action, Citizens Action Coalition of Indiana, Hoosier Environmental Council, Valley Watch, Ohio Valley Environmental Coalition, West Virginia Environmental Council, Clean Air Council, Izaak Walton League of America, National Wildlife Federation, Indiana Wildlife Federation, League of Ohio Sportsmen, Sierra Club and Natural Resources Defence Council.

quite novel at the time because its concern with the environment and health matters went beyond the traditional, narrow, constitutionally defined concerns of local government.

More importantly, the actions by the City of Toronto had an impact on the political and judicial process of a foreign country. The official at the EPO states that the brief was influential in assisting the parties in the United States to articulate their concerns about pollution from coal fired plants, in that "they could articulate the risks for health on not only American residents living downwind of old coal-fired power plants in the U.S., but could also demonstrate tremendous concern by Canadians living downwind. It demonstrated long-range transport issues, especially for ozone and particulates" (Environmental Protection Office 2008). Furthermore, according to the EPO official, the multi-year consultations and preparations before the filing of the brief "enabled collaboration and knowledge exchange with American NGOs and legal colleagues, and greater publicity of the impact of transboundary pollution on both sides of the border, through enhanced media attention in both countries" (Environmental Protection Office 2008). Finally, the EPO official argues that the publicity that was generated from the case put pressure on the Ontario Minister of the Environment to engage transboundary pollution and health issues. At the very first Shared Air Summit in 2005, a one-day meeting of experts and public officials from all levels of government across Canada and the United States hosted by the Ontario Minister of the Environment, Premier Dalton McGuinty announced that the province was seriously considering joining one of the fifty lawsuits dealing with air pollution that were taking place in the United States.

An indication that the strategy of using American political and judicial instruments to attain domestic aims by local governments in Canada was not an isolated incident but has become an embedded strategy, was the filing in November 2006 of a petition with the United States Environmental Protection Agency (EPA) to make a finding under section 115 (a) of the Clean Air Act (CAA) that air pollution emitted from 150 coal-fired plants located in seven midwestern states including Ohio, Illinois, West Virginia, Indiana, Kentucky, and Pennsylvania was endangering the health and welfare of Canadians (Ecojustice 2008a). The petition was organized by a Canadian nongovernmental organization called Ecojustice Canada, but its signatories included 13 Canadian cities and municipalities from Ontario and Quebec including the local or regional governments of Windsor, Essex, Halifax, Chateauguay, Peel, Chatham-Kent, Cornwall, Durham, Ajax, Laval, Goderich, Gatineau and Toronto which played a central role in the initiative. Mayor Miller of Toronto states, "Both cases [AEP and EPA] are examples where the city of Toronto chose to act independently of the position of national governments. In a post free-trade world, where so many institutions are linked together, it is logical for cities to spread out" (Miller 2010).

This was the first time that Canadian local governments had filed such a petition across the border directly with the EPA. It also signaled a change in local government consciousness and strategy. Rather than engaging American political and judicial actors as individual cities, local governments realized their strength in numbers and formed a temporary alliance. The EPA petition brought forth by 13 Canadian municipalities garnered intense media scrutiny and again brought transboundary pollution to the fore of the domestic political agenda on both sides of the border.

THE INSTITUTIONALIZATION OF LOCAL GOVERNMENT CROSS-BORDER COOPERATION

Perhaps the most remarkable recent development in the basin is that local government participation in cross-border environmental management has moved beyond these ad hoc and isolated activities and such participation is becoming increasingly institutionalized. The most important example is the emergence of the Great Lakes and St. Lawrence Cities Initiative[3]. In 2003, at the direction of Chicago Mayor Richard Daley, the Cities Initiative came into existence, becoming the first transnational municipal network operating across the border between Canada and the United States.

Transnational municipal networks are institutionalized forms of cross-border cooperation between local governments. Kern and Bulkeley (2009) observe that transnational municipal networks have three defining characteristics. First, they are more than a loose collaborative network of member-cities. They are self-governing institutions that operate from "non-hierarchical, horizontal and polycentric" (Kern and Bulkeley 2009) structures which include a secretariat, a board of directors, and a general assembly. Second, its member-cities are autonomous and can exit the organization at their choosing. Third, member-cities vote on and implement decisions taken by the network. Whereas Kern and Bulkeley also note that "such networks are often characterized as a form of self-governance", it is argued here that a more precise definition of transnational municipal networks emphasizes self-governance as a sine qua non of transnational municipal networks as a concept. The network's policies and strategies are driven by the member-cities that comprise the network, rather than by any other state or non-state actor.

As of 2010, the Cities Initiative consists of 73 mayors from American and Canadian local governments located within eight states and two provinces bordering the Great Lakes including Illinois, Indiana, Michigan, Minnesota, Ohio, New York, Pennsylvania, Wisconsin as well as Ontario and Quebec (see Figure 5.1 for map and Table 5.1 for list of member-cities).

[3]http://www.glslcities.org/

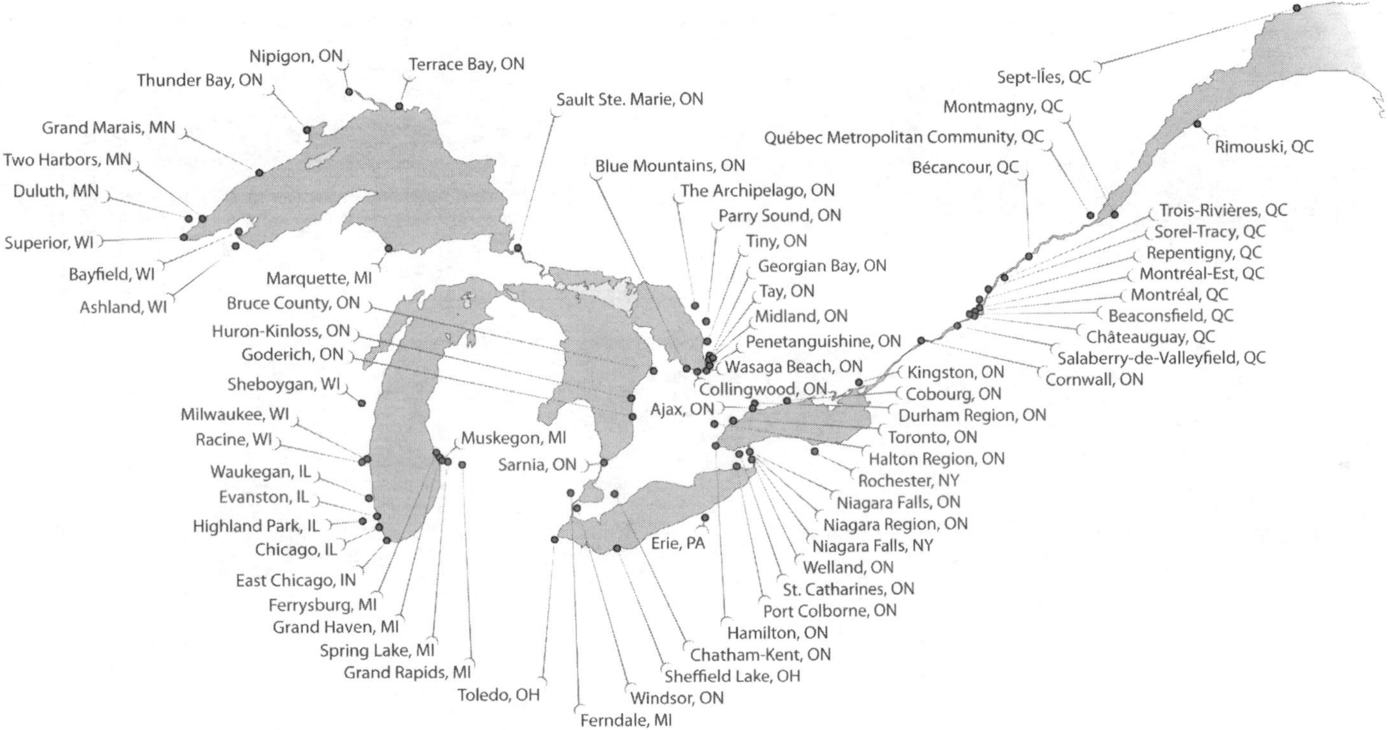

Fig. 5.1 GLSLCI Member-City Map 2010
Source: map by Matthew Campbell
www.mustardstudios.com

Table 5.1 GLSLCI Member-City List, 2010

ONTARIO	ONTARIO	Salaberry-de-Valleyfield	OHIO
Ajax (town)	Port Colborne	Sept-Iles	Sheffield Lake
The Archipelago (twnp)	St. Catharines	Sorel-Tracy	Toledo
Blue Mountains (town)	Sarnia	Trois-Rivieres	WISCONSIN
Bruce County	Sault Ste. Marie	ILLINOIS	Ashland
Carling Township	Tay (twnp)	Chicago	Bayfield
Chatham-Kent	Terrace Bay (twnp)	Evanston	Milwaukee
Cobourg	Thunder Bay	Highland Park	Racine
Collingwood	Tiny (twnp)	Waukegan	Sheboygan
Cornwall	Toronto	INDIANA	Superior
Durham Region	Wasaga Beach (town)	East Chicago	NEW YORK
Georgian Bay (twnp)	Welland	MICHIGAN	Niagara Falls
Goderich (town)	Windsor	Ferndale	Rochester
Halton Region	QUEBEC	Ferrysburg	PENNSYLVANIA
Hamilton	Beaconsfield	Grand Haven	Erie
Huron-Kinloss (twnp)	Becancour	Grand Rapids	
Kingston	Chateauguay	Marquette	
Midland (town)	Montmagny	Muskegon	
Niagara Falls	Montreal	Spring Lake (village)	
Niagara Region	Montreal-Est	MINNESOTA	
Nipigon (twnp)	Quebec Metropolitan Community	Duluth	
Parry Sound	Repentigny	Grand Marais	
Penetanguishene (town)	Rimouski	Two Harbors	

The emergence of this viable transnational municipal network in the Great Lakes challenges our present understanding of the limits of local government in bilateral environmental cooperation. The mere existence of the Cities Initiative directly contradicts the findings of a Policy Research Institute (PRI) report sponsored by the Government of Canada which concluded that local government cross-border cooperation is sparse, that it is predominantly bilateral in nature (i.e. between individual or very small groups of cities), and that it lacks institutionalization (Policy Research Institute 2005). It challenges previous findings that local actors in North America are unable to institutionalize their cross-border relations. For example, Brunet-Jailly (2004) concludes that a "competitive-city paradigm" permeates cross-border relations in North America predicated on intense market pressures that precludes local governments from establishing and sustaining institutionalized forms of cross-border cooperation. He argues that "local actors primarily compete" (Brunet-Jailly 2004) and that in North America "intra-metropolitan regional competition exists in principle". Consequently, Brunet-Jailly concludes that no cross-border institutions exist involving Canadian and American local governments. Norman (2009), in her analysis of transboundary water governance along the United States-Canada border, concludes that "although local actors are genuinely attempting to engage in transboundary governance, they encounter limited success due to inadequate resources and restricted capacity" and that "local actors are less able to transcend the border than their nation-state counterparts". In contrast to these findings, the Cities Initiative is the clearest manifestation of the increased policy capacity of American and Canadian local governments, and provides evidence that local governments are able to transcend market forces, inter-municipal competition, and sparse resources in developing institutionalized cross-border relations.

The emergence of the Cities Initiative introduced a completely new layer to the transnational environmental cooperation already taking place at different levels of government around the Great Lakes, joining an already dense web of networks including the Council of Great Lakes Governors, the Great Lakes Legislative Caucus, and the Great Lakes Commission, which operate at the regional level, as well as the International Joint Commission (IJC) and the North American Free Trade Agreement's Commission for Environmental Cooperation (CEC), a bi-national and tri-national agency respectively, that operate at the supranational level above the level of the nation state. These sub-national and supranational networks operate beside more traditional forms of bilateral cooperation between the federal, state and provincial governments in both countries, such as the Great Lakes Fishery Commission, as well as beside non-state actors, for example, non-governmental organizations such as Great Lakes United.

TRANSNATIONAL MUNICIPAL NETWORKS VERSUS PROTO-NETWORKS

We can illustrate the distinct nature of transnational municipal networks by briefly looking at an organization called the International Association of Great Lakes and St. Lawrence Mayors that had already institutionalized cross-border relations between Canadian and American local governments around the Great Lakes Basin before the Cities Initiative was formed. As the following section demonstrates, the International Association cannot be considered a TMN because its governance structure was composed of and is dependent upon actors from industry as well as provincial and state governments. Here we will see at play the critical element of self-governance to the definition of transnational municipal networks. The member-cities of the International Association were not self-governing therefore at best it can be considered a proto-network.

The International Association of Mayors was established in 1987 in Quebec City by the St. Lawrence Economic Development Council (SODES) which endowed this new organization with a near-total economic orientation focused almost exclusively on issues related to the shipping industry. SODES is an organization that promotes the interests of the shipping and maritime transportation industry, which is reflected in its governing structure. In 2010, for example, the executive was chaired by the Vice-President of Canada Steamship Lines and the Vice-President of the Societe des Traversiers du Quebec, which is an organization representing ferry services. Its Secretary was an associate of the law firm Langlois Kronstrom Desjardins, which specializes in maritime and admiralty law, and its Treasurer was Vice-President of Fednav Ltd., which is an international shipping company. Mayor Ross, of Superior, Wisconsin, who was a member of both the International Association and the Cities Initiative noted that the former organization was mostly composed of small, port cities, and he noticed in his first annual meeting that there were "a good number of shipping presentations by shipping organizations" and that the International Association was "more balanced towards industry, and less balanced towards environmental", a balance he determined to be "80/20" economic versus environmental (Ross 2009).

The Co-Secretariat of the International Association of Mayors was composed of the SODES and also the Great Lakes Commission. The Great Lakes Commission is a public agency formed in 1955 by an interstate compact between eight Great Lakes states and two associated provinces Quebec and Ontario principally in order to oversee the expanding maritime industry resulting from the construction of the St. Lawrence Seaway. Although broadening its scope of interests, the traditional concern of the Commission has been navigation and shipping (Great Lakes Commission 2005). Consequently, the International Association of Great Lakes and St. Lawrence Mayors was a creature, first and foremost, of industry, and second, of the states and provinces with shipping interests.

A former member of the Great Lakes Basin Commission states that the International Association of Mayors "did not participate in the Great Lakes community, it was not a force, it was not even an actor, it just existed, and it did not do anything except for proposing resolutions that were put forth by the Great Lakes Commission" (Great Lakes Basin Commission, personal interview 2009). In essence, the International Association of Mayors existed to legitimize policies of the Great Lakes Commission.

Chicago was the first big city to join the organization which until that time had been composed primarily of small port cities. According to the former member of the Great Lakes Basin Commission, "when Mayor Daley found out how ineffective and trivial it was, he transformed it" (Great Lakes Basin Commission, personal interview 2009). Meeting in Chicago in July 2003, the International Association of Great Lakes and St. Lawrence Mayors and the newly founded Great Lakes and St. Lawrence Cities Initiative agreed in principle to merge the two organizations. Many of the member cities of the International Association joined the Cities Initiative and assumed positions on the Board of Directors which was comprised solely of representatives of local government. The merger was completed in 2005. The headquarters was moved from Quebec City to Chicago. Mayors Badawey and Ross, who were members of both organizations, noted that after the merger an even orientation between economic and environmental concerns exists within the Cities Initiative (Badawey 2009, Ross 2009).

A EUROPEAN PHENOMENON

Transnational municipal networks have been studied most extensively, one could even say exclusively, in the European context where they were first observed. The first such network, the Council of European Municipalities and Regions (CEMR), was formed in 1951 and is today the most important network composed of over 50 national associations of towns, municipalities and regions across 37 countries representing over 100,000 member cities and regions. This network addresses every political issue from the local to the global. In the 1980s, Europe witnessed a significant expansion of transnational municipal networks and the emergence of different types. These new networks were based not on an aggregation of national municipal associations, but on direct membership of cities, such as the Eurocities network, with 120 members in over 30 countries. More recently, these general purpose networks have been supplemented by the emergence in the 1990s of regionally defined, direct membership organizations such as the Union of Baltic Cities, with over 100 cities in ten countries, the Alliance of the Alps, with 250 local authorities, as well as direct membership organizations centered around specialized issues, such as those dealing with climate change including the Climate Alliance, Cities for Climate Protection, and Energie-Cites totaling some 1,400 member cities across the continent (Heinelt and Niederhafner 2005, Kern and Bulkeley 2009).

The exclusive focus on transnational networks in the European context and as a European innovation has a basis in the long pedigree of transnational municipal mobilization stretching back to the 13th century in the form of the Hanseatic League. The Hanseatic League was an alliance of trading cities along the coast of northern Europe, stretching from the Baltic Sea to the North Sea, governed by its own set of laws and institutions that provided order and protection from outside threats. Continuing that legacy in the post-second-world-war era, the first "twin cities" were established in Europe and have since mushroomed to somewhere in the neighborhood of 30,000 twinning arrangements (Ewen and Hebbert 2007).

There is a body of literature that points to the distinctive character of the development of European social organization (Crouch 1999), involving a dramatic move away from formal, enclosed styles of government (John 2001) towards a system of multilevel governance, especially under the European Union (John 2000), and the adoption of greater pluralism in power relationships (Benington 1994) which favors the autonomy of cities (Bagnasco and Le Gales 2000). These observations undergird subsequent research demonstrating that networks of municipal government have flourished under the European Union and its multilevel governance framework (Le Gales 2002).

However, even in this European context of advanced multilevel governance, Kern and Bulkeley (2009) state that,

> while the role of local authorities in European integration has been discussed more frequently in acknowledgement of the growing multi-level character of European governance (John 2000, Le Gales 2002), the emergence of TMNs [transnational municipal networks] has been widely neglected. Most studies which mention such networks do not focus on them (John 2000, Marshall 2005), but tend to consider them solely in terms of their role as lobbying agencies (Bennington and Harvey 1998, Ward and Williams 1997) and pay little attention to their internal dynamics.

Research into the internal and external dynamics of these networks has been limited to three studies (Heinelt and Niederhafner 2005, Kern 2007, Kern and Bulkeley 2009). Hence, this chapter expands a shallow body of literature on transnational municipal networks in general, and also introduces the first purely North American perspective into the discourse (Kusmierczyk 2010).

INTERNAL STRUCTURES

The bureaucratization of the Cities Initiative follows the established "non-hierarchical, horizontal and polycentric" pattern typical of transnational municipal networks (Kern and Bulkeley 2009; see Fig. 5.2). Three standard levels or groups of actors emerged: (i) an international *secretariat* located in Chicago (ii) the *executive* composed of chairpersons, a board of directors comprising the mayors of member-cities selected to those posts, and a general assembly of all the member cities who gather annually to vote on resolutions, and (iii) the individual *member-cities* themselves (Kern and Bulkeley 2009).

As in the European transnational municipal networks, the Cities Initiative has established sub-networks to deal with specific issues, although these are less formal than the separate governance structures operating in European TMNs which deal with specific national territories or policy areas. For example, the Climate Alliance and Energie-Cites have decentralized structures with national satellite offices spread throughout various European countries. The Union of

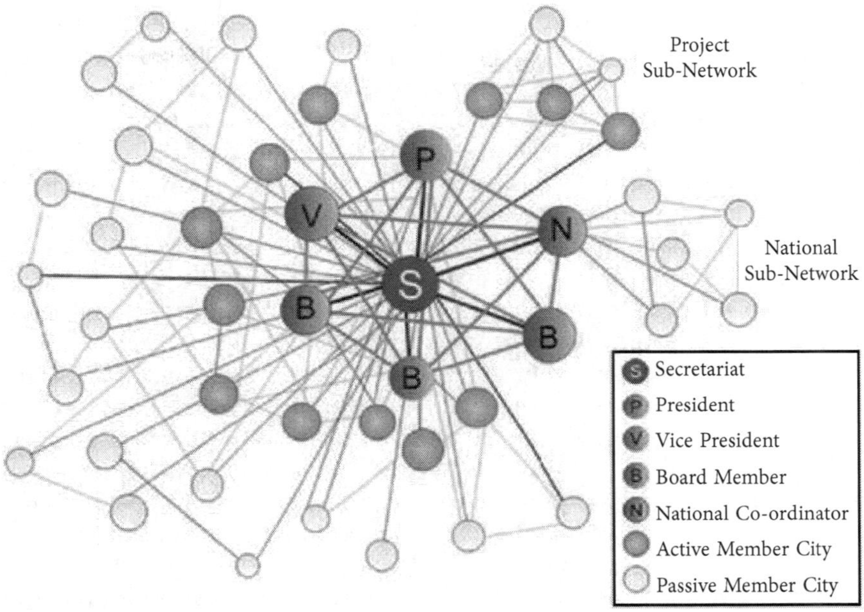

Fig. 5.2 Structure of Transnational Municipal Networks
Source: Kern and Bulkeley 2009

Baltic Cities has its headquarters in Gdansk, Poland while the Environment and Sustainable Development Commission is situated in Turku, Finland and its Energy Commission in Oskarshamn, Sweden. In contrast, although the Cities Initiative has representatives in Ottawa, Montreal, and Quebec City, no dedicated permanent office exists in these cities in order to coordinate regional or policy specific issues in Canada.

The internal complexity of the European transnational municipal networks simply reflects a larger resource base. Although revenue from membership dues is similar for both organizations, European transnational municipal networks receive substantial funding support from public sources such as the European Commission, while no such external funding is available for North American networks. For example, the Cities Initiative collected approximately US$ 306,000 from its member-cities in 2008 (GLSLCI 2008c), while the Union of Baltic Cities (UBC) plans to collect EUR 243,135 in 2010 (UBC 2010). The amounts are comparable. However, in addition to membership dues, the UBC's Environment and Sustainable Development Commission alone received EUR 308,560 from the European Union for various environmental projects in the basin in 2009 (UBC 2010). This does not include external funding from other important partners such as the Nordic Council of Ministers, the Finnish Ministry of the Environment, or the Baltic Sea Unit at the Swedish

International Development Cooperation Agency. The Cities Initiative, on the other hand, received its one major external grant in 2007 from a private foundation for US $500,000 that was dispensed over a period of three years. The major difference is that whereas the Cities Initiative likely cannot count on continuous access to such large grants from that particular private funding source, European networks such as the UBC repeatedly return to, and draw from, a dedicated funding source from the EU that is larger and has a longer time horizon. For example, from 2000-2006, the UBC drew from the Baltic Sea Region INTERREG III B Neighborhood Program which makes available EUR 196 million for transnational projects in the basin[4]. The Baltic Sea Region Program 2007-2013 was established through the EU Structural Fund program, providing EUR 236.6 million for transnational projects.[5] More importantly, in 2009, the European Union created the EU Strategy for the Baltic Sea Region, its first macro-regional policy for the basin. Although no new lines of funding were established, the strategy focuses the work of existing financial institutions in the region on priority areas, with environmental sustainability identified by the European Council as being the foremost challenge (European Commission 2009). For example, the already existing Cohesion Policy, which will disburse EUR 50 billion in the region over the period 2007-2013, is a key instrument of the Baltic Sea strategy, and the Common Fisheries Policy—contributing another EUR 1.25 billion—is another important source of funding particularly for the marine research community (European Commission 2009).

Nonetheless, despite limited resources, informal regional meetings of Cities Initiative members do occur, for example, as when Toronto hosted a meeting of Ontario member cities in 2007 in order to discuss regional issues. In addition, certain board members of the Cities Initiative have regional designations based on four geographical quadrants—Lower Lakes, Upper Lakes, St. Lawrence/Quebec, and Great Lakes/Ontario—and a responsibility to coordinate and represent issues pertaining to those sub-regions. At the moment these regional designations have not coalesced into an organized sub-network, although an unsuccessful initiative was launched in 2009 to establish a parallel network of Quebec municipalities along the St. Lawrence River called the Forum Des Municipalites Riveraines au fleuve Saint-Laurent. The Forum, which was spearheaded by the mayors of Montreal, Sorel and Beaconsfield, was rejected by the Cities Initiative as an unnecessary duplication of existing governing structures.

The executive of the Cities Initiative is composed of the Board of Directors made up of 16 mayors, an elected chair serving a one-year term, a vice-chair and a secretary-treasurer. Mayor Daley of Chicago and Mayor Miller of Toronto each held a Founding Chair position on the Board of Directors during their time in office. It remains unclear whether they will retain this position

[4]www.bsrinterreg.net

[5]http://eu.baltic.net/Funding.113.html?

once they leave office. The Board elects new members to the executive. The Board meets mid-winter to assess progress and to bring new issues to the table. These issues are presented to the General Assembly at the Annual Meeting in the summer. At the Annual Meeting, and with help of the Executive Director, common policy positions, resolutions, and a general agenda are proposed which are debated, voted on and adopted by the Board of Directors. These common positions or resolutions are then presented and voted on by the member cities attending the General Assembly at the annual conference. In addition to fulfilling internal governing functions, the members of the Board of Directors also represent the Cities Initiative in its external relations by testifying before legislative bodies in federal and state capitals, speaking at conferences, taking part in regional stakeholder meetings, publishing editorials in newspapers, or engaging in advocacy. For example, on 30 July 2008, Mayor George Heartwell of Grand Rapids, Michigan, testified on behalf of the Cities Initiative as its vice-chair in front of a United States Senate Committee deliberating on the "Water Compact", essentially a groundbreaking piece of water diversion legislation.

The local governments comprising the Cities Initiative vary in size representing large cities such as Chicago (pop. 2,837,000) and Toronto (pop. 2,500,000), regional municipalities such as Niagara (pop. 427,000), towns such as Wasaga Beach (pop. 15,000) and appropriately named townships like Tiny (pop. 10,750). It also includes one village called Spring Lake (pop. 2,514) and a geographical curiosity in the collection of island and bays covering some 600 square kilometers that forms the township of The Archipelago (pop. 576) on the Georgian Bay. The cities are geographically dispersed, stretching from the most northern city, Nipigon, at the top of Lake Superior, as far east as Rimouski, Quebec, on the St. Lawrence River, to Sheffield Lake on the southern shores of Lake Erie, and Duluth on the western tip of Lake Superior.

However, as is often the case in European networks, there is a clear territorial pattern within the Cities Initiative. In Europe, language is often the "critical issue" in setting territorial patterns (Kern and Bulkeley 2009). For instance, Climate Alliance, the largest transnational municipal network dealing with climate change issues, has faced problems recruiting members from countries other than Germany and Austria because its working language is German (Kern and Bulkeley 2009). The working language in the Cities Initiative is English, but this has not presented a difficulty as evidenced from the fact that nine member cities come from Quebec, though it is possible that some smaller, rural French speaking towns may well have found language to be a barrier to membership.

Territorial patterns have emerged in a distinct clustering effect with two groups of cities concentrated around the most powerful members—Chicago on Lake Michigan and Toronto on Lake Ontario including a cluster of cities north of Toronto along the southern shore of Georgian Bay. This pattern reflects the influence that these two major cities have had on the recruitment process and their gravitational power as global cities (Sassen 1991). Recently, Foreign Policy ranked Chicago and Toronto within their top global cities index (Foreign Policy 2008). Recognizing the influence of these two global metropolises, Denis Lapointe, the Mayor of Salaberry-de-Valleyfield, a Quebec town with a population of 30,000, said that "we realized that with the capacity to pressure government of the mayor of Chicago and Toronto, things have changed, we were finally listened to" (Lapointe 2009). The clustering effect around Toronto and Chicago may be due to the fact that smaller local governments within the vicinity of these two cities have even more incentive to collaborate with Toronto and Chicago because they have a better chance of having their localized issues taken up and placed on the national agenda. In addition to a patterned dispersal, there is also a marked Canadian bias in the membership. Forty-seven of the 73 members—or 64.5%—come from Canada, with Ontario alone enlisting nearly half—31 members. However, the Board of Directors is divided equally between eight Canadian and eight American mayors. The cities represent both rural and urban interests, different geographical concerns based on their proximity to a particular Great Lake, varied socio-economic statuses, different cultural-linguistic backgrounds, and they voluntarily join for a number of reasons including potential access to information, funding, and political support.

Cities that join transnational municipal networks distinguish themselves from other cities in that they already tend to be highly motivated and active in a certain policy field, to the degree that Kern and Bulkeley (2009) argue that TMNs are "primarily networks of pioneers for pioneers". Nonetheless, even transnational municipal networks can be further divided into active and passive members, with the latter forming the majority. As a result, the typical polycentric governance structure in these networks also appears in the Cities Initiative. In its 2007 Annual Report, the Great Lakes and St. Lawrence Cities Initiative acknowledged that it had purposefully undertaken the "development of a core membership base" (GLSLCI 2008a) which was strategically cultivated because "establishing a central group of committed members has helped the organization focus its efforts around key issues throughout the Basin".

CITIES INITIATIVE EXPANDING LOCAL GOVERNMENT POLICY CAPACITY

The Cities Initiative has been the primary vehicle for the expansion of local government policy capacity within the Great Lakes Basin. It is important to note that the Great Lakes Regional Collaboration Strategy, the large multilevel, multi-stakeholder network mentioned earlier, incorporated local government participation not on an ad hoc basis but through the inclusion of the Great

Lakes and St. Lawrence Cities Initiative as a member of its Executive Committee and as the sole representative of regional local government interests. When state governors and state legislatures began deliberations in 2005 on the creation of a Great Lakes Water Compact Council to regulate out-of-basin water diversion and other water quantity issues, the Executive Director of the Cities Initiative was the sole representative of regional local governments on its Advisory Committee. In March, 2005, Mayor Daley and Mayor Miller, as the founding chairs of the Cities Initiative, took part in a joint address before the Great Lakes Congressional Breakfast in Washington, D.C. It was the first time that a Canadian mayor had addressed this annual gathering of congressmen, governors, mayors, and other stakeholders. Miller notes, "I met a number of congressmen, Rahm Emanuel among them. I can call him up now if we need to reach the Obama administration" (Miller 2010). The Executive Director of the Cities Initiative was also appointed by President Bush in 2006 to the joint United States and Canada Great Lakes Fishery Commission which has been managing fisheries in the basin since 1954. The Executive Director of the Cities Initiative also represented the local government perspective on the Agreement Review Committee (ARC) set up by the Bi-national Executive Committee, working alongside officials from the Environmental Protection Agency and Environment Canada to formulate ideas on improving the Great Lakes Water Quality Agreement. As a result, the Committee adopted a principle recognizing that local governments play a key role in the development and the implementation of the Great Lakes Water Quality Agreement.

The work of the ARC may appear symbolic except when seen in the light of recent developments. United States Secretary of State Hillary Clinton and Canadian Minister of Foreign Affairs Lawrence Cannon agreed on 13 June 2009, that both countries would undertake a formal review of the Great Lakes Water Quality Agreement for the first time in more than two decades. This announcement was based in part on the recommendations made by the Agreement Review Committee (ARC). If those recommendations relating to local governments are adopted by the new Great Lakes Water Quality Agreement, it will mark the first formal recognition of the role of cities in environmental management around the Great Lakes and surely a sign of their growing importance.

The most significant example of this institutional imbedding was the signing of the Canada Ontario Agreement Memorandum of Cooperation (COA MOC) on 17 July 2008, between the Ontario Government and the Great Lakes and St. Lawrence Cities Initiative. The memorandum commits the Cities Initiative to facilitate a new process of collaboration between the mayors of Ontario's local governments and the signatories from the three provincial government ministries including the Ministry of the Environment (MOE), the Ministry of Natural Resources (MNR), and the Ministry of Agriculture, Food and Rural

Affairs (MAFRA) thereby establishing a new forum for provincial-municipal cooperation regarding the environmental management of the Great Lakes.

This agreement is significant for bringing local governments closer to the COA process. The COA is the primary mechanism by which Ontario assists Canada in meeting some of its commitments under the Great Lakes Water Quality Agreement. The COA establishes both the shape of the programs and the size of the budget dedicated to those programs. The timing of the memorandum of cooperation is also very significant. Since 1971, the COA has been renegotiated numerous times. Therefore, the signing of the COA memorandum indicates that local governments will have, for the first time, through this new forum, a strategic voice in shaping the discussions within the most important intergovernmental institution in environmental policy on the Canadian side of the Great Lakes.

After signing the agreement, John Garretsen, then Ontario Minister of the Environment, announced that "today we *formally* recognize the importance of engaging municipalities as key partners in sustaining the health and vitality of the Great Lakes" (City of Toronto 2008). The then Minister of Natural Resources, Donna Cansfield, stated that the memorandum "marks the beginning of an exciting new era of collaboration among municipal, provincial and federal governments" (City of Toronto 2008), while Leona Dombrowsky, then Minister of Agriculture, Food and Rural Affairs, stated that "this agreement reinforces the collaborative approach of the Government of Ontario" (City of Toronto 2008).

The work of the Cities Initiative was central to attaining greater inclusion of municipal governments in the COA process, and it played a key role in facilitating the finalization process which led up to the signing of the COA MOC. As early as 16 February 2007, in a letter to the Manager of the Great Lakes Office, Land and Water Policy Branch at the Ontario Ministry of the Environment, the Executive Director of the Cities Initiative called on the provincial government to give the Cities Initiative a greater role in future renegotiations of the COA than the heretofore normal mechanisms which had been reserved for NGOs, business, academic institutions and cities. In the letter the Executive Director implied that these were inadequate mechanisms. Rather, he argued that it was essential that the future renewals of the COA "involve municipal governments as full partners" (GLSLCI 2007c). The Cities Initiative played a key role in convincing the provincial government to expand the COA process to include local government.

Furthermore, during the six-month run-up to the signing of the memorandum, the Cities Initiative helped organize a Great Lakes Municipal Working Group which organized and prioritized municipal issues into common positions. The Municipal Working Group was made up of nine mayors from the Great Lakes and St. Lawrence Cities Initiative, 16 senior officials from municipal governments including chief administrative officers, managers of various departments including the environment, transportation, public utilities, and water infrastructure, three directors from the secretariat of the Cities Initiative and the senior policy advisor from the Association of Municipalities of Ontario. On this last point, it is quite important that the Association of Municipalities of Ontario, the chief provincial association of municipal governments, supported the fact that the Cities Initiative was the key player in breaking new ground not only on behalf of the members of the Cities Initiative, but on all municipal governments in Ontario, in its relationship with the Province and, indirectly, the federal government (GLSLCI 2008g).

While working together to prioritize its set of common positions, the members of the Municipal Working Group were simultaneously working within a Joint-Municipal-Provincial Committee which included 12 senior officials from the MOE, MNR, and the MAFRA, as well as participants from the Ministry of Tourism, Ministry of Health and Long-Term Care, and the Ministry of Economic Development, to identify areas of common interest and to produce recommendations that the province could potentially act upon both within and outside the parameters of the COA. From this new collaborative process emerged a report published in May 2009 produced by the nine mayors from the Great Lakes and St. Lawrence Cities Initiative who participated in the Municipal Working Group entitled At the Shoreline: A Mayors' Collaborative Action Plan to Protect the Great Lakes, which presented a five-point plan directly to the cabinet ministers representing the MOE, the MNR, and the MAFRA (GLSLCI 2009a).

FROM EMBEDDED TO EMBOLDENED NETWORKS OF LOCAL GOVERNMENT

During the Annual Board Meeting of the Cities Initiative in Trois-Rivieres, Quebec, in 2009, while congratulating the members for embedding the network deeper into the governance structures through the COA MOC and the Regional Collaboration, a senior official of External Affairs for the City of Chicago stated that the network can "declare victory on getting seats, but now we must exercise those seats" (GLSLCI 2009b), thus signaling a subtle shift from a focus on embedding and empowering the network to a focus on effectively employing their new influence.

A document handed out at the Board Meeting contained a section on the "Future Direction" of the Cities Initiative which stated there were two overarching goals for the network in the next five year period, 2009-2014. The first is to continue to better integrate environmental, social, and economic concerns in the policies and agendas of local government, in what has been termed by the members the Triple Bottom Line. The second goal is to expand and deepen collaboration with other orders of government (GLSLCI 2009e).

Towards this second goal, on 5 May 2009, a month before the Annual Board Meeting, several members of the cities initiative presented the Ontario Ministers of the Environment, Natural Resources, and Agricultural, Food and Rural Affairs their report titled At the Shoreline which included a five-point action plan (GLSLCI 2009a). Building on the success of the COA memorandum of cooperation, the first and most important point called for the establishment of "a new collaborative relationship among federal, provincial and municipal governments to reinvigorate and reorient Great Lakes protection" (GLSLCI 2009a). In light of the fact that no such intergovernmental forum exists for cooperation and collaboration between all levels of government on Great Lakes issues, the mayors of the Cities Initiative called for the establishment of a federal-provincial-municipal Great Lakes Table (GLSLCI 2009a). Towards that goal, the first ever tripartite meeting was held in October 2009, thus fulfilling the vision of former Prime Minister Paul Martin who stated that, "We cannot as a nation, remain isolated inside the old silos that prevent partnership between the orders of government" (quoted in Sutcliffe 2007).

The establishment of the COA memorandum of cooperation between provincial and municipal governments, combined with the fact that plans have been put forth for the establishment of a federal-provincial-municipal forum in the form of the Great Lakes Table, demonstrates that the Cities Initiative has been an active catalyst not only for the expansion of the multilevel governance system around the Great Lakes but, more accurately, they have been an active catalyst for the expansion of their own policy capacity and role within an expanding multilevel governance system around the Great Lakes, thereby lending support to Jeffery's (2000) argument that local governments are more than passive recipients of authority and resources bequeathed from higher levels of government. As in Europe, the expansion of local government policy capacity "reflects and constitutes" (Kern and Bulkeley 2009) the growth of multilevel governance around the Great Lakes Basin.

THE PRACTICAL SIGNIFICANCE OF INCREASED LOCAL GOVERNMENT POLICY CAPACITY

There are several compelling reasons, besides updating the record, for bringing local government bilateral environmental cooperation into the spotlight. First and foremost, it is becoming apparent that environmental challenges will not be solved at the international level or through the cooperation of federal governments alone. As Betsill and Bulkeley (2006) argue with regards to climate change, "it is increasingly clear that nation-states will be unable to meet their international commitments for addressing climate change without more explicit engagement with sub-national action" adding that especially in the area of airborne pollutants, for example, "the local is the most appropriate political jurisdiction for bringing about any necessary reductions in these emissions".

Cities potentially have the greatest impact in addressing most environmental challenges, not just climate change, for the simple fact that three-quarters of the developed world now lives in urban regions, which makes them major consumers of energy and emitters of pollution but also sites with the greatest potential for positive change (United Nations 2000). In their study, "The Carbon-Neutral Individual", Vandenbergh and Steinemann (2007) calculate that individual Americans, through everyday transportation and household energy use, contribute roughly one-third of all carbon emissions in the United States, a proportion that exceeds even total industrial emissions. They conclude that tremendous reductions in pollution can be achieved through small behavioral changes of individual citizens that have a monetary benefit and requires little effort, what they term "low-hanging fruit". Since six-out-of-ten Americans live in the 100 largest metropolitan centers, local governments, as the government closest to individual citizens, are arguably the best placed governments to influence their behavior and make a dramatic positive environmental impact by enacting by-laws, implementing urban planning, organizing information and outreach programs with regards to "low-hanging fruit" such as recycling, engine idling, home insulation and heating, etc.

Cities not only have influence over their inputs and outputs through their control over land-use planning, transportation supply, energy management, and wastewater treatment, but they have extensive expertise to offer federal governments and international regimes in environmental conservation, prevention, mitigation and remediation (Bulkeley and Betsill 2003, Vandenbergh and Steinemann 2007). With regards to other contributable resources, in terms of dollars-and-cents, American and Canadian local governments around the Great Lakes cumulatively spend US $15 billion annually on water quality management in the basin including, for example, capital costs for wastewater treatment and restoration projects (GLSLCI 2008a). Simply put, it is time for a reappraisal of local government policy capacity in cross-border environmental cooperation around the Great Lakes so that our perceptions are commensurate with their contributions.

Another compelling reason for analyzing local government is that their increased participation in environmental governance around the Great Lakes has the potential to improve accountability within water governance structures, adding an extra layer of voluntary oversight to an otherwise deteriorating regulatory system in the United States and Canada. The problem with accountability is acknowledged on both sides of the border. As early as 2001, a report by the Canadian Commissioner of Environment and Sustainable Development on the state of Great Lakes governance stated that the Canadian federal government "decided to rely on others, and when others failed to deliver, it did not assume the lead" (Office of the Auditor General of Canada 2001) and that "limited use of federal powers, weakness in basic management and accountability and the politics of federal-provincial relations have all played a part" in the failure to meet the goals set forth in the Great Lakes Water Quality Agreement. Shortly thereafter, in 2003, a report prepared by the United States General Accounting Office on the Great Lakes highlighted the fragmentation and lack of accountability in the American governance structure by observing that there were 148 federal and 51 state restoration programs and that "the myriad of current strategies and coordination efforts makes it difficult to determine which organization is in charge" (United States General Accounting Office 2003).

American and Canadian observations about the gross inadequacies in their water governance structures were synthesized and given their most forceful expression in the 13th Biennial Report on Great Lakes Water Quality published by the International Joint Commission (IJC) in 2006, which for the first time focused entirely on the problem of accountability. The IJC report concluded that "there is now a compelling need to…improve accountability so that persistent and emerging challenges to the Lakes can be effectively and definitively addressed" (International Joint Commission 2006), arguing that the objectives of the Great Lakes Water Quality Agreement are "indefinite and outdated, and there are few limits and thresholds or schedules for action", that the "precise roles and responsibilities for achieving the Agreement's objectives are unspecified", that the monitoring system is "still insufficient as well as uncoordinated across jurisdictional boundaries, and fails to provide the comprehensive, focused information required to assess progress under the Agreement", and that requirements for reporting "have become outdated and…left undone". The holes in the governance structure identified by the report leads to the general conclusion that "we have no collective sense of the progress we are making" in managing the Great Lakes environment and "the public has neither an accounting of specific progress toward identified goals nor confidence that any reports tell the full story".

The accountability gap is not restricted to the Great Lakes. A recent article published in the New York Times on 13 September 2009, entitled "Toxic Waters", found that violations of the Clean Water Act have risen at a steady pace across the United States, and that between 2004-2009, factories and plants have violated the Clean Water Act 506,000 times by over 23,000 companies (Duhigg 2009). The report states that, "tap water in parts of the Farm Belt, including cities in Illinois, Kansas, Missouri and Indiana, has contained pesticides at concentrations that some scientists have linked to birth defects and fertility problems" (Duhigg 2009). One in 10 Americans, meaning 31 million people, have been exposed to drinking water "that contains dangerous chemicals or fails to meet a federal health benchmark" (Duhigg 2009). Sixty percent of polluters in violation of EPA standards were deemed to be in "significant noncompliance" which means that the discharges of cancer causing agents such as copper, lead, zinc, chlorine and selenium into water systems were not being measured or reported.

More troubling, and reflective of the moth-eaten regulatory system in the country, less than 3% of Clean Water Act violations resulted in fines. In an internal memo sent by the current EPA administrator Lisa Jackson to her enforcement deputy, she noted that "in many parts of the country, the level of significant noncompliance with permitting requirements is unacceptably high and the level of enforcement activity is unacceptably low" (Duhigg 2009). The significance of local government mobilization effectively as voluntary or informal oversight agents is a critical development in the light of recent comments by Minnesota Congressman James Oberstar that "the EPA and states have completely dropped the ball. Without oversight and enforcement, companies will use our lakes and rivers as dumping grounds—and that's exactly what is apparently going on" (Duhigg 2009). In such a context, the mobilization of local government as additional informal oversight agents, as catalysts for the improvement of accountability structures within the water governance system, takes on a new imperative.

CONCLUSION

Cross-border environmental cooperation between local governments is increasingly becoming a vital instrument for managing the waters of the Great Lakes Basin and a vital component in managing the bilateral environmental relations between Canada and the United States which were once the exclusive domain of federal governments. Their growing involvement in the Regional Collaboration, the Canada-Ontario Agreement (COA) and through a recently established, and heretofore non-existent, trilateral intergovernmental forum are but a few examples that testify to the democratization of bilateral environmental relations around the Great Lakes Basin. As the first bi-national review of the Great Lakes Water Quality Agreement in two decades moves forward, it represents an ideal opportunity to formally embed local governments within the governance structures in recognition of their growing participation in, and contribution to, the management of a shared and treasured ecosystem.

REFERENCES

Allee, D.J. 1993. Subnational Governance and the International Joint Commission: Local Management of United States and Canadian Boundary Waters. Natural Resources Journal 33.

Alper, D. 1996. The Idea of Cascadia: Emergent Transborder Regionalisms in the Pacific Northwest-Western Canada. Journal of Borderland Studies 10 (2).

Alper, D. 1997. Transboundary Environmental Relations in British Columbia and the Pacific Northwest. The American Review of Canadian Studies 27 (3).

Anderson, E. Personal interview. June 18, 2009.

Badawey, V. Personal interview. June 18, 2009.

Bagnasco, A. and P. Le Gales [eds.]. 2000. Cities in Contemporary Europe. Cambridge University Press. Cambridge, UK.

Bennington, J. and J. Harvey. Transnational Local Authority Networking within the European Union: Passing Fashion or New Paradigm? *In*: D. Marsh [ed.]. 1998. Comparing Policy Networks. Open University Press. Buckingham, UK.

Betsill, M.M. 2001. Mitigating Climate Change in US Cities: opportunities and obstacles. Local Environment 6 (4).

Betsill, M.M. and H. Bulkeley. 2006. Cities and the Multilevel Governance of Global Climate Change. Global Governance 12 (2).

Betsill, M.M. and H. Bulkeley. 2007. Looking Back and Thinking Ahead: A Decade of Cities and Climate Change Research. Local Environment 12 (5).

Brunet-Jailly, E. 2004. Comparing Local Cross-Border Relations Under the EU and NAFTA. Canadian American Public Policy 54.

Bulkeley, H. and M. Betsill. 2003. Cities and Climate Change: Urban Sustainability and Global Environmental Governance. Routledge, New York, USA.

City of Toronto. 2000. Air Pollution Burden of Illness in Toronto. http://www.toronto.ca/health/burdenof.pdf

City of Toronto. 2004. Air Pollution Burden of Illness in Toronto: 2004 Summary. http://www.toronto.ca/health/hphe/pdf/air_and_health_burden_illness.pdf

City of Toronto. 2007. U.S. Court Decision on Transboundary Air Pollution. http://www.toronto.ca/legdocs/mmis/2008/hl/bgrd/backgroundfile-9849.pdf

City of Toronto. 2008. News Release July 17. http://www.toronto.ca/greatlakes-stlawrence/pdf/GLSLCI_English_Release_17July2008.pdf

Crouch, C. 1999. Social Change in Western Europe. Oxford University Press, Oxford, UK.

Duhigg, C. 2009. Toxic Waters. New York Times: September 12, 2009.

Ecojustice. 2008a. Ecojustice letter to EPA July 30.

Ecojustice. 2008b. Phone interview November 27.

Environmental Protection Office, City of Toronto. 2008. Personal correspondence December 1, 2008.

Ewen, S, and M. Hebbert. 2007. European Cities in a Networked World during the Long 20th Century. Environment and Planning C: Government and Policy 25 (3).

Foreign Policy. 2008. The 2008 Global Cities Index.

Frame, T.M. and T. Gunton, and J.C. Day. 2004. The Role of Collaboration in Environmental Management: An Evaluation of Land and Resource Planning in British Columbia. Journal of Environmental Planning and Management 47 (1).

Great Lakes Commission. 2005. http://www.glc.org/advisor/report/2005/2005AR-final.pdf

Great Lakes Basin Commission. 2009. Personal interview. October 1.

GLSLCI. 2006a. David Ullrich testimony before Senate Environment and Public Works Committee, March 16.

GLSLCI. 2007a. Mayors Call for Less Pollution From BP Refinery to Lake Michigan, July 18.

GLSLCI. 2007b. We must all protect, not contaminate, lake. Chicago Tribune, July 20.

GLSLCI. 2007c. Ullrich letter to Carolyn O'Neill. February 16.

GLSLCI. 2007d. Ullrich testimony in front of Chicago City Council. August 2.

GLSLCI. 2008a. Annual Report 2007-2008.

GLSCI. 2008b. Board of Directors Meeting July 16.

GLSLCI. 2008c. Consolidated Financial Reports.

GLSLCI. 2008d. Personal interview with official. January 13.

GLSLCI. 2008e. Personal interview with official. June 9.

GLSLCI. 2008f. Personal interview with official. July 18.

GLSLCI. 2008g. Summary of Discussion: Canada-Ontario Agreement of Cooperation Municipal Great Lakes Working Group Meeting #1. http://www.glslcities.org/documents/MWGMtg1Oct28SummaryFinal111008.pdf

GLSLCI. 2009a. At the Shoreline: A Mayors' Collaborative Action Plan to Protect the Great Lakes.

GLSCI. 2009b. Board of Directors Meeting June 17, 2009.

GLSLCI. 2009c. General Assembly, June 19, 2009.

GLSLCI. 2009d. Letter from John Piskura April 29.

GLSLCI. 2009e. 2009 Annual Members Meeting – Agenda.

Heartwell, G. Personal Interview. June 18, 2009.

Heinelt, H. and S. Niederhafner. 2005. Cities and Organized Interest Intermediation in the EU Multi-level System. Discussion Paper Nr. SP IV 2005-103: Social Science Research Center Berlin. Berlin, Germany.

International Joint Commission. 2006. 13th Biennial Report on Great Lakes Water Quality.

Jeffery, C. 2000. "Sub-national mobilization and European integration: Does it make any difference?" Journal of common market studies 38(1).

John, P. 1994. UK Subnational Offices in Brussels: Regionalisation or Diversification? Regional Studies 28 (7).

John, P. 2000. The Europeanization of Sub-national Governance. Urban Studies 37 (5-6).

John, P. 2001. Local Governance in Western Europe. Sage Publications Inc. London, UK.

Kern, K. 2007. When Europe Hits City Hall: The Europeanization of Cities in the EU Multi-level System. Biennial Conference of the European Studies Association. Montreal, Canada.

Kern, K. and T. Loffelsend. 2004. Sustainable Development in the Baltic Sea Region: Governance beyond the Nation State. Local Environment 9 (5).

Kern, K. and H. Bulkeley. 2009. Cities, Europeanization and Multi-level Governance: Governing Climate Change through Transnational Municipal Networks. Journal of Common Market Studies 47 (2).

Kusmierczyk, I.W. 2010. Water Over the Bridge: Examining Transnational Municipal Networks of American and Canadian Local Governments in the Context of Canada-U.S. Bilateral Environmental Relations within the Great Lakes Basin. PhD Dissertation, Vanderbilt University, Nashville, Tennessee.

Lapointe, D. Personal interview. June 18, 2009.

Latouche, Daniel. 1988. State Building and Foreign Policy at the Subnational Level. *In:* I. Duchacek, D. Latouche and G. Stevenson [eds.] Perforated Sovereignties and International Relations: Trans-Sovereign Contacts of Subnational Governments. Greenwood Press. Westport, Connecticut, USA.

Le Gales, P. 2002. European Cities: Social Conflicts and Governance. Oxford University Press. Oxford, UK.

Marks, G. Structural Policy in the European Community. *In*: A. Sbragia [ed.]. 1992. Europolitics: Institutions and Policymaking in the "New" European Community. The Brookings Institute. Washington DC., USA.

Marks, G. An Actor-Centred Approach to Multilevel Governance. *In*: C. Jeffery [ed.]. 1997 The Regional Dimension of the European Union. Frank Cass.London, UK.

Marks, G. and L. Hooghe. Contrasting Visions of Multi-level Governance. *In*: I. Bache and M. Flinders [eds.]. 2004. Multi-level Governance. Oxford University Press.Oxford, UK.

Marshall, A. 2005. "Europeanization at the Urban level: local actors, Institution and the dynamics of multi-level interaction." Journal of European public policy 12(4).

Miller, D. Personal interview. January 11, 2010.

Munton, D. and J. Kirton. 1988. Province-State Interactions Project: Final Report. Ottawa: Canadian Department of External Affairs and International Trade.

Munton, D. and J. Kirton. 1996. Beyond and Beneath the Nation-State: Province-State Interactions and NAFTA. International Studies Association Annual Conference. San Diego, California, USA.

Norman, E.S. 2009. Navigating Bordered Geographies: Water Governance Along the Canada-United States Border, PhD Dissertation, University of British Columbia. Vancouver, Canada.

Norman, E. and K. Bakker. 2005. Drivers and Barriers of Cooperation in Transboundary Water Governance: a Case Study of Western Canada and the United States. The Walter and Duncan Gordon Foundation. www.watergovernance.ca/PDF/Gordon_Foundation_transbandary_Report.pdf

Norman, E. and K. Bakker. 2009. Transgressing Scales: Water Governance Across the Canada-U.S. Borderland. Annals of the Association of American Geographers 99 (1).

Office of the Auditor General of Canada. 2001. Report of the Commissioner of Environment and Sustainable Development.

Parish, S. Personal interview. June 18, 2009.

Policy Research Institute. 2005. The Emergence of Cross-Border Regions: Interim Report. PRI. Ottawa.

Policy Research Institute. The Emergence of Cross-Border Regions Between Canada and the United States: Reaping the Promise and Public Value of Cross-Border Regional Relationships. *In*: A. Downs [ed.]. 2008. PRI Project: North American Linkages. PRI. Ottawa, Canada.

Priscoli, J.D and A.T. Wolf. 2009. Managing and Transforming Water Conflicts. Cambridge University Press. Cambridge, UK.

Rabe, B.G. 1997. The Politics of Ecosystem Management in the Great Lakes Basin. American Review of Canadian Studies 27(3).

Rabe, B.G. 1999. Federalism and Entrepreneurship: Explaining American and Canadian Innovation in Pollution Prevention and Regulatory Integration. Policy Studies Journal 27 (2).

Ross, D. Personal interview. June 18, 2009.

Sassen, S. 1991. The Global City. Princeton University Press. New York, London, Tokyo, Princeton.

Schwartz, A.M. Great Lakes: Great Rhetoric. *In*: J. Lemco [ed.]. 1992. Tensions at the Border: Energy and Environmental Concerns in Canada and the United States. Praeger. New York, USA.

Schwartz, A.M. 2000. The Canada-U.S. Relationship at the Turn of the Century. The American Review of Canadian Studies 30 (2).

Smith, P.J. 2002. Cascading Concepts of Cascadia: A Territory or a Nation? International Journal of Canadian Studies 25.

Sutcliffe, J. 2007. Local Government in a Multi-level setting: lessons from England and Ontario. Regional and Federal Studies 17(2).

Trepanier, P. Speaking at GLSLCI General Assembly. June 19, 2009.

UBC. 2010. Personal correspondence. October 26.

United Nations. 2000. Human Development Report. Oxford: Oxford University Press. http://hdr.undp.org/en/media/HDR_2000_EN.pdf

United States General Accounting Office. 2003. Great Lakes: An Overall Strategy and Indicators for Measuring Progress Are Needed To Better Achieve Restoration Goals.

Valiante, M. 2007. The Role of Local Governments in Great Lakes Environmental Governance: A Canadian Perspective. University of Michigan Journal of Law Reform 40 (4).

Vandenbergh, M.P. and A.C. Steinemann. 2007. The Carbon-Neutral Individual. New York University Law Review 82 (6).

VanNijnatten, D.L. 1999. Participation and Environmental Policy in Canada and the United States: Trends Over Time. Policy Studies Journal 27 (2).

VanNijnatten, D.L. 2003. Analyzing the Canada-U.S. environmental relationship: a multi-faceted approach. American Review of Canadian Studies 33 (1).

VanNijnatten, D. Ground-Level Ozone: A Multi-Faceted Approach. *In*: P.L. Prestre and P. Stoett (eds.). 2006. Bilateral Ecopolitics: Continuity and Change in Canadian-American Environmental Relations. Ashgate Publishing Limited. Burlington, USA.

Ward, S. and R. Williams. 1997. From hierarchy to networks sub-central government and EU Urban environment policy. Journal of common market studies 35(3).

CHAPTER 6

The Economic Impact of Remedial Action Plans: Case Studies from Ontario

Atif Kubursi

Econometric Research Ltd. and McMaster University,
Ontario, Canada
E-mail: kubursi@mcmaster.ca

The waterfront and watersheds of Ontario are important economic assets that first attracted settlement to the area and continue to draw industry and employment. The economic functions of the waterfront and watersheds have changed over time, however, and are changing again now.

At the time of European settlement, the waterfront and watersheds provided transportation and bountiful resources of fish, wildlife, timber, agricultural land and water power. Later, they provided good locations and transportation routes for manufacturing. With natural resource depletion and waste production from these uses and from an expanding population, the environmental quality and ecological productivity of the waterfront and watersheds declined. Water, soil and sediments were contaminated, and terrestrial and aquatic habitats were degraded.

Toronto, like many other cities, is now in the later stages of a transition from a manufacturing-based economy to a knowledge/service-based economy. The manufacturing that remains is more knowledge/skill-intensive and less resource-intensive. Current industries are less dependent on resources and bulk transportation; rather, they are attracted and retained by good environmental amenities and strong cultural, educational and research communities.

In this transition, the waterfront and watersheds serve increasingly important economic functions as environmental amenities. *Industrial and commercial activities* are attracted and retained by restored waterfront and watersheds and the provision of "green infrastructure" (Royal Commission on the Future of the Toronto Waterfront 1992). *Residential developments* in Toronto feature waterfront and ravine locations. This new economic development depends on, and perhaps even anticipates, environmental restoration, enhancement and protection of the waterfront and watersheds. In turn, this development also increases people's interaction with, and awareness of, the waterfront and watersheds, and thereby increases the direct benefits that can be achieved from restoration, enhancement and protection.

Just as the benefits of restoration, enhancement and protection of urban watersheds are being recognized, there is also increasing recognition of the negative consequences of continuing urban sprawl. Although the population of the Toronto area has grown fourfold over the past 50 years, this has occurred by geographic expansion of the urbanized area rather than by increases in density. With a recent trend toward migration of industry and commerce to the outskirts, there is renewed concern that Toronto might follow the "hole-in-the-donut" pattern of American cities. Numerous policy documents have acknowledged the negative long-term implications of these trends for sustainable development, and advocated a greater emphasis on compact urban form that can be served by public transport. The restoration, enhancement and protection of the watersheds and waterfront are essential to this purpose, by providing the "green infrastructure", amenities and services that can attract industrial, commercial and residential intensification to already developed areas. Our analysis follows this view by assuming that this form of development in particular, and continuing, sustainable development in general, ultimately depend on, and are made possible by the restoration, enhancement and protection of the watersheds and waterfront.

On this basis, here we evaluate in a comprehensive and systematic way the potential economic development and direct benefits from restoration, enhancement and protection of the waterfront and watersheds of Metropolitan Toronto and Region (as defined in the Metropolitan Toronto and Region Remedial Action Plan (1989) to include all of the watersheds flowing into the Metropolitan Toronto waterfront). This evaluation draws on visions of the integration of economic development and environmental restoration that have been articulated by several bodies, including the Royal Commission on the Future of the Toronto Waterfront (1990, 1992), the Metropolitan Toronto and Region Conservation Authority (1994), the Office of the Greater Toronto Area (1992), and the municipalities in their official plans and other policy documents. An essential element of all of these visions is restored and enhanced waterfront and watersheds that can serve both as waterways and greenways for recreation, and as habitats for diverse, self-sustaining fish and wildlife populations.

EVALUATION METHOD

Our evaluation method, which was developed in a previous study (VHB Hickling et al. 1993), is based on the recognition that watershed restoration is an economic activity that is necessarily part of the overall economic development path of an area. On one hand, watershed and waterfront restoration foster growth and technical progress in the environmental protection and restoration industries, which are increasingly important sectors in the global economy. On the other hand, they also provide the amenities sought by the trade, commercial and service sectors (including hospitality and tourism), which are also major growth sectors.

Mindful of these relationships, the evaluation method sets out:

1. elements of watershed and waterfront regeneration that suggest general directions of potential development, taking account of complementarities or "synergies" among environmental restoration and forms of economic development; and
2. an accounting system for assessing development options with regard to five categories of economic benefits and impacts.

Figure 6.1 depicts how initial environmental restoration investments are catalysts that make possible follow-on investments, which together lead to economic benefits and impacts (as indicated by the boxes connected by heavy arrows). The dashed feedback lines (above and connecting the boxes) are intended to show that planning of initial restoration investments and follow on investments is an *iterative* process in which actions can be modified according to the effects that are predicted or realized from these actions.

The vertical dotted lines in Fig. 6.1 connecting each of the three boxes above with "structural change" beneath is intended to suggest that the investments and benefits occur in relation to broader structural changes in the economy and society (e.g. the growth of the service sector and the "aging" of population profiles).

In the Great Lakes Basin, watershed restoration is generally associated with five complementary economic development activities:

1. *Initial restoration investments*, such as Remedial Actions Plans (RAPs), are being undertaken to restore water quality and habitat (Hartig 1988, Hartig and Thomas 1988).
2. *Watershed enhancement* is providing open parks and greenways for recreation, and as travel corridors for humans and wildlife (United States National Park Service 1991).
3. *Waterfront development* is occurring to make better use of waterfronts for amenity, cultural, commercial and residential uses (Centre for the Great Lakes 1991).
4. *Business promotion* is being pursued to enhance the economic base, especially in the new knowledge/service-based fields.

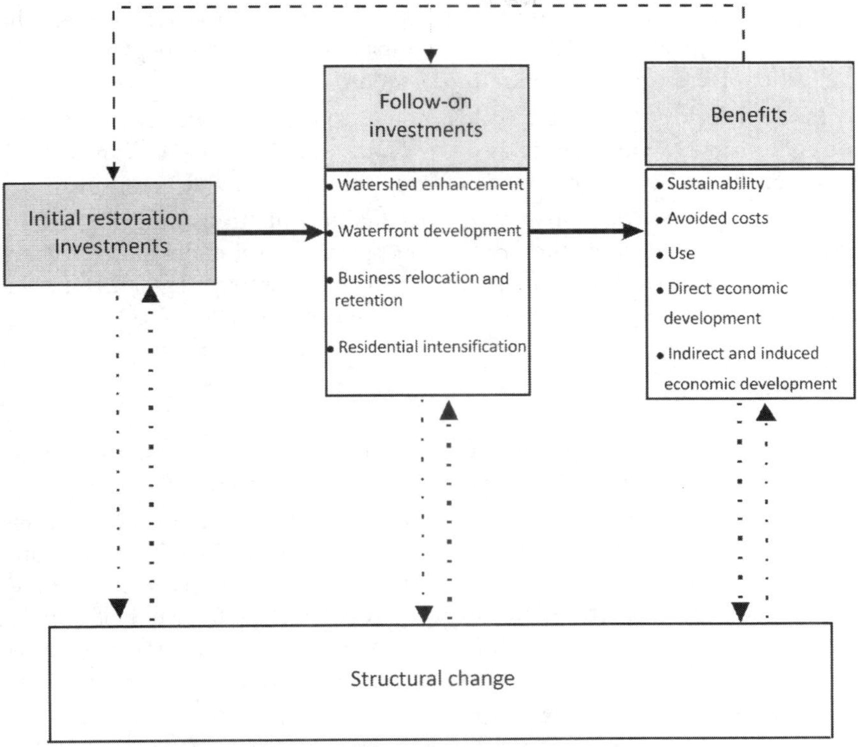

Fig. 6.1 The Benefits of Restoration Investments

5. *Residential intensification* is being encouraged to provide conveniently located residences, and reduce public service and commuting costs (Savelli 1991).

As depicted in Fig. 6.1, we can distinguish five economic categories of *benefits* and *impacts* for evaluating an envisioned outcome.

Sustainability Benefits

Sustainability benefits are the benefits of maintaining and enhancing essential life-support functions of the environment to serve future needs. Here, we distinguish sustainability benefits according to four categories of environmental functions (de Groot 1986):

Regulation functions are those functions of natural ecosystems that regulate and maintain essential ecological processes and life support systems (e.g. watershed protection and water catchment, assimilation of conventional pollutants, biodiversity and habitat provision).

Carrier functions relate to the capacity of natural ecosystems to provide space and a suitable substrate or medium for human activities (e.g. habitation, cultivation, recreation, and nature conservation).

Production functions are the ecological functions that provide resources that are consumed by humans, such as oxygen, water, food and energy (e.g. air and water quality, and sustainable yields of harvestable fish, wildlife and plants).

Information functions are the environmental functions contributing to cognitive development, cultural expression, spiritual inspiration and recreation (e.g. improved attitudes/perceptions, educational/scientific nature-based activities, and cultural activities based on natural/historical heritage).

Avoided Costs

Investments in watershed restoration allow savings of costs that would otherwise be incurred. These savings can occur in the following main areas:

- the costs of illness and of health care, to individuals, businesses and the public sector can be reduced because of reduced pollution, better health and fitness (as a consequence of enhanced recreational amenities), and reduced social and psychological stress (as a consequence of a more buoyant economy with less involuntary unemployment); the (unit) costs of water services and sewage treatment can be reduced by water conservation, and upgrading of sewage facilities;
- residential intensification can allow more efficient use of new and existing infrastructure, yielding savings in the provision of services such as community and recreation centres, fire, police, postal service, schools, etc.

Use Benefits

Use benefits from watershed restoration arise from the enhanced current uses that are made possible by the restoration and follow-on investments. These benefits are associated with:

- enhanced wildlife (consumptive and non-consumptive uses);
- enhanced fisheries (commercial and non-commercial benefits);
- other enhanced recreational amenities;
- less pollution and improved health; and
- aesthetic improvements.

Direct Economic Development Potential

Direct economic development potential is the potential follow-on economic investment and activity as a consequence of initial restoration investments.

Whether or not these economic impacts are seen as benefits generally depends on the circumstances in which they occur. For example, increased employment may be viewed as a benefit if there are sufficient numbers of unemployed workers available. In the dynamic context of structural economic change, economic impacts may be seen as benefits if they promote structural changes that enhance future economic prospects.

Direct economic development impacts arise from increased capital investments by both private investors and government, as well as increased annual expenditure associated with new economic activities. Both private investment and private annual expenditure are assumed to be incremental expenditure which is *not* offset by reductions in other expenditure. This is consistent with the view that restoration attracts new investment and expenditures that would otherwise not be made within the jurisdictions of interest. As a result these expenditures generate new tax revenues (there is an increase in the tax base) which can be used to finance increased public expenditure, including that for the watershed restoration and enhancement.

Indirect and Induced Economic Impacts

Both initial restoration investments and follow-on investments and activities lead to indirect and induced economic impacts, as impacts circulate through the economy. *Indirect* economic development impacts are the new expenditures of firms that supply the requirements that arise as a consequence of initial restoration investments and follow-on investments and activities. *Induced* economic development impacts are the new expenditures from the increased income that is generated by direct and indirect economic impacts.

An input/output model using Statistics Canada economic data is used to assess the indirect and induced impacts in Metropolitan Toronto and in Ontario as a whole. The model calculates full economic impacts, including the effects on income, employment and tax receipts by each level of government.

In principle, all of these benefits and impacts could be quantified in monetary terms. In practice, there are substantial gaps in the current state of knowledge so that reliable quantification of some benefits is not yet possible.

Initial Restoration Activities for Metropolitan Toronto Watersheds

The remedial options to restore water quality, wildlife habitat and greenways identified for the Metropolitan Toronto Area of Concern fall into 10 categories of investment opportunities:

- stormwater control
- combined sewer systems
- sanitary sewer and treatment plants
- site remediation
- habitat restoration and protection
- access improvement
- land acquisition
- public education
- policy, law and planning
- research and monitoring

The estimated total level of investment is almost Can$1.5 billion in capital investments and Can$17 million in annual operating expenditures over the life of the restoration activities. This estimate is considered conservative since the capital and operating requirements for many of these activities have yet to be quantified.

A VISION OF SUSTAINABLE REDEVELOPMENT BASED ON WATERSHED RESTORATION

The vision for sustainable redevelopment enunciated in numerous policy and planning documents for Metropolitan Toronto and Region reflects several discernible themes:

Healthy natural environment

- Fresh air
- Sparkling waters
- Good earth
- Abundant wildlife

Health communities

- Urban intensification
- Communities for all people
- Transit options
- Green-space connections
- Vibrant culture
- Keeping good health
- Sustainable farms

Healthy economy

- Smart resource use
- Green business
- Attractive place to visit and live

The unfolding of the vision will differ for each part of Greater Toronto. Historical development and the facts of geography have given each community its own unique character.

Etobicoke

On the Etobicoke waterfront mixed use redevelopment will occur mostly on lands formerly devoted to industrial uses as well as the motel strip. Developments include the creation of a waterfront trail and park system, some on new lakefill, and some on open space acquired for the purpose.

Central Waterfront

In the Central waterfront a key focus will be the reconnecting of the city to the lake through sensitive redevelopment, pedestrian and transit access improvements, and making the waterfront itself a more diverse and enjoyable mix of greenspace and urban places. Within the Central waterfront, the variation of existing conditions gives the opportunity for the creation of a diverse and vibrant environment. Conditions range from the near wilderness of the Leslie Street Spit, through the contaminated under-used industrial areas of the Lower Don Lands and the very urban character of Harbourfront to the parks in the west and on Toronto Island with its unique car-free community.

Scarborough

On the Scarborough waterfront, access to and along the lake is the primary focus of greening efforts. A lakefront trail and parks system will greatly improve access to and enjoyment of the waterfront. There is redevelopment potential in the Centennial Industrial District, and in the historical Port Union Village, creating a lakefront gateway to the city.

The Rouge River

Unlike the other watersheds of Greater Toronto, the lower Rouge retains significant wild areas. They will be protected and managed through the creation of a park stretching from the lake to Steeles Ave. The Rouge Park Management Plan sets out as a vision:

> "The Rouge Park will be a special place of outstanding natural features and diverse cultural heritage in an urban-rural setting, protected and flourishing as an ecosystem in perpetuity. Human activities will exist in harmony with the natural values of the park. The park will be a sanctuary for nature and the human spirit" (Ministry of Natural Resources 1994).

Urban Mid-watersheds

Large areas of the watersheds of Greater Toronto have been urbanized. As such, the problems they face and the vision for their restoration are similar. The most comprehensive current urban watershed vision is the *Forty Steps to a New Don* (Don Watershed Task Force 1994). The report's three general principles: i) protect what is healthy; ii) regenerate what is degraded; and iii) take responsibility for the Don, focus on water quality, natural functions and habitats, and community access and support for the health of the river. Within these urban watersheds, opportunities for urban intensification have been identified in Official Plans. The scale of intensification varies from small infill developments in residential areas to the development of large "Metropolitan Centres" in combination with new rapid transit lines.

Rural Mid-watersheds

Outside of the Oak Ridges Moraine, the largest watershed areas still in a mostly rural state, are in the Rouge River. The upper reaches of the Humber watershed, and pockets of the Don River and Mimico and Etobicoke Creeks also remain rural. Here the vision emphasizes retaining as much as possible of these areas free of urban development, protecting natural areas and "greening" agricultural practices. Where development does take place in the rural area, great care will be taken to preserve natural features and functions, including stormwater infiltration, and will favour mixed use, higher density urban form-oriented to transit rather than the automobile. The Markham Centre development on the Rouge Valley is an example of this approach.

Oak Ridges Moraine

The headwaters of the Humber, Don and Rouge Rivers rise in the Oak Ridges Moraine. The Oak Ridges Moraine Area Strategy for the Greater Toronto Area (Oak Ridges Moraine Technical Working Committee 1994) envisions protecting and enhancing natural habitats, ensuring clean abundant groundwater flows and preserving landscapes as the primary goals for the area. Development will be limited to expansions and intensification of existing urban areas, and even there, protection of the ecological integrity of the moraine is paramount. A trail system along the Oak Ridges Moraine will link to trails up the major river valleys from the Waterfront Trail, and to the Bruce Trail in the west and the Ganaraska Trail in the east.

SUSTAINABLE REDEVELOPMENT POTENTIAL IN THE VISION

Principles of sustainable redevelopment consistent with the vision, and a table of redevelopment opportunities in accord with these principles are set out in Table 6.2. These opportunities include an estimated 148 thousand new residential units in the Metro Toronto watersheds. People will be attracted to these new residential developments by the increased access to recreation amenities, such as greenways and waterfront trails, services and employment opportunities. The expected new capital expenditure from this residential intensification are about Can$19 billion.

Watershed restoration will also result in an expanded level of economic activity occurring in and around the restored areas. A greater emphasis will be placed on providing services and green activities that complement the restored watershed, resulting in almost 10 million cubic meters of additional office, commercial, retail and industrial space in the area. The local benefits include increased income and new employment opportunities.

New public transit lines will also be developed to serve the increased residential population, commuters to new business developments, and improve access to leisure and recreation facilities in the Metropolitan Toronto watersheds. The total expenditure on these new transit facilities are an estimated Can$3 billion. The expansion of public transit is an integral component of developing demand for new residential development and business activity.

The total quantifiable new capital expenditures in the vision for the Metropolitan Toronto watersheds are an estimated Can$32 billion. There will be about 321,000 new jobs directly involved in the operations of the new commercial, retail, office and industrial facilities accounted for in the vision.

DIRECT BENEFITS IN THE VISION

Sustainability Benefits

The Metropolitan Toronto Area of Concern has suffered from indiscriminate urbanization for 200 years. The vision for the area includes the restoration of a healthy watershed and the protection of important heritage and habitat areas. The current status of environmental functions in Metropolitan Toronto watersheds and their expected status after restoration are detailed in Table 6.2. Sustainability benefits are not quantified, but some of the key potential benefits in each category of environmental function are as follows:

Regulation Functions

- Rehabilitated watershed areas
- Improved diversity, distribution and protection of indigenous aquatic and wildlife populations
- Reduced releases of contaminants into watershed

Production Functions

- Improved air and water quality
- Increased fish stocks, and fish consumption
- Increased commercial and recreational fishing
- Improved energy efficiency

Carrier Functions

- Decreased number of beach closings and recreational access to green and open spaces
- Stable agricultural community
- Shift to green industries and the practice of industrial ecology
- Regeneration and improved health of aquatic and terrestrial areas

Information Functions

- Better linkages and understanding between communities and local ecosystems
- Enhanced research into and understanding of the remediation and maintenance of the watershed
- Improved preservation of cultural and heritage sites
- Better public access to green spaces

Avoided Costs

Current economic development patterns are based on extensive land use including devolution from core development nodes, unmixed land use, lack of diversity of building structures. Watershed restoration can reduce costs to local governments and other public agencies: by preserving green-space and avoiding extensive land development, municipal governments reduce costs for public services such as sewers, roads, school facilities, fire and police protection and waste management services (National Parks Services 1991). The restoration of a watershed can also help control water, air and noise pollution by natural means, reducing future public sector expenditures in pollution control and mitigation.

The total quantifiable additional savings (avoided costs) from restoration activities in the Metropolitan Toronto Area of Concern are about Can$1,500 million in capital expenditures and Can$110 million in operating expenditures. These savings neglect many avoided costs that remain to be quantified. The economic impacts of avoided costs are not estimated because they are assumed to be offset by decreases or increases in other spending.

Use Benefits

Use benefits from watershed restoration arise from the enhanced uses that are made possible by the restoration and follow-on direct economic development investments. Use benefits include the benefits of better quality water supply, better recreational amenities and aesthetics, and the private benefits of better health status that also follow from these improvements.

Improved recreational benefits are a major objective of watershed restoration activities in the Metropolitan Toronto watersheds. Accordingly, it is expected that improvements in the accessibility and quality of ecosystems and aquatic and terrestrial habitats will result in increased passive recreation use (notably in the new park in the Rouge River Valley and Toronto, Scarborough and Etobicoke river systems, and on the waterfronts).

With improved water quality, there will also be more recreational fishing and other water-related activities. In Metropolitan Toronto, local recreational benefits are expected to arise from the use by local residents of recreational activities related to the restoration of the area, e.g. walking, hiking, sailing and fishing.

The total quantifiable additional use benefits from restoration activities in the Metropolitan Toronto Area of Concern are about Can$133 million. Estimated local use expenditures are about Can$17 million. Some use benefits have not been quantified. The economic impacts of user expenditures are not estimated because they are assumed to be offset by decreases or increases in other spending.

ECONOMIC IMPACTS OF WATERSHED RESTORATION AND ENVISIONED REDEVELOPMENT

Impact of Initial Watershed Restoration Activities

The quantifiable economic impacts of watershed restoration are summarized in Table 6.3. In examining this table it is important to bear in mind the distinction between economic benefits and economic impacts. Economic impacts refer to the macro-economic changes which occur as a result of watershed restoration. Whether these impacts are seen as benefits will depend on the circumstances in which they occur. For example, increased employment may be viewed as a benefit if there are sufficient numbers of unemployed workers available. However, if labour markets are already tight, then increased employment can only occur at the expense of other activities and should be viewed as a cost.

Initial restoration expenditure (both capital and operating) are assumed to be financed by reductions in both local and provincial government spending, determined according to the extent to which these governments finance the expenditures. As a result the economic impacts of these expenditures are the result of a reallocation of existing local spending to watershed restoration, which involves a change in the type of local expenditure but not in the amount of local expenditure, together with a reallocation of existing provincial government expenditure to watershed restoration, which involves both a transfer of expenditure across regions and a change in the type of expenditure. Since the reallocation of expenditure across regions leads to more spending in the local areas, there are positive/negative impacts on local income, employment and tax revenue, in the Area of Concern.

In addition to these onetime effects, initial restoration requires annual expenditure in the local regions. These annual expenditures also involves a reallocation of expenditure to the local regions from other regions in the province, as well as a change in the type of expenditure. In Table 6.1 we provide a summary of the capital and annual operating expenditure by sector (e.g. residential, commercial, public transit infrastructure etc.). These expenditures drive the impact results.

Impact of Potential Economic Development Activities

The largest economic impacts associated with watershed restoration are those which arise from direct economic development investment, which represent new incremental spending which is not assumed to be offset by reductions in spending elsewhere in the province (Table 6.1). These impacts represent new activities which add to overall economic activity (see Table 6.2). The use benefits and use costs of remedial actions are presented in Table 6.3.

Table 6.1 Summary of potential economic development investments (1993$)

Development activity	*Capital investment ($million)*	*Annual expenditures ($million)*
Residential development and intensification	18,970.3	n.a.
Commerical, retail, office activity	10,546.7	320,740 (person years employment)[a]
Industry activity	n.a.	n.a.
Public transit infrastructure	2,865.6	n.a.
Total	32,382.6	n.a.
Source:	Appendix B.	
Notes:	n.a.: not available. a: New jobs, including new industrial employment.	

Table 6.2 Potential economic development opportunities in the Metro Toronto AOC (1993$)

Location/category/project	*Capital/onetime expenditures (million$)*	*New job creation*	*Comments*	*References*
The Waterfront – Etobicoke				
Residential intensification				
Motel Strip – Etobicoke	high density 2,960 units; $303.6 million		2960 units on 16.2 ha mixed use area permitted.	Motel Strip Secondary Plan, Dec. 1993.
Parklawn/Lakeshore – Etobicoke	high den. 3,700 units, med. den. 600 units; $384.4 million		4,300 additional residents is objective, 2,000 new units permitted, total area not given; excludes sanitary sewer system costs.	Park Lawn Road/Lake Shore Boulevard Secondary Plan, May 1992.
Etobicoke – Central Western Lakeshore redevelopment	low den. 73 units, med. den. 2,699 units, high den. 1,263 units; $459.1 million		105 ha total – on 17.2 ha parcel seek minimum of 1,500 units and maximum of 2,200 units, 1,000 for other parcels; excludes sanitary sewer system costs.	Central/Western Lakeshore Area Secondary Plan, Aug. 1993.
Retail and commercial activity				
Motel Strip – Etobicoke	62,000 m^2 of commercial gross floor area; $68.2 million	2,070 new jobs	62,000 m^2 of commercial gross floor area can be accommodated (15,000 m^2 retail, rest office).	Motel Strip Secondary Plan, Dec. 1993.

Table 6.2 contd. ...

Table 6.2 contd. ...

Location/category/project	*Capital/onetime expenditures (million$)*	*New job creation*	*Comments*	*References*
Parklawn/Lakeshore – Etobicoke	113,445 m^2 of commercial space without transit improvements; $125.7 million	5,350 new jobs	5,350 additional jobs is objective. More will be possible after transit improvements, IBI Group projects 13,400 jobs in total in the area.	Parklawn Secondary Plan, 1992. GO Transit and Min. of Transportation. GO Transit 3rd Party Contributions. prepared by IBI Group, Oct. 1994.
Etobicoke – Central Western Lakeshore redevelopment	Retail: 8,670 m^2 Office: 16,600 m^2; $27.7 million	640 new jobs	640 job opportunities in Long Branch Mixed Use area	Malone, Given, Parsons. Fiscal impacts of the recommended land use designations: Central/Western Lakeshore Area Secondary Planning Study.
Industry				
Parklawn/Lakeshore – Etobicoke			Land reserved for additional industrial development adjacent to Christies – no targets given	Park Lawn Road/Lake Shore Boulevard Secondary Plan, May 1992.
Etobicoke – Central Western Lakeshore redevelopment			Some light industry in mixed use area of Long Branch area (no target in Secondary Plan, province objects to mixing industrial with other uses)	Central/Western Lakeshore Area Secondary Plan, Aug. 1993.

	Transit					
		Motel Strip – Etobicoke	see Parklawn/ Lakeshore		Shared with Parklawn/ Lakeshore area	Motel Strip Secondary Plan, Dec. 1993
		Parklawn/Lakeshore – Etobicoke	$35.7 million for GO station includes $25.5 million for track work		New transit facilities planned – Mimico GO Gateway facility, waterfront light rail transit	Park Lawn Road/Lake Shore Boulevard Secondary Plan, May 1992.
		Etobicoke – Central Western Lakeshore redevelopment	Not available		Consideration given to possible relocation and enhancement of Long Branch GO station	Central / Western Lakeshore Area Secondary Plan, Aug. 1993.
The Central Waterfront						
	Residential intensification					
		Lower Donlands – Toronto	high den. 863 units; $128.4 million		Gooderham and Worts project proposes 863 new units on 6 ha. The future of the Ataratiri site is currently uncertain – the 7,000 units planned on 32 ha do not seem realizable.	Development Potential and other benefits from restoration, enhancement and protection of the lower Don Watershed, Nov. 1994.
		Railway Lands – Toronto	high den. 6,768 units; $621.8 million		6,768 units, majority in railway lands west area	Fax from Robert Gosse – City of Toronto Planning Department

Table 6.2 contd. ...

Table 6.2 contd. ...

Location/category/project	*Capital/onetime expenditures (million$)*	*New job creation*	*Comments*	*References*
Garrison Common – Toronto	8,800 units, all med. and high density; $1,059.9 million		WRT initially called for 17,600 units, now given lack of funding for rail relocation and expectation that southern part will be more focused on the trade centre, estimate has been cut in half.	Waterfront Regeneration Trust. 1993. Garrison Common Implementation Plan. Personal communication, Mike Ufford, City of Toronto Planning Department
Retail and commercial activity				
Lower Donlands – Toronto	1,191,636 m^2 commercial/industrial floorspace; $1,311.6 million	39,720 new jobs	Projections include redevelopment in the Port Industrial area (67% of total floor space) and Eastern Ave./Lesliegate, Gooderham and Worts, and Ataratiri.	Development Potential and other benefits from restoration, enhancement and protection of the lower Don Watershed, Nov. 1994.
Railway Lands – Toronto	1.078 million m^2 non-residential floorspace; $1,186.5 million	35,930 new jobs	1.078 million m^2 non-residential floorspace, majority in railway lands east, primarily office towers, considerable mixed use allowed.	Fax from Robert Gosse – City of Toronto Planning Department
Garrison Common – Toronto	718,000 m^2 of commercial space on private land; $790.3 million	23,930 new jobs	1st 5 years, 57,000 m^2 of non-residential space on private land, 2nd 5 years – 150,000 m^2 of additional and renovated non-residential space (private lands); then	Waterfront Regeneration Trust. 1993. Garrison Common Implementation Plan

					additional 511,000 m^2 of non-residential space in following years. In addition the plan calls for the construction of the 700,000 m^2 National Trade Centre which is already under way.	
	Transit					
		Lower Donlands – Toronto			Extension of waterfront LRT from Bay St. east to Woodbine through Port Industrial Area proposed beyond Phase 1 of Metro Rapid Transit planning.	Metro Toronto Official Plan, June 1994.
		Railway Lands – Toronto			Well served by existing structure, Metro OP calls for transit corridor along Front St.	Metro Toronto Official Plan, June 1994.
		Garrison Common – Toronto	Rail relocate $114.2 million includes GO station (WRT reports no funding available). Front St. to Gardiner extension $88.3 million, Phase II Front St. to Dufferin extension $32.3 million.		Extension of LRT through CNE in two phases; relocation of GO station; extension of Front St. to Dufferin St. in two phases.	Waterfront Regeneration Trust. 1993. Garrison Common Implementation Plan

Table 6.2 contd. ...

Table 6.2 contd. ...

Location/category/project	*Capital/onetime expenditures (million$)*	*New job creation*	*Comments*	*References*
The Waterfront – Scarborough				
Residential intensification				
Port Union Community	440 med. and 1,760 low density units; $645.9 million		New residential community on vacant and under-used industrial land, some neo-traditional features such as laneways, centred around pedestrian oriented mixed use zone and village square, provides connection to lake, and waterfront trail, close to Rouge Hill GO station	Port Union Secondary Plan, Wayne Quinn – Scarborough Planning Department.
Retail and commercial activity				
Port Union Community	About 9,000 m^2 non-residential floorspace expected; $9.9 million	300 new jobs	Mixed use zone encourages retail and commercial uses at grade along Port Union Rd. and around Village Square	Port Union Secondary Plan, Wayne Quinn – Scarborough Planning Department
Transit				
Port Union Community	Relocation of GO station $10.9 million		Changes to Rouge Hill GO station, including relocation permitted, but relocation not likely.	Port Union Secondary Plan, Wayne Quinn – Scarborough Planning Department Ministry of Transportation

					Improvement in TTC bus service to area, including service on new interior arterials expected. Improvement to Port Union Rd. interchange at 401 includes new on-ramps	and GO Transit. GO Transit 3rd Party Contributions. Prepared by IBI Group, Oct. 1994.
Urban Watersheds						
	Residential intensification					
		City Centre – Etobicoke	high den. 3,300 units; $275.8 million		Additional 3,300 apartment units to be constructed over 10-20 years in Kipling-Islington subway areas. Pedestrian orientation, protection of and linkage to Mimico Creek open space. More development possible with improvement in public transit. Excludes sanitary sewer system costs.	Etobicoke Official Plan, Office consolidation, April 1993, pp. 125-134.
		Metro Toronto Main St. initiative	20,000 high. den. units; $1,671.4 million		Assumes about 900 units a year built until the year 2011 along Metro arterials; excludes sanitary sewer system costs.	1993 Report by Brenda Bernards of Metro Planning.

Table 6.2 contd. ...

Table 6.2 contd. ...

Location/category/project			*Capital/onetime expenditures (million$)*	*New job creation*	*Comments*	*References*
		Scarborough City Centre	13,000 high density units; $1,086.4 million		Site specific housing densities range from 62-366 units/ha; excludes sanitary sewer system costs.	Scarborough Official Plan, Oct. 1994.
		Agincourt Centre (Sheppard - Kennedy)	1,300 high and medium density units; $135.8 million		Another 700 residential units are possible after the subway is built; excludes sanitary sewer system costs.	
		York City Centre	4,845 high density units; $461.1 million		Mixed use city centre and new residential areas to be integrated with transit and Black Creek open space system. Southern part of development area slightly to heavily contaminated industrial land. Requires clean-up prior to new development. Cost and area not available.	City of York Official Plan, 1994
		East York Centre	700 high density units; $58.5 million		Redevelopment of existing business park to become focal point of the Borough. Encourage more mixed uses including high	East York Official Plan, Feb. 1994

					density residential, and commercial. Development limited by transportation bottlenecks – 700 units assumed based on increased capacity provided by HOV lanes. Excludes sanitary sewer system costs.	
		North York Centre (Uptown and Downtown)			Mixed use high density development on Yonge subway line. Total residential population to increase to 27,000. Uptown (Finch subway stn.) to have greater proportion of residential to commercial.	North York Official Plan, 1992
		York University			Mixed use high density development combined with new subway line.	Metro Official Plan, June 1994
		Queen St. Corridor – Brampton	7,900 med. and high density units (assumes high scenario); $825.3 million		Main St. and adjacent area intensification from Hwy 10 east to Bramalea City Centre; exlcudes sanitary sewer system costs.	Brampton Central Commercial Centre, Final Report, Feb. 1991. Personal communication, David Waters, Brampton Planning Department

Table 6.2 contd. ...

Table 6.2 contd. ...

Location/category/project	*Capital/onetime expenditures (million$)*	*New job creation*	*Comments*	*References*
Bramalea Rd Gateway	high den. 5,000 units, med. den. 140 units; $435.4 million		Declining industrial area to be redeveloped with mixed commercial/ residential uses. Exlcudes sanitary sewer system costs.	Bramalea Rd South Gateway Planning Concept Study, Draft final Report, Nov. 1994. Personal communication, David Waters, Brampton Planning Department.
Retail and commercial activity				
City Centre – Etobicoke	376,000 m^2 of office space; $413.8 million	12,530 new jobs	Additional 376,000 m^2 of office space to be constructed over 10-20 years in Kipling-Islington subway areas. More development possible with improvements in public transit.	Etobicoke Official Plan, Office consolidation, April 1993, pp. 125-134.
Metro Toronto Main St. initiative	New commercial space potential not available		New larger buildings on Main St. will have some increased commercial/ retail space, but change not expected to be large	Brenda Bernards, Metro Planning Department, personal communication.
Scarborough City Centre	1,600,000 m^2 of commercial space; $1,761.0 million	53,330 new jobs	major centre in Metro OP	Metro Official Plan, June 1994
Agincourt Centre (Sheppard – Kennedy)	500,000m^2 of commercial space; $550.3 million	16,670 new jobs	intermediate centre in Metro OP	Metro Official Plan, June 1994

		York City Centre	305,500 m^2 of commercial space; \$336.2 million	10,180 new jobs	Most commercial space in mixed use area at intersection of Eglinton and Black Creek Dr.	
		East York Centre	70,800 m^2 of office space; \$77.9 million	2,360 new jobs	Redevelopment of existing business park to become focal point of the Borough. Encourage more mixed uses including high density residential, and retail/commercial. Development limited by transportation bottlenecks.	East York Official Plan, Feb. 1994
		North York Centre	1.2 million m^2 commercial space; \$1,320.8 million	40,000 new jobs	Major mixed use commercial/residential development – focus of commercial use to be downtown (Sheppard Subway).	Metro Official Plan, June 1994
		York University	510,000 m^2 commercial space; \$561.3 million	17,000 new jobs	designated intermediate centre in Metro OP	Metro Official Plan, June 1994
		Queen St. Corridor–Brampton	Retail space 170,000 m^2; new office/commercial space 300,000 m^2; new institutional 100,000 m^2; \$627.4 million	19,000 new jobs	Mixed use redevelopment of existing Main St in Brampton	Brampton Central Commercial Centre, Final Report, Feb. 1991.

Table 6.2 contd. ...

Table 6.2 contd. ...

Location/category/project	*Capital/onetime expenditures (million$)*	*New job creation*	*Comments*	*References*
Bramalea Rd Gateway	146,000 m^2 office space 21,600 m^2 retail space; $184.5 million	5,590 new jobs	Mixed use office, retail residential development on former industrial lands.	Bramalea Rd South Gateway Planning Concept Study, Draft final Report, Nov. 1994. Personal communication, David Waters, Brampton Planning Department
Transit				
City Centre–Etobicoke			More development possible with improvement in public transit, including expansion of subway to Sherway	Etobicoke Official Plan, Office consolidation, April 1993, pp. 125-134.
Metro Toronto Main St. initiatives			Use existing TTC infrastructure	
Scarborough City Centre	RT expansion: $421 million (75% province, 25% Metro)		Extension of Scarborough RT to Sheppard and Markham Rd. Sheppard subway will eventually terminate at the City Centre, but not as part of Phase I.	Metro Official Plan, June 1994
Agincourt Centre (Sheppard–Kennedy)			Will eventually be served by the Sheppard Subway, but phase I does not extend to Kennedy. Proposed relocation of existing GO station and new cross town GO line to intersect existing GO line.	Metro Official Plan, June 1994

		York City Centre	Eglinton subway: $732.6 million 25% Metro, 75% province		Proposed Eglinton Subway ends at Centre, joins to a new proposed.	City of York Official Plan, 1994
					GO station, to create major "Gateway". Secondary plan also calls for widening of Black Creek Dr. to six lanes	
		East York Centre			Secondary plan aims for improved TTC service, GO connection, Leslie St. extension with busway or LRT (latter not likely for environmental and cost reasons).	East York Official Plan, Feb. 1994.
		North York Centre	Sheppard subway: $935.6 million; 25% Metro, 75% province		First phase of Sheppard Subway begins at North York Centre	Metro Official Plan, June 1994
		York University	Subway to York: $495 million; 25% Metro, 75% province		Extension of Spadina Subway line will end at York University	Metro Official Plan, June 1994
		Queen St. Corridor–Brampton			HOV or bus lanes on Queen St., integration of existing bus service with Queen St. service, integration of proposed intra regional service on Hwy 410; improved GO train service	Brampton Central Commercial Centre, Final Report, Feb. 1991.

Table 6.2 contd. ...

Table 6.2 contd. ...

Location/category/project	*Capital/onetime expenditures (million$)*	*New job creation*	*Comments*	*References*
Bramalea Rd Gateway			Hwy 407, improvements to GO train service	Bramalea Rd South Gateway Planning Concept Study,
			and to existing station, interregional transitway service and local Brampton transit service integrated with GO	Draft final Report, Nov. 1994. Personal communication, David Waters, Brampton Planning Department
Rural Watersheds				
Residential intensification				
York Region Housing & Employment Targets, some intensification			By 2021 population expected to double to 1.1 million and employment to increase from 248,000 to 625,000. 20% of new housing to be intensification and infill.	York Region Official Plan, Oct. 1994.
Markham Centre Plan	low den. 1,130, med den. 4,861, and high den. 9,343 units; $1,906.2 million		36,000 residents on 506 ha (150 ha parkland), 15,344 residential units, 8 elementary schools, 2 secondary schools. Developed over 30-50 years. Total cost $6-10 billion	Notes on the Markham Centre Study recommended master plan, June 2 1994. "Markham plan focuses on bringing livability to burbs" Toronto Star, June 11, 1994.

		Cornell (Ninth Line)-Markham	low den. 6,216, med. den. 2,279, and high den. 1,865 units; $2,035.6 million		27,000 residents on 625 ha, 10,360 units	"Town centre plan impresses" Toronto Star, June 9, 1994, Section NY. Cornell Official Plan Amendment and Secondary Plan. July 1994
		Angus Glen–Markham	low den. 725, med. den. 300, and high den. 225 units; $488.3 million		Neo-traditional grid streets with laneways for 3,730 on 107 ha. Includes Golf course cited as a model for the GTA by the MTRCA for ecologically sensitive development	
		IBM Golf Course–Markham	low den. 667, med. den. 467, and high den. 97 units; $486.5 million		Neo-traditional grid streets with laneways for 3,863 residents. Valley lands donated to city will be ecologically restored.	
		Urban Village I, II & GO study area–Vaughan	Low den. 15,730, med. den. 10,690, and high den. 7,300 units; $5,500.9		33,720 housing units proposed on about 2,150 ha. Nodal development with mixed use concentrations along arterials and at district centres. (Note: Block 12 and parts of 47/40 have been removed from the totals to reflect reviewers criticisms of OPA 400.)	Official Plan Amendment 400 Stan Holiday

Table 6.2 contd. ...

Table 6.2 contd. ...

Location/category/project	*Capital/onetime expenditures (million$)*	*New job creation*	*Comments*	*References*
Commercial and retail activity				
Markham Centre Plan	687,500 m^2 employment space, 92,900 m^2 retail space; $858.9 million	26,010 new jobs	687,500 m^2 employment space, 92,900 m^2 retail space in mixed use zones, mostly	Notes on the Markham Centre Study recommended master plan, June 2 1994. "Markham plan focuses on bringing
			organized around central pedestrian-friendly corridor. Developed over 30-50 years.	livability to burbs" Toronto Star, June 11, 1994.
Cornell (Ninth Line)–Markham	46,000 m^2 of commercial space; $50.6 million	1,530 new jobs	Retail space provided in neighbourhood centres and along main street central core, which is a mixed use area.	"Town centre plan impresses" Toronto Star, June 9, 1994, Section NY. Cornell Official Plan Amendment and Secondary Plan. July 1994
Angus Glen–Markham	18,100 m^2 of commercial space; $19.9 million	600 new jobs	Two mixed use commercial areas on 8 ha.	Markham OP Amendment No. 19, July 1994
IBM Golf Course–Markham	not available		Local commercial centre planned on 0.51 ha	Markham OP Amendment No. 25, Oct. 1994.
Urban Village I, II & GO study area–Vaughan	240,000 m^2 commercial space allowed in district and neighbourhood centres; $264.2 million	8,000 new jobs	Development includes 3 levels of mixed use centres, plus mixed use residential areas and yet to be determined commercial uses in the	Vaughan Official Plan Amendment 400, Paul Robinson & Stan Holiday, Vaughan Planning Department

					GO study area. Potential total gross floor area for all these areas has not been estimated.	
	Transit					
		Markham Centre Plan			Proposed new GO station, Hwy 407, transitway along Hwy 407	Notes on the Markham Centre Study recommended master plan, June 2 1994. "Markham plan focuses on bringing livability to burbs" Toronto Star, June 11, 1994.
		Cornell (Ninth Line)-Markham	3.1 km "transit supportive corridor"		Hwy 407, Markham By-pass, 2 intersecting busways, transit station in central core of Cornell and one at 407, links to rapid transit	"Town centre plan impresses" Toronto Star, June 9, 1994, Section NY. Cornell Official Plan Amendment and Secondary Plan. July 1994
		IBM Golf Course–Markham			Hwy 407, integration of local bus service with transitway along 407	Markham OP Amendment No. 25, Oct. 1994.
		Urban Village I, II & GO study area–Vaughan			Hwy 407, HOV lanes along Hwys 7 and 400, transit in Hwy 407 corridor, relocation of Maple GO station to Urban Village II, additional transit services on arterials, and higher order transit (latter is long-term)	Vaughan Official Plan Amendment 400, Paul Robinson & Stan Holiday, Vaughan planning department.

Table 6.3 Direct use benefits and expenditures from watershed restoration

Direct use category	*Use benefit ($1993)*	*Use expenditures ($1993)*
Watershed/waterfront recreation activities	10,286,100	1,907,400
Boating	34,590,270	2,479,620
Fishing	23,166,000	3,433,320
Swimming	33,893,640	5,149,980
Skating	31,152,000	3,814,800
Total	133,088,010	16,785,120
Source:	Appendix E.	

CONCLUSION

The economic development benefits arising from initial restoration activities were estimated for a few locations on the Great Lakes in Ontario using a regional input output model. The other benefits were estimated using a collection of methodologies outlined in this chapter.

The estimated economic development benefits of restoration activities are based on initial data supplied by Environment Canada, Environment Ontario and Harbour Commissions. Only those economic benefits arising from easily quantifiable expenditures andinvestment activities are included in the model.

The results of the estimates are summarized in Tables 6.2 and 6.3 above. It is important to bear in mind the distinction between direct economic benefits and economic development benefits. Direct economic user benefits refer to the value of impacts on the individual welfare from new and increased use of the watershed. On the other hand, economic development benefits (impact effects) refer to the social (macro) changes that occur as a result of watershed restoration. Whether these impacts are seen as benefits depends on the circumstances in which they occur. Employment impacts can be seen as a benefit if unemployment is a problem. If full employment is achieved, additional jobs come from other existing jobs and could end up raising the cost of production for all.

Restoration expenditures have direct impacts on local incomes and revenues for local governments. These are typically positive and fully appreciated by the communities. Besides, these impacts can be very large and represent vital interests to many sectors in the local economy.

REFERENCES

Ashworth, W. 1986. The late, Great Lakes: an environmental history. Collins. Toronto, Canada.

Centre for the Great Lakes. 1987. Fact Sheets: Great Lakes Areas of Concern, Windsor, Ontario, Canada.

Centre for the Great Lakes, and the Centre for the Great Lakes Foundation. 1989. The Environment and Economic growth: A success story. Great Lakes Facts, Windsor, Ontario, Canada.

Centre for the Great Lakes. 1991. Water Works 1991: a survey of Great Lakes/St. Lawrence Waterfront Development. Center for the Great Lakes. Chicago, USA.

Ciccheti, C. and V.K. Smith. 1991. Recreation benefit estimation and forecasting: Implications of the identification problem. Water Resources 8(4).

City of Toronto Planning and Development Department. 1990. City Plan 91.

Crombie, D. 1991. Regeneration – Toronto's Waterfront and the sustainable city: Final Report. Royal Commission on the Future of the Toronto Waterfront, Toronto.

De Groot, R.S. 1986. A Functional Ecosystem Evaluation Method as a Tool in Environmental Planning and Decision Making. Nature Conservation Department, Agricultural University. Wageningen, The Netherlands.

Dornbusch, D.M and S.M. Barrager. 1973. Benefit of Water Pollution Control on Property Values. United States Environmental Protection Agency. Office of Research and Monitoring. Washington, D.C., U.S.

Garland, G. 1991. Greater Toronto Region Waterfront: Community Overview. Royal Commission of the Future of the Toronto Waterfront. Toronto, Canada.

Gertler, M.S. 1991. Toronto: The State of the Regional Economy. Royal Commission of the Future of the Toronto Waterfront. Toronto, Canada.

Gorrie, Peter. 1987. Cleaning Up Hamilton Harbour. Canadian Geographic 107(3): 34-45.

Government of Canada. 1990. Canada's Green Plan. Ministry of Supply and Services. Ottawa, Canada.

Gray, C.B. and D. Rathkey (eds). 1988. Remedial Action Plan Research Needs. Windsor. International Joint Commission, Council of Great Lakes Research Managers, October. Ontario, Canada.

Hartig, J. 1988. Great Lakes Program whose Time has Come! The Great Lakes: Living with North America's Inland Waters, American Water Resources Association. November: 45-51.

Hartig, J. and R. Thomas. 1988. Development of Plans to Restore Degraded Areas in the Great Lakes. Environmental Management. 12(3) 327-347.

IJC. 1991. Review and evaluation of the Great Lakes Remedial Action Plan Program 1991. Great Lakes Water Quality Board, Report to the International Joint Commission, Windsor, Ontario, Canada

Inwood, K. and T. Sullivan. 1992. Comparative Perspectives on Nineteenth Century Growth: Ontario and the Great Lakes Region. Discussion Paper 1992-06. University of Guelph. Department of Economics. Ontario, Canada.

Jacobs, J. 1961. The Death and Life of Great American Cities. Vintage Books. New York, USA.

Jacobs, J. 1969. The Economy of Cities. Random House. New York, USA.

Jacobs, J. 1982. Cities and the Wealth of Nations: Principles of Economic Life. Random House. New York, USA.

Metro Toronto Remedial Action Plan Team. 1988. Metro Toronto Remedial Action Plan. Stage 1. Environmental Conditions and Problem Definition. Environment Canada, Environment Ontario, Ministry of Natural Resources and Metropolitan Toronto and Conservation Authority.

Meyer, S.M. 1992. Environmentalism and Economic Prosperity: Testing the Environmental Impact Hypothesis. MIT. Cambridge, Massachusetts.

Ontario Round Table on Environment and Economy. 1992. Restructuring for Sustainability. Ontario Round Table on Environment and Economy.

Pearce, D. and A. Markandya. 1989. Environmental Policy Benefits: Monetary Valuation. OECD. Paris, France.

Savelli, S. 1991. An Ecosystem Approach to Natural areas, Inner-city and Waterfront Development in the Hamilton Harbour Watershed. Environment Canada, CCIW.

United States National Parks Services. 1991. Economic Impacts of Protecting Rivers, Trails and Greenway Corridors. Rivers, Trails and Conservation assistance Program. Washington DC, USA.

VHB Hickling, Econometric Research Limited. LURA Group and Michael Michalski Associates. 1993. Development Potential from Restoration Enhancement and Protection of Watersheds in the Great Lakes Basin. Ottawa: Hickling Corporation.

Section 2

Principles of Integrated Water Resources Management

CHAPTER 7

A Brief Introduction to Integrated Water Resources Management

Abigail Cruickshank and Velma I. Grover

24048 109th Avenue, Maple Ridge, British Columbia, V2W 124

916-981 Main St. W, Hamilton, ON, L8S 1A8, Canada

INTRODUCTION

The Great Lakes Water Quality Agreement (1987) specifies that the governments of the United States and Canada take an "ecosystem approach" for lake management and restoration of the areas of concerns under the Remedial Action Plans (RAPs). The Hamilton Harbour RAP (the first in Canada) defines this approach as:

> *"a basin-wide approach to planning, research and management ... including emphasis on holistic, coordinated management, a systems approach for managing flows of materials and energy and [a] ... combined social-economic-environmental approach to problem solving"*
>
> (LURA, 1986 as quoted in Kellogg, 1998).

The intention of a holistic or ecosystem perspective is to look beyond the man-made political boundaries, involve all the stakeholders concerned in coming to a solution and have a multi-disciplinary approach. In the water management arena, advocacy for a systems approach led to the emergence of Integrated Water Resources Management (IWRM).

IWRM as a philosophy for managing freshwater resources is not a new concept. Many examples embodying the principles of IWRM were present prior to the 1992 Earth Summit in Rio de Janeiro where it was formally adopted. In Canada, two precursors to IWRM were the "comprehensive river basin planning" organized by the Canadian federal government and some provincial governments and the Ontario Conservation Authorities (Mitchell, 2006). Both initiatives used the river basin as a management unit, focused on developing partnerships and participation, and actively sought to incorporate environmental considerations into planning and development.

On the global agenda, integrated management was born as a result of observable shortcomings in traditional water management. Prior to IWRM, paradigms were top-down, sectoral, technical and driven by supply and demand. The main sub-sectors, water supply, sanitation, irrigation and energy, would predict their needs, in isolation, and develop infrastructure to meet the void (Savenije and Van der Zaag 2008). There was a lack of interagency coordination and little regard was given to the finite amount of water in the basin or to the environmental and social impacts of large-scale development (Merrey, 2008). Single focused management systems were noticeably insufficient to deal with increasingly complex problems of poor water quality, overexploitation of water resources, and ecosystem degradation. Heathcote (1998 as quoted in Roy et al. 2009) provided the example of pollution control efforts that solely focused on point sources such as sewage treatment plants and industrial discharges. In contrast, an integrated, ecosystem or watershed approach would enable the additional consideration of non-point sources of pollution.

There is no universally accepted definition for IWRM, however, the most widely quoted is the one provided by the Global Water Partnership (GWP), in which, IWRM is referred to as:

> *a process which promotes the coordinated development and management of water, land and related resources, in order to maximize the resultant economic and social welfare in an equitable manner without compromising the sustainability of vital ecosystems*
>
> (GWP 2000)

At its most basic IWRM is a guiding conceptual framework for water managers. It emphasizes the understanding that water resources are an integral part of an ecosystem, a natural resource and a social and economic good. This chapter provides a brief review of IWRM, including defining the concept, providing a framework and guide to implementation and a discussion of the barriers and challenges faced by managers when trying to operationalize the philosophy.

WHAT IS IWRM?

IWRM is a process of achieving sustainable development, allocation and monitoring of water resource use in the context of social, economic and environmental objectives. It is not an end in itself but a means of achieving the often conflicting goals of efficiency, equity and ecological integrity. The 3 E's, described below, formulate the main objectives of IWRM (Postel 1992, Cap-Net 2008):

Efficiency (to make water resources go as far as possible) – Water is a finite, scarce and vulnerable resource that needs to be managed efficiently. Institutional arrangements and management decisions should promote conservation and cost recovery.

Equity (in the allocation of water across different social and economic groups) – Access to water in adequate quality and quantity is a basic human need and right. Water is a finite resource but also fundamental to survival and therefore needs to be managed, protected and equitably distributed amongst the users (Savenije and Van der Zaag 2008).

Ecological Integrity (to protect the water resources base and associated ecosystem) – As ecosystem services support life on earth and regenerate freshwater, this objective recognizes the need to ensure enough water is protected so as not to undermine the integrity of ecosystems.

These three objectives were all expressed in the Rio Summit's Agenda 21 Chapter on Freshwater, where IWRM was first conceptualized as being:

> *(B)ased on the perception of water as an integral part of the ecosystem, a natural resource and a social and economic good, whose quantity and quality determine the nature of its utilization. To this end, water resources have to be protected, taking into account the functioning of aquatic ecosystems and the perenniality of the resource, in order to satisfy and reconcile needs for water in human activities. In developing and using water resources, priority has to be given to the satisfaction of basic needs and the safeguarding of ecosystems. Beyond these requirements, however, water users should be charged appropriately*
>
> (Snellen and Schrevel 2004)

IWRM offers a common-sense "way of thinking" approach to the management of freshwater resources. There is no one authoritarian framework advocated by the approach instead IWRM is founded on a set of guiding principles. The International Conference on Water and the Environment held in Dublin prior to the Earth Summit was tasked with outlining priority issues related to freshwater. The result of the conference was a set of four principles that would become the compass orientating reformation in water management. The four principles dubbed the Dublin Principles are described in Box 7.1.

The broad conceptualization of IWRM has opened the approach to a number of criticisms. Merrey (2008) linked it to a religious text providing "a set of unquestioned assumptions and assertions about how water resources should be developed and managed". He argues that IWRM is impossible to implement and asserts that it "fails as a guide to practical action". Within the field of water management, IWRM has been described as a "nirvana concept" that idealizes the possibility of simultaneously satisfying the water needs of all sectors (including those of ecosystems) (Molle 2008).

To translate the principles into action, the GWP developed a general conceptual approach for implementing IWRM (Fig. 7.1). The framework

Box 7.1 The Dublin Guiding Principles

Principle No. 1: Fresh water is a finite and vulnerable resource, essential to sustain life, development and the environment

Since water sustains both life and livelihoods, effective management of water resources demands a holistic approach, linking social and economic development with protection of natural ecosystems. Effective management links land and water uses across the whole of a catchment area or ground water aquifer.

Principle No. 2: Water development and management should be based on a participatory approach, involving users, planners and policy-makers at all levels

The participatory approach involves raising awareness of the importance of water among policy-makers and the general public. It means that decisions are taken at the lowest appropriate level, with full public consultation and involvement of users in the planning and implementation of water projects.

Principle No. 3: Women play a central part in the provision, management and safeguarding of water

This pivotal role of women as providers and users of water and guardians of the living environment has seldom been reflected in institutional arrangements for the development and management of water resources. Acceptance and implementation of this principle requires positive policies to address women's specific needs and to equip and empower women to participate at all levels in water resources programmes, including decision-making and implementation, in ways defined by them.

Principle No. 4: Water has an economic value in all its competing uses and should be recognized as an economic good

Within this principle, it is vital to recognize first the basic right of all human beings to have access to clean water and sanitation at an affordable price. Past failure to recognize the economic value of water has led to wasteful and environmentally damaging uses of the resource. Managing water as an economic good is an important way of achieving efficient and equitable use, and of encouraging conservation and protection of water resources

(Source: The Dublin Statement. International Conference on Water and the Environment: Development issues for the 21st century. 26-31 January 1992, Dublin, Ireland).

focuses on simultaneously developing three basic pillars: positive Enabling Environment, clear Institutional Framework and realistic Management Instruments. Under the three pillars, the GWP proposes 13 key areas for potential change (Box 7.2.).

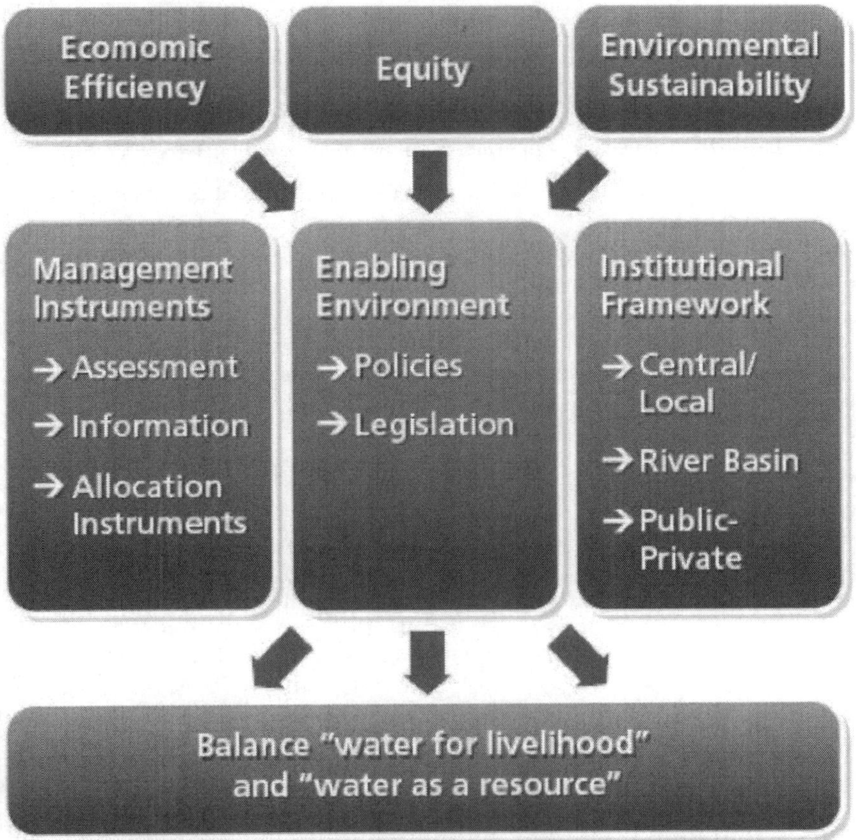

Fig. 7.1 GWP conceptual framework for IWRM (Pahl-Wostl and Sendzimir, 2005)

Enabling Environment

A positive enabling environment created by appropriate legislation and policies will ensure that management and decisions on allocation and conservation comply with the Dublin principles. In essence, the enabling environment defines the "rules of the game" and ensures that there are social and financial capacities to put the policies into practice. The government is viewed as an "enabler" promoting and providing arenas for participatory dialogues and cross-sectoral integration, as well as addressing water use rights, cost sharing and responsibilities of the various institutions and stakeholders (GWP 2000). Water policies need to outline national objectives and identify how management of water resources fits into an overall sustainable development strategy. The legislative framework then supports the achievement of policy goals by defining

Box 7.2 The 13 Key IWRM Change Areas (GWP, 2004)

The enabling environment

1. Policies – setting goals for water use, protection and conservation.
2. Legislative framework – the rules to follow to achieve policies and goals.
3. Financing and incentive structures – allocating financial resources to meet water needs.

Institutional roles

4. Creating an organizational framework – forms and functions.
5. Institutional capacity building – developing human resources.

Management instruments

6. Water resources assessment – understanding resources and needs.
7. Plans for IWRM – combining development options, resource use and human interaction.
8. Demand management – using water more efficiently.
9. Social change instruments – encouraging a water-oriented civil society.
10. Conflict resolution – managing disputes, ensuring sharing of water.
11. Regulatory instruments – allocation and water use limits.
12. Economic instruments – using value and prices for efficiency and equity.
13. Information management and exchange– improving knowledge for better water management.

relevant laws such as ownership, permits to use or to pollute, protection and conservation (GWP 2004). Policies and their legislative foundation are not sufficient in themselves. Governments need to provide financial mechanisms, information sharing and build capacity to ensure that stakeholders and managers have access to enough means to participate effectively and implement the plan of action.

Institutional Roles

Formulation and implementation of policies and legislation is completely dependent on institutional development. IWRM promotes institutional frameworks that are centrally/locally focused, inclusive of both top-down and bottom-up participation and effective at integration and coordination across government agencies but also companies, NGOs, community based organization and communities at-large. However, the philosophy also emphasizes that frameworks are not one-size-fits all and therefore institutional arrangements and roles need to be context specific, considering local customs and practices, geographic settings, political structures, financial and human resources of each group and stage of development (GWP 2004). GWP (2000) notes that "flawed demarcation of responsibilities between actors, inadequate co-ordination mechanisms, jurisdictional gaps or overlaps, and the failure to match responsibilities, authority and capacities for action are all major sources of difficulty with implementing IWRM". Reform in institutional arrangements should begin with two areas of change: creating an organizational framework and developing institutional capacity. Creating an organizational framework essentially seeks to answer the question – how should water users be brought together to make decisions on water management? The first step is to identify the gamut of existing institutions (from transboundary to basin level, regulatory to local authorities and civil society organizations), recognize ways in which organizations currently coordinate and how this can be scaled-up and develop capacity for organization to undertake great management and integration responsibility.

Management Instruments

A range of activities occur under IWRM including allocating and monitoring water resources, designing and maintaining infrastructure, creating incentives for efficient use, managing contamination, and protecting the resources. Management instruments are the methods used in these activities and the tools to make tradeoffs between the three objectives of economic efficiency, equity and ecological integrity. Finding the right balance requires making difficult choices for courses of action, and access to and implementation of a variety of both hard/structural and soft/institutional measures. Methods and tools available within IWRM include water resources assessments, education, allocation and conflict resolution, regulatory instruments such as polluter-pays principle and technological option sets (Box 7.2).

IMPLEMENTING IWRM

Those who have been successful at implementing IWRM have not followed a generic set of step-by-step instructions. Nor did they set out with the goal of simply putting IWRM into practice and assuming all water related problems would be solved. Instead groups addressed specific water related issues by subscribing to an IWRM philosophy, applying key points and linking solutions and methodology to a larger water management and sustainable development arena. Essentially, managers need to remember to view situations holistically, recognizing the interconnections of water and land resources and understand the wider implications of their decisions. The idea of 'integration' is fundamental. Integration in terms of linking the biophysical resource base across space and time, connecting stakeholders, and managing the various water uses and values to create a shared vision, are all key strategies employed by successful IWRM case studies.

Cross-scale Linkages

An IWRM approach calls for integrating all elements in a river basin: surface water, groundwater and marine environments; land and water; and human activities. Water, sediments, nutrients and energy flows connect the air, soil, vegetation and water bodies across a landscape. This flow and subsequent usage through different land uses, vegetation covers and development, determines the distribution, quantity and quality of the resource (GWP 2000). The resources are not bound by political, man-made boundaries but vary spatially across geographic scales and temporally throughout the day, the year and between years.

The river basin is often promoted as the most appropriate boundary for water management because it is based on a natural demarcation. The river basin collects precipitation, filters and stores water and connects the flows of water. This management unit urges managers to take into consideration "green water" in addition to "blue water". "Blue water" is the water available in rivers, lakes and aquifers, and has been the traditional focus of management. "Green water" is the water flow that supports plant production in forests, grasslands, rainfed agriculture and wetlands (Moberg and Galaz 2005). Falkenmark and Rockstrom (2006) explain the importance of managing the flows of both by stating:

> *..it is not only a question of water allocation among irrigation, industry, and municipalities but involves difficult decisions for balancing green and blue water for food, nature, and society. It will change the role of water-resource planners and managers. Water resources planning and management will have to incorporate land-use activities consuming green water and its interaction with blue water, generating surface runoff and groundwater recharge.*

Integration of upstream and downstream sections of a watershed is also facilitated through the use of the river basin management unit. As water naturally flows to the lowest point, upstream land uses and water usage fundamentally affects downstream water quality and availability. Land use changes upstream such as deforestation for agriculture might alter surface water flows downstream or groundwater recharge rates. Such changes could impact economic activity such as through reduced irrigation or result in the loss of wetland ecosystems. High pollutant loading upstream will detrimentally affect downstream environments and will require coordinated management.

A successful example of applying an ecosystem approach was observed in the case of Lake Erie. Of all the Great Lakes, Lake Erie is the shallowest, most biologically productive and the most populated basin. From the 1950s to the 1970s, the lake experienced accelerated eutrophication and toxic substances contamination (Herdendorf and Rockwell 1983). It was often referred to as the "Dead Sea of North America" because the decaying algal blooms resulted in hypoxia and anoxia greatly impacting fisheries populations (Fitzpatrick, n.d). When the International Joint Commission established that high phosphorus levels were the cause of eutrophication, action for load reduction was immediately put in place. At the time, it was believed that phosphorus-based detergents were the major contributor, which resulted in a basin wide usage ban. Municipalities also implemented controls on domestic and industrial discharges (De Pinto et al. 1999). However, combined these controls were not sufficient to achieve target phosphorus levels. Managers had to look beyond point sources and consider the impacts of non-point source pollution. As 67% of the basin is agricultural land, programs were put in place to encourage farmers to implement Best Management Practices (BMPs). These included practising conservation tillage and improved fertilizer management. As a result annual phosphorus loads were reduced from 28,000 tonnes in 1968 to 11,000 tonnes in 1985 (OAG, 2001). The Water Quality Laboratory at Heidelberg College (1980) estimated that non-point sources of phosphorus accounted for over 51% of lake loading, while point sources and atmospheric loading combined only amounted to 42%. Effective management of Lake Erie expanded well beyond the lake, to include the tributaries and land uses of all 34 sub-watersheds and relied on considerable cooperation between nations and numerous municipalities.

Integrating Stakeholders

Systems approaches such as IWRM advocate involving stakeholders and linking government agencies. The creation of mechanisms and forums for integration and participatory dialogues are the key to making tradeoffs and decisions regarding complex circumstances. Users and stakeholders who have participated in the formulation of decisions are more likely to work cooperatively, facilitate implementation and comply with conditions (GWP 2004). Sadoff and Muller (2009) state that "good management practices that are developed in user communities are more likely to be sustainable than rules imposed from outside by formal organisations". Encouraging participation, which can balance insufficiencies in top-down and bottom-up approaches, may require capacity building, education and the provision of economic resources. In contrast to consultation, IWRM calls for real participation meaning that stakeholders at all levels should be provided an opportunity to have an impact on decisions. The realization of this might require certain levels of support to different entities such as NGOs with limited resources. Stakeholders need to be provided with the tools necessary to be informed and available in order for involvement to be effective and helpful (GWP 2000).

Beyond, stakeholder involvement is the urgency for cross-sectoral integration (Fig. 7.2). Most governments have at least four ministries that in some capacity manage/use water resources. Each agency has its own mandates and needs but

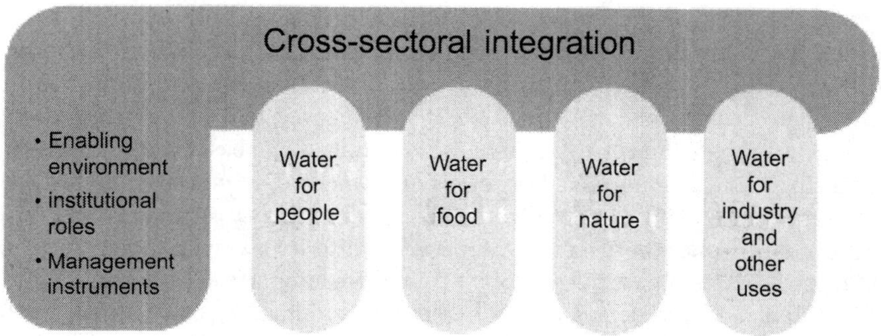

Fig. 7.2 IWRM and its relations to sub-sectors (GWP, 2000)

at the crux is the undeniable fact that there is a finite "pool" of water. Population increases are putting pressure of all uses of water, resulting in inter-ministerial co-operation becoming more and more essential in order to effectively and appropriately allocate the resource and maintain ecological integrity.

Building partnerships with strong communication channels will allow governments to more easily manage change, make decisions that coincide with wider sustainable development plans and implement agreed upon plans of actions. The Hamilton Harbour RAP provides a great example of the process of developing cross-sector linkages and facilitating stakeholder participation (see Chapter 12 in the book for further details). The issue addressed by the RAP was poor water quality resulting from municipal sewage treatment plant discharges, urban and rural non-point sources and industrial discharges (Hamilton is Canada's largest steel producer with plants lining the south shore of the harbour). According to Kellogg (1998), three elements contributed to the system-wide sources of water pollution being addressed, the implementation of joint research projects on waste treatment technologies and the restoration of a marsh area. The three variables were the RAP coordinator's goal for interdepartmental collaboration and stakeholder participation, the skills and expertise of the facilitator and the vision and technical capacity of the stakeholder committee members.

Prior to coordination for the creation and execution of the RAP, data related to water quality was spread across several ministries, each interested in a specific aspect. Therefore the coordinator's first attempt to induce cross-sectoral linkages was to organize a report summarizing all existing water quality studies on the harbour (Kellogg 1998). The RAP coordinator further commited to an ecosystem approach by hiring a facilitator to form a stakeholder committee. The facilitator sought only those who had worked on long-standing environmental issues in the community. This caveat ensured that members had an existing knowledge of issues and would be capable contributors. The stakeholder committee composed of representatives from education, community groups, environmental and conservation groups, municipal government, businesses, industry, sailing and sports clubs as well as ministry staff redefined the scale to be addressed by the RAP from the harbour to the entire watershed, thereby showing the value of stakeholder processes in identifying wider connections. Kellogg (1998) noted that the facilitator assisted stakeholders to develop shared perceptions and full consensus decisions, which in turn eased the uptake of remedial actions, for example the steel companies agreed to assist in the restoration of a harbour inlet.

Shared Visions

The ultimate task in the management of water is protection and strategic allocation for social and economic needs without compromising the health of ecosystems. These tasks are extremely difficult when taking into consideration the numerous levels of uncertainties, the number of competing uses and the abundance of water users. In solving these issues, IWRM is not an "end" goal. IWRM provides a means and requires the process to be iterative as circumstances change and new factors need to be integrated. The "end" or goals for each watershed are determined by the values at play. Every group, household and individual has different values or concerns (e.g., flood control, fishing, recreation, conservation) and therefore has different priorities for water management. Adoption, compliance and engagement in management can be facilitated by creating shared goals or shared visions for the landscape.

The Fraser Basin Council (FBC) in British Columbia formed due to concerns over industrial and agricultural pollutants, over-fishing, urbanization and the belief that management priorities needed to be addressed on the Basin level. The FBC is a collaboration of four levels of government, interest groups, First Nations, the private sector and civil society. The council acts as an impartial facilitator, working to tackle basin-wide issues, overcome conflict and make informed decisions. The society acts primarily to bring together governments, private sectors and NGOs to find solutions and tackle regional and basin wide problems. Through the council's sustainability indicators and "state of the basin" report, all levels interacting in the basin can develop a shared perspective on the state of the environment, agriculture and food security, consumption and waste, social and economic well-being, land use, transportation and population and health. These indicators and reports help create a level playing field where stakeholders are able to come to mutual understandings and from there find workable solutions.

As a basin-wide organization, the society also outlines and puts into practice tools for change. Mitchell (2005) provided the example of FBC managing population growth by encouraging local governments to encourage more compact communities through regulatory tools, Official Community Plans, zoning bylaws and density bonusing. As a result over 80% of the growth from 1986 to 1996 was high-density neighbourhoods.

The three examples provided showed positive instances of IWRM in practice. The various mechanisms employed prove the variety of ways and means of implementing the philosophy. As with every approach IWRM is not without its challenges and limitations.

BARRIERS FOR IMPLEMENTATION

The different physical, social, economic, political and cultural conditions across scales call for IWRM to be adapted to local contexts. This can be challenging, as there is not a clearly outlined set of strategies to be implemented. The above guidelines and examples provided different outlooks on how IWRM philosophy can be put into practice and showed evidence of the myriad of ways IWRM is interpreted. Addressing some of the barriers to implementation can also provide significant insight.

Definition of Integration

One of the problems with IWRM is the definition of "integration". It can mean different things to different people. For instance, for some it can mean "technical integration" – where scientific descriptions of the environment being studied are reported in a compatible manner and where each report should be useful to other groups involved. There can be "procedural integration" – where an agreed set of protocols is used for all the aspects of the IWRM to try and make all the information accessible to others in a standard or known format or "imposed integration" where one or a few agencies drive the process and define the scope, methods, format and reporting of the various aspects of the study. Lastly, there can be "reporting integration", where the various aspects are summarized, analyzed and reported by an appointed group or unit (and they integrate the various aspects).

Jurisdictional Boundaries

The challenge of integrating multiple jurisdictions within a river basin can be overwhelming. Rarely do political boundaries align within watersheds nor does authority over the range of water related concerns align along the same level of governance. Finding an appropriate means of integration can be a significant barrier to moving forward with IWRM. Management issues, such as flow levels of transportation versus pollution hot spots, operate at different scales and require different levels of integration (Morin 2009). Decisions regarding establishing appropriate forums and scales of management for one problem may in turn impede or facilitate issues at other levels.

The creation of a new River Basin Organization is often associated with an IWRM approach. This not only constitutes a considerable challenge for most countries but can also impede the process of IWRM. Butterworth et al. (2010) note that the considerable financial and human resources devoted to establishing catchment agencies does not always ensure acceptance or legitimacy. In many cases, it is argued that working with existing local arrangements can be more effective. The above examples proved that IWRM can be applied at many scales from integrated management of areas of concerns (Remedial Action Plan for Hamilton Harbour) to basin wide initiatives (Fraser Basin Council).

Drowning in Complexity

Integration of all connecting variables and stakeholders within a watershed are the key concepts promoted by IWRM. This task can be a significant barrier to effective implementation, as it is easy for managers to get bogged down by too many variables and participants. Mitchell (2006) distinguishes between two interpretations of systems or ecosystem approaches – the comprehensive approach and integrated approach. To complete a comprehensive report on the status and interconnections of a river basin takes a considerable amount of time and resources, often bottle necking any plans of action. Mitchell (2006) notes that "it was unnecessary to examine every component and every relationship, since each was not significant in accounting for variability in system behaviour". The integrated approach in contrast advocates focusing on key components and relationships. Identification of the key factors can be a challenge and managers are encouraged to draw on experience within the basin by facilitating participation.

Other Common Problems with IWRM Implementation

Another problem related to IWRM arises because of the various disciplines involved. This leads to issues such as a lack of a common vocabulary and understanding of IWRM. Another problem is that the possible solutions are rated differently by the groups belonging to different disciplines, the role of the local community is considered differently by the technical groups; the lead group often determines the agenda despite many issues occurring at the local level and experts operating with a broader viewpoint. The key thing is to strike a balance between the technical community and local participation and between top-down (where the delivery mechanism is not available) and bottom-up approach (which is so local that it cannot be replicated at any other place or cannot be scaled up) (Fig. 7.3). Scaling up is not just replication of technologies or approaches, but the expansion of principles and knowledge, such that people build capacity to make better decisions and influence decision-making authorities. Thus, scaling-up has power and development dimensions (Grover et al. 2005).

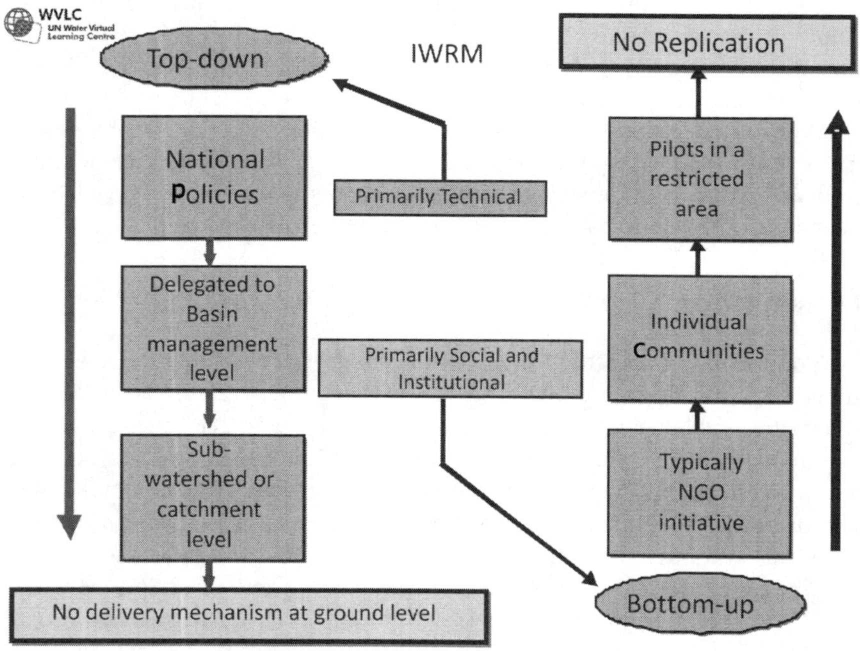

Fig. 7.3 Gaps between bottom-up and top-down approach (Grover et al. 2005)

CONCLUSION

An IWRM approach fosters mechanisms, governance and forums to solve water problems, avoid poor investment and expensive mistakes and enables strategic allocation of water resources (GWP 2004).

Moving beyond sectoral focus supports decisions that are less likely to have larger damaging impacts. IWRM incorporates such diverse management concerns as flood, drought, potable water and water for fisheries management etc. facilitated through inter-agency and stakeholder communication and cooperative decision-making. Bringing together the various interests allows for more comprehensive analyses of decisions. The decision purview is expanded from local to watershed levels, from short-term to long-term ramifications, and incorporates lost opportunities and economic implications.

IWRM principles outline that choices should not undermine ecological integrity. Many lessons have shown the high cost associated with remedial actions to revert ecosystems after inappropriate decisions were made. Plans of action and allocation of water resources is to be tied to sustainable development goals. This process enables strategic distribution and ensures that national goals underpinned by IWRM principles are at the forefront of decisions.

The approach advocated by IWRM is not without its challenges and barriers. Communities, nations and governments have to analyze their contexts and implement strategies that align with IWRM philosophy but are also tied to the political, social and economic realities of their situations. Management should not follow a prescription of actions and should not be overwhelmed with trying to integrate and put in place an ideal IWRM management system. Approaches need to be iterative, timely, and appropriate in order to have a beneficial impact.

REFERENCES

ADB. 2006. ADB Water Financing Program 2006-2010 – Helping to Introduce IWRM in 25 River Basins in the Asia-Pacific Region. http://www.adb.org/water/wfp/basin.asp

Allan, T. 2003. IWRM/IWRAM: a new sanctioned discourse? Discussion Paper No. 58. Water issues study group, University of London. London. UK.

Butterworth, J. and J. Warner, P. Moriarty, S. Smits, and C. Batchelor. 2010. Finding practical approaches to Integrated Water Resources Management. Water Alternatives 3(1): 68-81

Cap-Net. 2008. Integrated Water Resources Management for River Basin Organizations [Training Manual and Operational Guide]. Capacity Building for Integrated Water Resources Management, UNDP. South Africa <www.cap-net.org>

DePinto, J. and V.J. Bierman, T.J. Feist, and J. Kaur. 1999. Conceptualization of an Aquatic Ecosystem Model for Integrated Management of Lake Erie Presentation – White Paper. IJC – Council of Great Lakes Research Managers. Ohio, United States.

Falkenmark, M. and J. Rockstrom. 2006. The New Blue and Green Water Paradigm: Breaking new ground for water resources planning and management. Journal of Water Resources Planning and Management 132(3): 129-132.

Fitzpatrick, J.J. (n.d). A History of Eutrophication Modeling in Lake Erie. IJC - Council of Great Lakes Research Managers. Ohio, United States.

GLC. 2000. Lakewide Management Plans: An ecosystem approach to protecting the Great Lakes. Great Lakes Commission. Michigan, United States.

Grover, V.I. and C. Mayfield, R. Daley, M.S. Babel, KifleKhasai, and P. Ofori-Danson. 2005. Integrated Water Resources Management - its development and water virtual learning center, Proceedings of MTERM International Conference, June 2005 AIT, Thailand.

GWP. 2000. Integrated Water Resources Management, TAC Background Paper No.4. Global Water Partnership. Stockholm, Sweden.

GWP. 2004. Catalyzing Change - A handbook for developing integrated water resources management and water efficiency strategies. Global Water Partnership. Stockholm, Sweden.

GWP. 2009. Investing in infrastructure: The value of an IWRM approach Policy Brief No. 7. Global Water Partnership. Stockholm, Sweden.

Heathcote, I. 1998. Integrated Watershed Management: Principles and Practices. John Wiley and Sons, Inc. New York, USA.

Herdendorf, C.E. and D.C. Rockwell 1983. Lake Erie Water Quality 1970-1982: A Management Assessment. U.S. Environmental Protection Agency. Illinois, United States.

Jonker, L. 2007. Integrated Water Resources Management: The theory-praxis-nexus, a South African perspective. Physics and Chemistry of the Earth 32: 1257-1263.

Kellogg, W. 1998. Adopting an ecosystem approach: Local variability in remedial action plans. Society and Natural Resources 11(5): 465-491.

Koudstaal, R. and F.R. Rijsberman and H.H.G. Savenije. 1991. Water and Sustainable Development. International Conference on Water and the Environment (ICWE). *In:* Keynote Papers by ICWE Secretariat, c/o World Meteorological Organization, Geneva, Switzerland.

Medema, W. and B.S. McIntosh, and P. Jeffrey. 2008. From Premise to Practice: a Critical Assessment of Integrated Water Resources Management and Adaptive Management Approaches in the Water Sector. Ecology and Society 13(2): 29.

Merrey, D.J. 2008. Is normative integrated water resources management implementable? Charting a practical course with lessons from Southern Africa. Physics and Chemistry of the Earth 33: 899-905.

Mitchell, B. 2005. Integrated Water Resource Management, institutional arrangements, and land-use planning. Environment and Planning 37: 1335-1352.

Mitchell, B. 2006. IWRM in Practice: Lessons from Canadian Experiences. Journal of Contemporary Water Research and Education 135: 51-55.

Moberg, F. and V. Galaz. 2005. Resilience: Going from Conventional to Adaptive Freshwater Management for Human and Ecosystem Compatibility. Policy Brief No. 3. SIWI, Sweden.

Molle, F. 2008. Nirvana concepts narratives and policymodels: Insight from the watersector. Water Alternatives1(1): 131-156.

Moriarty, P. and J. Butterworth, and C. Batchelor. 2004. Integrated Water Resources Management and the domestic water and sanitation sub-sector. Thematic overview paper. IRC International Water and Sanitation Centre. Delft, The Netherlands.

Morin, A. 2009. Strengthening Integrated Water Resource Management in Canada [Discussion Paper]. Policy Research Initiative. Ottawa, Ontario, Canada.

OAG. 2001. The Commissioner's Perspective 2001 Commissioner of the Environment and Sustainable Development to the House of Common. Office of the Auditor General of Canada. Ottawa, Ontario, Canada.

Pahl-Wostl, C. and J. Sendzimir. 2005. The relationship between IWRM and Adaptive Water Management. NeWater Working Paper No 3.

Postel, S. 1992. Last Oasis, Facing Water Scarcity. W.W. Norton, New York, USA.

Sadoff, C. and M. Muller. 2009. Water Management, Water Security and Climate Change Adaptation: Early Impacts and Essential Responses. TEC Background Paper No, 14. Global Water Partnership. Elanders, Germany.

Savenije, H.H.G. and P. Van der Zaag 2008. Integrated Water Resources Management: concepts and issues. Physics and Chemistry of the Earth 33: 290-297.

Schreier, H. and R. Bestbier, S. Brown, K. Hall, S. von Westarp, and L. Elliot. 2003. Water and International Development [CD Rom]. Vancouver, BC: Institute for Resources and Environment, University of British Columbia, Vancouver, BC., Canada.

Schreier, H. and K. Hall, S. Brown, L. Lavkulich, and P. Zandbergen. 2003. Integrated Watershed Management [CD Rom]. Institute for Resources and Environment, University of British Columbia, Vancouver, BC., Canada.

Shepherd, G. 2004. The Ecosystem Approach: Five Steps to Implementation. IUCN, Gland, Switzerland and Cambridge, UK.

Snellen, W.B. and A. Schrevel. 2004. IWRM: for sustainable use of water. Fifty years of

international experience with the concept of integrated water management. Background document to the FAO/Netherlands Conference on Water for Food and Ecosystems.

UNEP. (n.d.). Proceedings of International Symposium on Building Partnerships between Citizens and Local governments for Sustainable Lake Management. Freshwater Management Series No. 3. United Nations Environment Programme. Hamilton, Ontario, Canada.

UN-Water and the GWP. 2008. Roadmapping for Advancing Integrated Water Resources Management (IWRM) Processes.

www.unwater.org/downloads/UNW_ROADMAPPING_IWRM.pdf

CHAPTER 8

Lessons Learned from Implementing Low Impact Development in the Credit River Watershed

Christine Zimmer,[1] Phil James,[2] Kyle Vanderlinden,[3] Robb Lukes[3] and Tracy Patterson[4]

[1]Manager of Water Resources and Restoration, Credit Valley Conservation Authority, 1255 Old Derry Road, Mississauga , Ont L5N 6R4

[2]Water Resources Engineer, Credit Valley Conservation Authority, 1255 Old Derry Road, Mississauga. Ont L5N 6R4

[3]Water Resources Specialist, Credit Valley Conservation Authority, 1255 Old Derry Road, Mississauga. Ont L5N 6R4

[4]Principal, Freeman & Associates, 701 Rossland Road E #325, Whitby, Ont, L1N 9K3

INTRODUCTION

For many years, the International Joint Commission (IJC) has identified the discharge of water-borne pollutants from urban land use as a principal influence on the deterioration of water quality in the Great Lakes (IJC 1978, 1987, Kok and Shaw 2005). In the early 1990s, the US Environmental Protection Agency (US EPA) ranked urban stormwater runoff as the second largest source of impairment in rivers (Lee and Jones-Lee 1994). The National Water Research Institute (NWRI) of Canada has also identified urban runoff as one of 15 major threats to drinking water quality in Canada (Environment Canada 2005). In 2003, Gaffield et al. found that urban expansion remains one of the major threats to water quality in the US and is linked to chronic and acute illnesses. Gaffield et al. (2003) concluded that reducing stormwater runoff and associated non-point source pollution was a potential viable component of an integrated strategy to protect public health at the least cost.

Bannerman and Dodd (1992) found that streets, sidewalks and driveways contribute the largest amount of urban pollutants. With streets contributing 65-75% of the total loadings of suspended solids, total phosphorus, copper and zinc. This raises concerns given current trends to widen streets from the 1950s standard road widths of 6 m to current widths of 8 m (City of Mississauga Standard local road 2002).

Depending on the age of subdivision development, a mixture of stormwater practices can be seen across neighborhoods in Ontario. Urban stormwater management in Ontario can be characterized by three eras, starting with stormwater directed through sewers to receiving waterways without treatment (pre-1970s) to control of peak flows or flooding (1970s to early 1980s), to current development standards that are designed to remove up to 80% suspended solids from runoff and manage peak flows (Watt et al. 2003). In light of current erosion and water quality concerns, a fourth emerging era is that of Low Impact Development (LID), which is designed to mimic natural hydrology to improve water quality, maintain groundwater supply and reduce erosion.

Across Ontario, most urban centers were built between the 1950s and 1970s, meaning that they do not meet today's stormwater standards and are also due for replacement. The City of Toronto has targeted $ 1 billion dollars (Cdn) over the next 25 yr to overcome its stormwater deficit. The City of Mississauga, like its neighbors to the east (Toronto) and west (Hamilton), was mostly built prior to stormwater management requirements (Maunder 2010). Currently only 15% of the City of Mississauga receives water quality treatment through end-of-pipe stormwater management facilities. There is, however, the potential to increase this to 25% through retrofitting of existing dry detention facilities and construction of new end-of pipe facilities (Maunder 2010). In order for the City to have a significant impact on water quality, contemporary LID treatment approaches will have to be implemented. Given municipalities' already strained fiscal situation, we are rapidly approaching a tipping point on the infrastructure deficit, one that will seriously harm both our quality of life and our competitiveness and productivity (Bitti 2010).

Due to infrastructure deficits and land constraints, solutions for cities like Toronto and Mississauga to treat surface water quality, minimize erosion, reduce basement flooding liability and build resiliency for climate change is through the implementation of LID retrofitting in existing urban areas and as a requirement for infill and redevelopment. Municipalities on the growing urban fringe must be proactive in incorporating LID into greenfield development to avoid costly stormwater management retrofits in the future. For smaller municipalities that rely on groundwater supply, LID is a cost effective alternative to pumping water from the Great Lakes to meet water demands as LID provides an opportunity to maintain groundwater supply while addressing growth needs.

LOW IMPACT DEVELOPMENT (LID)

Stormwater management systems in use today have been designed to convey flows from a single large design storm event (typically storm events that occur on average once every 2 yr or up to 100 years) away from residential areas. Designing control systems for a single extreme event does not imply that they will perform effectively under other scenarios (CWP 2000a, Paul and Meyer 2001, Marsalek 2002, MOE 2003). The 2-100 yr storm event contributes only a small fraction of the total average annual pollutant load due to their infrequent occurrence (O'Connor et al. 2004). Thus, water quality is not adequately addressed by current stormwater management design methods and remains a concern (PGDER 1993, Marsalek 2002, MOE 2003).

Current flow control standards have their origin in reducing property damage while having little focus on ecosystem protection (PGDER 1993, CWP 2000b, Watt et al. 2003, Marsalek 2004). For preserving stream integrity, experience has demonstrated the importance of a stormwater system designed for more frequent, smaller storm events (i.e. 6 mon to 2 yr storm event) (PGDER 1993, US EPA 2000, NAHB Research Center 2003, Coffman 2004, Guillette 2004a, Pitt et al. 2005). Using decentralized site-based source controls, LID structural practices can be designed for more frequent events and therefore it is a more effective ecosystem approach than conventional end of pipe stormwater management ponds (PGDER 1993, US EPA 2000, Coffman 2001, NAHB Research Center 2003, Guillette 2004b).

LID is an innovative stormwater management approach designed to mimic the natural water balance by:

- protecting natural areas and their functions;
- optimizing urban form by minimizing paved surfaces; and
- managing rainfall at its source using distributed stormwater controls such as green roofs, rainwater harvesting, rain gardens, and conveyance controls such as perforated pipe, and street swales, etc.

Conventional development is designed to convey runoff generated from paved surfaces into low points in the topography where large end-of-pipe stormwater facilities are located in order to store and slowly release runoff. LID by contrast involves an integrated planning approach to incorporate upfront natural landscape features, urban form and distributed source, conveyance and end-of-pipe stormwater controls to better mimic the natural hydrologic cycle. This approach provides flexibility for new development, urban retrofits, and redevelopment/revitalization projects.

When planning for urban development, consideration should also be given to limiting the amount of impervious area by reducing road widths, use of shared driveways, constructing sidewalks on one side of the road, avoiding curb and gutter roads, installing pervious driveways, parking lots and sidewalks where possible, reducing front lawns, and discouraging street layouts such as cul-de-sacs due to their turning radius which increase impervious area (US EPA 2000, PGDER 1999, Coffman 2001, NAHB Research Center 2003, Guillette 2004b, Pitt et al. 2005).

In addition to optimizing the urban form, distributed source and conveyance controls are integrated into the urban fabric. Table 8.1 provides a summary of LID structural controls.

Table 8.1 LID Structural Controls (CVC/TRCA 2010)

LID Practice	*Description*
Rainwater Harvesting	Rainwater harvesting is the process of intercepting, conveying and storing rainfall for future use, typically landscaping or other non-potable uses.
Green Roofs	Green roofs consist of a thin layer of growing medium and vegetation installed on top of a conventional flat or sloped roof and stores rainwater until it is evapotranspired by the plants, evaporates or slowly drains away.
Roof Downspout Disconnection	Simple downspout disconnection involves directing flow from roof downspouts to a pervious area that drains away from the building and preventing it from entering the storm sewer.
Soakaways, Infiltration Trenches and Chambers	Soakaways, infiltration trenches and chambers are underground stormwater infiltration practices that store runoff until it can be infiltrated into the native soils.
Bioretention	Bioretention is a vegetated shallow depression, typically with an engineered soil media, that temporarily stores, treats, and infiltrates runoff.
Vegetated Filter Strips	Vegetated filter strips are gently sloping, densely vegetated areas that treat runoff as sheet flow from adjacent impervious areas
Permeable Pavement	Permeable pavements, an alternative to traditional impervious pavement, allow stormwater to drain through them and into a stone reservoir where it is infiltrated into the underlying native soil or temporarily detained.
Enhanced Grass Swales	Enhanced grass swales are shallow, broad, open channels, sometimes incorporating check dams, used to convey, filter, and attenuate stormwater runoff.
Dry Swales	A dry swale can be thought of as linear bioretention or as an enhanced grass swale that incorporates an engineered soil bed and optional perforated pipe under drain.
Perforated Pipe	Perforated pipe systems can be thought of as long infiltration trenches that are designed for both conveyance and infiltration of stormwater runoff

While the USA has a long history of applying LID, Fig. 8.1 provides an overview of LID implementation across southern Ontario.

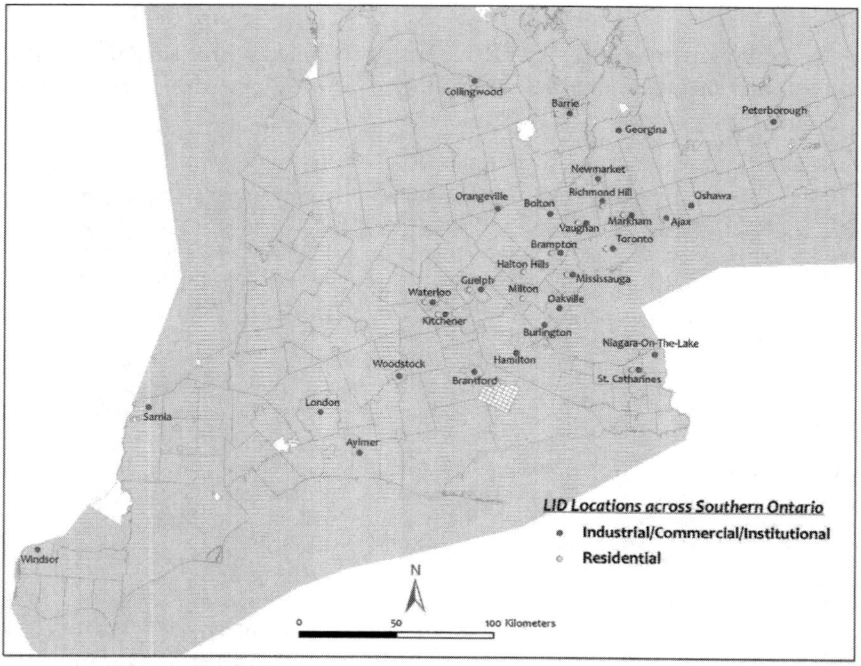

Fig. 8.1 LID implementation across southern Ontario

Color image of this figure appears in the color plate section at the end of the book.

Benefits

LID can provide numerous benefits over conventional stormwater management approaches such as:

- By managing runoff close to its source through LID BMPs, runoff is reduced to predevelopment runoff volumes and rates, thereby improving groundwater and surface water quality through enhanced infiltration and reduced impervious cover (PGDER 1993, US EPA 2000, CWP 2000a, Guillette 2004a, O'Connor et al. 2004).
- Reducing impervious cover provides an economic benefit due to lower capital and maintenance costs for infrastructure (Guillette 2004b, LIDC 2005, Pudget Sound Action Team-Washington State University Pierce County 2005). Experience has shown that LID designs provide cost

savings to developers and municipalities through reduced infrastructure and site preparation (PGDER 1993, US EPA 2000, Guillette 2004a, Pudget Sound Action Team-Washington State University Pierce County 2005). Case studies and pilot programs have shown a 10-30% in cost associated with site development, stormwater fees, and maintenance (Guillette 2004a, LIDC 2005, Pudget Sound Action Team-Washington State University Pierce County 2005). This reduction is achieved by reductions in clearing, grading, pavement, curbs, piping, land and infrastructure for pond construction (LIDC 2005).

- LID not only provides stormwater benefits, such as groundwater recharge and improved water quality, but also protects the function of natural features, reduces urban heat island effect, improves air quality, and reduces thermal stream pollution,
- LID connects residents to the environment by providing more open spaces for recreation and improved aesthetics of a community; and
- LID designs build resiliency into stormwater infrastructure for climate change (MOE 2010, IBC 2010)

LID Implementation Obstacles

A common set of barriers to adopting LID, both technical and social, has been identified by municipalities and the development community over the years. Many of the technical issues have been addressed in the last 10 yr by universities and research centers across North America. While some technical challenges remain, it is the social barriers that have shown to be the greater challenge. Typical social barriers that CVC encountered are outlined in Table 8.2.

In a literature review of municipal LID policies, the Low Impact Development Center (LIDC 2010) found these barriers to be common among the Great Lake States and Provinces. Section 3 provides context for these issues at a watershed level and how the Credit Valley Conservation (CVC) and its member municipalities are beginning to address them.

CREDIT VALLEY CONSERVATION (CVC) JURISDICTION

The Credit River watershed is nestled between the Humber River in Toronto and 16 Mile Creek to the west in Oakville. Part of the Greater Toronto Area, which for a decade has been one of North America's fastest growing areas, the Credit River Watershed, like its 650,000 residents, is diverse. The 1,000 km^2 watershed is covered by an extensive greenlands system made up of forests, valley lands, wetlands and meadows and the lower two thirds of the watershed is a mix of

Table 8.2 Perceived Barriers to LID Implementation

Perceived Barrier or Obstacle	*Description or Example*
Property owner acceptance of lot level stormwater management	The public may not perceive lot level LID practices, like rain gardens as a desirable amenity to preserve and maintain.
Contamination of groundwater due to maintenance and pollutant handling practices.	Lack of awareness with respect to proper handling/storage of pollutants Overuse of salt due to slip and fall liability concerns and economic incentives Overuse and improper application of fertilizers
Lack of funding sources for retrofitting and maintaining stormwater infrastructure	Most municipalities are unprepared for the maintenance costs of current stormwater management ponds let alone those associated with a network of LID practices.
Lack of LID knowledge among design professionals	Designing a distributed stormwater management system differs from designing a centralized system. In the past stormwater management was often the purview of only the engineer, but LID designs require collaboration with other project team members like architects, landscape architects, and planners.
Delayed approvals	LID requires upfront integrated planning/design of urban form and SWM, if planners, architects, engineers, landscape designers, public works, transit, are not at the table from the beginning. Opportunities may be missed or delays in approvals due to reviewers not being in agreement from the beginning Uncertainty as to how to design/size SWM infrastructure Non-conformance to current standards/bylaws Lack of knowledge/skill set on LID amongst practitioners Outdated watershed studies that do not reflect current conditions, current state of the art knowledge of stormwater techniques, and do not consider climate change impacts.
Perceived increased risk	Property owner not maintaining features causing LID failures (i.e., ponding, upwelling) Perceived public health risks due to standing water (West Nile virus)
Lack of LID knowledge among construction professionals	47% of LID practices as constructed deviate from the design plans, which often compromises the effectiveness of the practice (Hirshman and Woodworth, 2010). Unlike conventional linear construction, construction of LID sites often requires non-linear thinking or building stormwater management practices at different stages of a project.

rural and urban. From the headwater streams, the Credit River travels 99 km to the highly urbanized shoreline of Lake Ontario and along the way the river passes through natural areas, farms and rural estates, heritage villages and new suburbs, industrial areas and long-established residential zones. It is in these urban areas that CVC has documented a great need for restoration and has focused a great deal of effort in bringing about environmental restoration in urban areas.

Urban Watershed Studies

Most cities are characterized by older infrastructure and limited available land to implement end-of-pipe stormwater management. This challenge has been complicated further with more intense rainfall events due to climate change and an increasing urban population. CVC sees these challenges as unique opportunities to implement long term sustainable stormwater management practices through integrated watershed studies, public consultation, LID implementation, public education and outreach and social marketing. The following section will detail CVC's efforts to implement sustainable stormwater management on a watershed scale, the challenges being faced and the lessons learned that can be used across the Great Lakes.

What did We Find?

Urban watershed studies within the CVC watershed involve scientific analysis, social outreach and stewardship. As a starting point, the environmental health of the urban watershed was determined by evaluating, monitoring and fieldwork data against provincial and federal targets for surface and groundwater quantity and quality, fisheries, forest and wetland cover. Through this evaluation, it was determined these urban watersheds were showing signs of degradation.

Given that many of the barriers to restoration activities are social in nature (see Section on LID Implementation Obstacles), CVC conducted a number of surveys to better understand watershed stakeholder's awareness of their local environment and their impact on their local environment to gain insight into what motivates the public to make behavioral changes. In November 2007 an initial baseline survey defined as Wave I (Fig. 8.2) found that:

- While people felt climate change was an important issue, they felt it was something that the government was responsible for solving; they did not see how their actions could mitigate climate change.
- Despite the fact that they lived near Rattray Marsh (one of the last remaining bar mouth bay coastal wetlands along the western shore of Lake Ontario), residents felt that urban areas didn't have any natural

features worth protecting, that the urban environment was a lost cause and that CVC should focus its resources on the rural areas.

- People admitted that they knew very little about where their water came from and where it went when it went down the drain.
- Residents stated that they were not aware that winter salt use, or summer lawn fertilizer use had an impact on their local environment; but were willing to learn better practices.

Upon compiling this survey information, CVC launched a public communication campaign called the Save the Leopard Frog (STLF) Campaign that focused on connecting people with their local environment through walking tours, educating residents on their impacts of winter salt use, fertilizer use and pool drainage. The STLF program was designed to be a non-threatening, community engaging campaign that highlighted local natural areas and easy ways in which residents could help protect them by subtle changes to everyday activities. In the fall of 2008, a follow up survey (defined as Wave II) was completed and positive changes were shown in growing awareness about issues that impact watershed health (Fig. 8.2). However, the survey also indicated that more needs to be done with communication in order to:

- Change consumer attitudes
- Tightly link the issues to core consumer values like security of family, health, finances, future, food and water.

In light of the follow up survey in the fall of 2008, the STLF campaign was revamped to connect with schools as they provide great demonstration areas for LID as well as unique opportunities to connect with the community. The STLF school presentations and field tours were designed to specifically provide children with a hands-on, interactive and educational opportunity to connect with their local environment. Along with the students, their parents and teachers became more interested and engaged in knowing and improving their local environment. This provided CVC with the opportunity to identify and design LID techniques on school properties and further engage parent/teacher associations and school children. Ribbon cuttings, BBQ events with parents, students and the community are great opportunities to showcase LID sites and teach residents what they can do on their own property (i.e. disconnecting downspouts and constructing rain gardens). In 2010, CVC designed and helped install two LID features at a school and intends to expand the program in 2011.

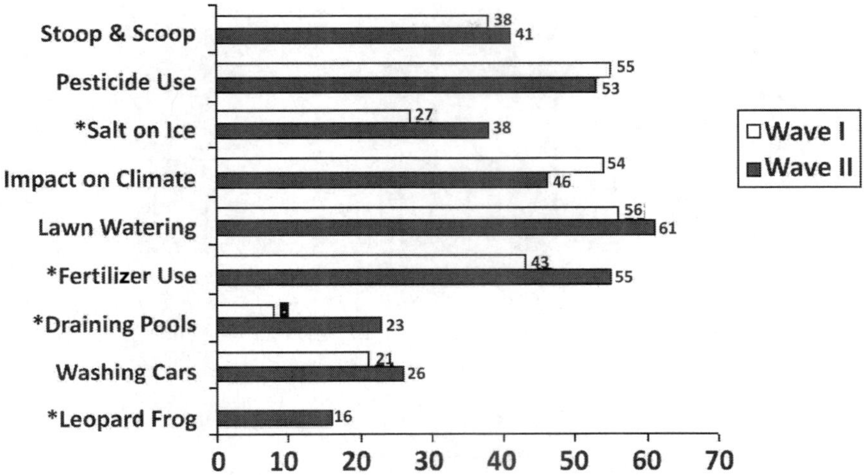

Fig. 8.2 Increased Awareness to Issues Impacting Watershed Health as a Result of CVC's Public Communication Campaign Measured in follow up surveys in Nov. 2008 (Wave II)

Market Research

As we build on the STLF program, delivering effective communication to address stakeholder concerns such as safety, nuisance conditions (i.e. mosquitoes, weeds, ponding water), and aesthetics is critical. As such CVC has drawn on market research studies conducted in municipalities across the Greater Toronto Area over a four-yr period to further develop its campaign.

The results of the market research, which are discussed below, has transformed CVC and member municipalities' approach to engaging residents on LID initiatives in their communities. The primary change has been the recognition of the strong emotional connection homeowners have to their house, surrounding property and, to a lesser extent, the neighborhood in which they live. Figures 8.3 and 8.4, taken from the City of Mississauga market research study, illustrate the emotional connection residents have with their home and its landscape.

Residents see their home as a safe sanctuary and the surrounding landscape as the image they project to the outside world and as a place to relax and spend time with family and friends. The words residents use to describe their home and landscape express deep emotions, not logic and rationale. Yet, outreach and communications from municipalities and conservation authorities and their consultants invariably try to inform and educate residents about the environmental, flood prevention and stormwater control aspects of LID.

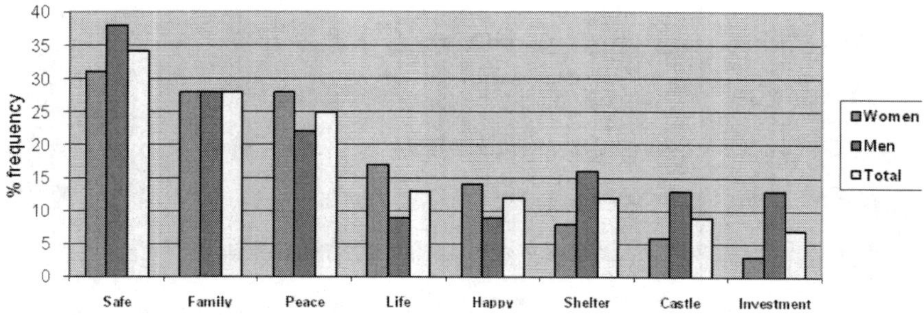

Fig. 8.3 Mississauga Resident Responses to "What Does Your Home Mean To You?"

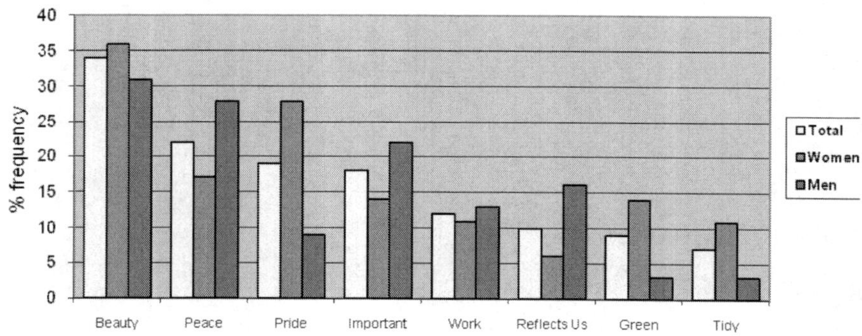

Fig. 8.4 Mississauga Resident Responses to "What Does Your Home's Landscape Mean to You?"

Residents have a very limited understanding of issues related to LID and stormwater management and watershed protection. In the market research completed for the Toronto Region Conservation Authority, residents were asked to "*name three plants* (not trees, or shrubs) *that are native to southern Ontario*". The top three answers provided by respondents in order were:

31% Trilliums

26% Dandelions

21% No answer

Residents were also asked "what is a watershed?", and the following is a summary of the primary responses:

35% No answer

30% Stores water – a natural holding tank

13% An area drained naturally by rivers, streams, etc.

12% Man-made rainwater storage

More often than not communication to residents seeks to educate them as to the use of native plants and the need to protect the watershed; however, the message is not being understood, or more likely, is failing to resonate with residents. It should be noted that all four surveys targeted single-family homeowners, the vast majority of whom have post secondary education with a significant portion having professional and/or post-graduate degrees. In other words, single-family homeowners represent the more educated demographics in the population.

The market research noted the disengagement rests in the fact that residents have busy lives; they are juggling work and family life and full schedules as stated by the respondents. Getting their attention means reaching them on an emotional level. Tapping those intrinsic motivations and engaging residents in a dialogue about the vision for their home landscape and their neighborhood must be the starting point for any LID initiative. Once individuals are emotionally engaged, they will not only be receptive to information on LID, but will actively seek out this information. Market research on stormwater management and water efficient landscaping, tested residents attitudes toward LID landscapes, their willingness to introduce LID elements and what they would encourage them to undertake making such changes pre- and post-viewing of a series of photos depicting these alternative landscapes. Post viewing of photos, the majority of respondents indicated that they would like their own home landscape to look like the landscapes depicted in the photos and indicated that informational resources on "how-to" achieve such landscapes would help them with the transformation process (TRCA 2008). Perceptions of LID and water efficient landscapes changed dramatically upon viewing photos. Negative misperceptions were dispelled and respondents were receptive to information and ideas about transforming their own home landscapes. In conjunction to the research, early findings from a pilot study (2010) by the Region of Peel testing this approach to social marketing – focusing on the aesthetic value of sustainable landscapes as opposed to the environmental, cost-saving and low maintenance aspects of these landscapes – in the community of Clarkson (1200 home community in Southern Peel Region) has resulted in uptake rates for landscape assessment (visits by landscape specialists providing advice on transforming one's landscape) 30 times the rate for the remainder of the Region.

Engaging residents on an emotional level involves building on their ideal for their home landscape and their neighborhood and addressing any concerns or misperceptions they hold about the visual impact of LID installations. In the study done for the TRCA, residents were asked to draw two full-color pictures – one of their "ideal" landscapes for their home and one of a "sustainable" landscape for their home (Fig. 8.5). Below is a sample of the drawings completed by respondents which are representative of the research sample. The drawing of the ideal landscape on the left depicts a front yard with flower

Fig. 8.5 Respondent's Drawings of an Ideal Landscape and Sustainable Landscape
Color image of this figure appears in the color plate section at the end of the book.

gardens, border shrubs and a flag stone walkway. In the same individual's depiction of a sustainable landscape, all color gardens are gone and a barren landscape remains. In conjunction, the flagstone walkway is no longer present and the detail on the garage doors has all but disappeared. This drawing is representative of 46% of respondents in the Mississauga research session and is consistent with responses from the other GTA market research studies that have been undertaken. With the exception of about 10 to 15% of home-owner respondents in three separate studies exploring LID and sustainable landscapes (TRCA-2006, Mississauga-2008 and Toronto-2009), residents held a negative perception of "naturalized", "sustainable" and "natural" landscapes. There was a perceived net loss of aesthetics and entertainment functions (swimming pools, BBQs, hot tubs, gazebos, decks/patios) in these landscapes. In other words, residents' perceptions of these landscapes as depicted in their drawings were a far departure from their ideal and therefore, undesirable.

Stakeholder consultation is an important process in generating support for LID. As part of identifying restoration projects, stakeholder consultation is important in gaining an understanding of residents' vision for their home landscape and streetscape.

Lessons Learned and Next Steps

In developing restoration and LID project, the following are lessons learned from the market research described below

- Neighborhood acceptance is important when screening potential restoration opportunities. When identifying demonstration site opportunities for new and existing development start small and learn lessons to build support for long-term initiatives. Selecting demonstration sites with a high probability of success will help with building confidence to implement across the watershed.
- Focus on behavior that residents can control and can understand and start small to build support for long term initiatives. Providing residents

with things they can do at home through subtle behavioral changes to everyday activities can help residents become champions of the health and long-term sustainability of their local watershed. Focus campaign communications around the positive benefits of change but don't ignore the negative aspects.

- Change will take time so build a long term model to grow and evolve the strategy over time. Measure success by building in measurement tools such as monitoring programs and surveys to identify what is and isn't working following the principles of Adaptive Environmental Management.
- When targeting residential areas focus on low and medium density as they are less transient.

Lakeview District Road Retrofit

The streets in the Lakeview District were identified 15 to 20 yr ago for updating from a rural to urban cross section as residents had been complaining about deep ditches that were safety hazards with standing water build-up. As part of the city's capital budget program, streets within the Lakeview District were scheduled for reconstruction works. The Lakeview District Road Retrofit (Lakeview Project) was identified as a demonstration site for implementing LID within the road right-of-way for the following reasons:

- The neighborhood homes are undergoing revitalization; the Lakeview Project provided an opportunity for the City of Mississauga and CVC to apply an innovative marketing approach for proposing LID while meeting the community's vision for their home landscape and neighborhood.
- Water quality in the receiving stream is a concern. By implementing LID within the road right of way, the City can improve water quality to receiving stream.
- The City had funding through their 10 yr capital budget.
- The Lakeview project offered a demonstration site for the City's Strategic Plan – Living Green Strategic Pillar, and the Mississauga Storm Water Quality Control Strategy update as well as CVC's Credit River Water Management Strategy Update

Using the information gained through the market research described in "market research section" above, a public meeting was held in June 2009 to discuss the proposed Lakeview Retrofit project with the community. The project was introduced to the community as a community beautification project that would also benefit the environment. A short overview of the proposed community improvement plan was provided and the majority of the time was for residents to walk around the room, view renderings of how the project would look and ask questions. The engineering design images were less important than the pictures that showed how the appearance of the neighborhood would be enhanced. This approach has also been used in Seattle, Chicago, Portland and other places and the response of residents has been overwhelmingly positive (CDOT, 2007; City of Seattle, 2010).

The residents were assured that their past concerns about the deep ditches within the right-of-way and standing water would be addressed through the improvement project. It was also stressed that the unique character of the neighborhood would be enhanced by minimizing the ditches and, where possible and with community input, redesigning them with attractive plantings. Landscape design specialists would be on the design team to work with homeowners to create designs that function and beautify the area.

To create a community vision for the project, several informal meetings were planned to update residents on design and landscape concepts and to address questions such as those listed below:

- Would the loss of ditches cause residents to park on the grass if the curbs are not constructed?
- Will you rip up my driveway?
- Will there be sidewalks?
- How will we deal with animals that like to live in the culverts and pipes?
- Is there some way we can create a guide for the snow ploughs?
- Will this impact emergency vehicle access and snow plowing operations?
- I have a ditch in my backyard. Will you also be dealing with ditches that are not in the road allowance?
- Does this project affect our taxes?
- What about ease of maintenance especially for seniors?

Research completed by Freeman & Associates (2008) was used to address the concerns residents expressed above. The following bullets detail the approach used to win support from residents to implement the Lakeview Project which are:

- Developing a script for staff to address questions listed above is critical to ensure consistency. As mentioned earlier, correct wording can make or break residents' adoption of a project. Having a script also provides staff with confidence and reassurance.
- Gaining trust and confidence by assuring residents that a great deal of research has gone into this initiative and draws on the best in stormwater management, environmental protection and community improvement from other jurisdictions around the world.
- That there will be a dedicated on-site construction co-coordinator available to answer questions and address your concerns.

- Less emphasis on environmental benefits and more emphasis on urban renewal, neighborhood rejuvenation, environmental legacy
- Visually attractive graphics less engineering/technical drawings
- Less formal presentation, more time for questions and interaction
- Find a champion resident and ask them to chair meetings, help to educate and market the improvement project.

In continuing the Lakeview Project, the following steps have been taken:

- Pre-monitoring construction monitoring was initiated in 2010 and will continue throughout construction and post-construction.
- Final design and approvals is scheduled for Winter 2011 with construction planned for Spring 2011
- CVC will be documenting the challenges faced during the construction of LID features so that constructions guidelines can be produced to assist contractors and municipal employees in future projects.
- Lessons learned will be posted on CVC's website to help others implement LID.

Elm Drive Road Retrofit Construction

Elm Drive is a local residential road located in the City of Mississauga just south of Square One. Elm Drive has a rural cross-section with open drainage features in the form of surface swales. As part of the Mississauga's 2010 capital budget program, Elm Drive was scheduled for pavement resurfacing and some minor improvements. The project included the installation of permeable pavers in the parking lay-bys and sidewalks within the road allowance. A bioretention planter system was installed within the adjacent boulevard to filter and store rainwater before slowly releasing it back to Cooksville Creek and Lake Ontario. Underground storage in the form of deep granular trenches were also included to provide additional storage.

As with all building projects, many unforeseen challenges arose during the construction process. One such challenge was building on private property as difficulty may arise due to the lack of proper documentation. The Elm Drive road retrofit was built on both public and private lands and during excavation both a fiber optic line and gas line were discovered. Both lines were not marked on any existing drawings so both lines had to be dealt with before construction had begun. These unexpected challenges may become more common as LID is implemented on private as well as public lands.

To help the contractor overcome these challenges, it is important that the engineer drawings translate clearly and accurately for construction. With the Elm Drive road retrofit, the contractor had several questions regarding the details of the engineer drawing as measurements, material descriptions and construction notes pertaining to sediment and erosion control were not thorough enough. Although the contractor obtained the necessary information from the engineering consultant, it will be important to have all necessary information on the design drawings in order to minimize any confusion or conflict for the contractor.

To help future LID construction projects, CVC is currently photographing construction of the Elm Drive Retrofit and documenting any questions raised or challenges encountered by the City and contractor. Proper documentation and understanding of municipal and contractor perspective with regards to LID will help in the creation of construction guidelines. The aim of these construction guidelines is to help minimize failure of LID features due to improper design, construction or materials used. The more comprehensive the guidelines can be for contractors and municipal inspectors, the better success rate of LID. Proper installation of LID features is crucial for LID success as failure will only lead to further barriers to its implementation.

POLLUTION PREVENTION (P2)

One of the main concerns regarding LID implementation is groundwater and soil contamination due to LID's reliance on water infiltration. Therefore, eliminating sources of pollution that drain into the LID feature will help ensure its success, but also reduce risk of groundwater contamination. Thus, developing a pollution prevention program (P2) along with LID implementation is critical. As part of the watershed study process, CVC and project partners completed pollution prevention surveys for residential, commercial, industrial, municipal and institutional uses to look for activities and practices that could result in pollutants entering the storm drainage system and receiving waterway and ultimately Lake Ontario. During site visits, potential threats to surface water were identified which could include outdoor fueling stations, vehicle washing stations, improper outdoor material storage, and improper waste storage or disposal of waste materials. An estimate of a site's pollution severity allows CVC to assess potential P2 activities needed at a site or general area. CVC launched pollution prevention demonstration sites to show both public and private sectors technologies common pollution sources and ways that small commercial industrial companies can eliminate these sources.

Pollution Prevention (P2): Demonstration Sites

The P2 projects were initiated in partnership with the City of Mississauga to help reduce threats to surface water flowing into the storm drainage network. As in other nearby urbanized watersheds, the storm drainage network in

Mississauga ultimately leads to Lake Ontario which is also the local municipal drinking water supply. CVC's P2 activities focused on building awareness through workshops and by implementing "in the ground" measures to reduce or eliminate threats to municipal drinking water supplies at the source. It is important to distinguish that these "in the ground" measures aimed to go above and beyond what was required of the property owners by the municipal storm sewer bylaw. The types of demonstration projects implemented were:

- Better dumpster management
- Improved outdoor material storage and waste handling
- Reduction of impervious surfaces
- Clean water buffers with soil amendments
- Improvement in snow storage and melt water treatment
- Fuel station secondary containment
- Loading/unloading secondary containment
- Emergency shut-off valves
- Information/Educational Signage

In addition, with the implementation of these "in the ground" measures CVC also aimed to achieve a range of supporting objectives through the education of property owners, tenants and outreach to the community which are:

- Empower stakeholders to take action and protect municipal drinking water supplies through implementation of on the ground projects.
- Increase public and community awareness of the importance of pollution prevention and making the connection between stormwater drainage and municipal drinking water supplies.
- Gain knowledge and experience that can be applied to future watershed studies pertaining to innovative pollution prevention and protecting surface water municipal drinking sources.
- Changing the business culture of participants to become environmentally and economically empowered local leaders.

Management Agreements

In order to address both landowner commitment and satisfy liability issues and as a condition of being a demonstration site in order to become eligible for funding, the landowner was required to sign a management agreement that included:

- Outline of project objectives
- Obligations of landowner
- Obligations of CVC
- Cross Indemnification clauses

- Insurance requirements
- Term of Agreement
- Termination clauses
- Amendments

The conditions of the funding were explained to the landowners and there was agreement to allow CVC to showcase their property, for property owners to attend CVC's P2 workshops, and allow restricted site visits to the public. Each landowner agreement required separate negotiation as concerns put forward by the property owners varied due to differences in the nature of their individual operations. Overall, the management agreements were a major step in the creation of the P2 demonstration sites and they were not without their challenges as two landowners declined from participating in the project based on legal advice from their solicitors.

Project Benefits

The pollution prevention projects provided a number of benefits to CVC, the City of Mississauga, the Region of Peel and the Ontario Ministry of Environment. Implementation of the pollution prevention demonstration sites provided the following:

- Supports the City of Mississauga's storm sewer bylaw, Storm Water Quality Strategy Update, and Green Development Strategy;
- Supports MOE with source water protection initiatives and guidelines;
- Supports the Air Quality Task Force Oakville – Clarkson Airshed;
- Supports Region of Peel's Sanitary Sewer Bylaw;
- Support and compliment LID initiatives;
- Supports Municipal, Region, and MOE in dealing with spill response and preparedness;
- Empower and educate businesses within the local ICI sector to become demonstration sites and provide leadership opportunities and green profile or participating businesses;
- Initiate local pollution prevention awareness and dialogue among local business operators;
- Create a cooperative partnership with participating businesses in protecting our water resources locally.

Lessons Learned

Many challenges were faced during the implementation of the pollution prevention demonstration sites. For many business operators, their initial reaction was one of fear of enforcement when both the municipality and conservation authority approached them. Many were unaware of how their facility stood in terms of legislative compliance and were reluctant to offer their site for close scrutiny by enforcement staff. To reduce any landowner fear or apprehension, it was very important that the intentions of the project were made clear during the initial meetings. Greater emphasis was placed on meeting the business operator or landowner in person rather than communicating the purpose of the project through email. At the same time, face to face communication also allowed CVC to work with the sometimes complex structure of a given business including working with property managers with multiple tenants and property maintenance personnel.

For most business operators, impacts to surface water or pollution prevention was just not on their agenda. There tended to be a general lack of knowledge of the local watershed and surrounding environment. The initial face-to-face meetings provided CVC and the municipality an opportunity to educate the property owner and tenants. Use of visual tools such as maps, aerial photographs, and concept drawings proved to be very beneficial in the education process as they helped to familiarize the property owners with their local watershed, its functions and how their businesses were connected to it. These facts are not always common knowledge and visualizations were important in explaining this technical information.

Once the project concept and plans had been finalized with the property/business owner, it was important that appropriate time be given for the permitting process. It was critical to work patiently through zoning and standard issues where there was uncertainty in the interpretation of the Ontario Building Code as it was silent regarding a particular P2 measure. Plans and designs needed to change so incorporating flexibility into the design was beneficial.

TOOLS FOR IMPLEMENTING LID ACROSS THE GREAT LAKES

Critics of LID cite the reliance on the property owner to maintain for example, rain gardens and rain barrels as a risk to municipalities (LIDC 2005); but this argument overlooks the fact that LID is based on a multi-barrier system which reduces the risk. Furthermore, non-structural practices of LID involve the reduction of paved surface that generate runoff (reduced road widths, curbless roads, buffer zones, grading strategies, open drainage systems, and conserving open space) which are not influenced by the behavior of the property owners (LIDC 2005).

In Maryland and Washington where a majority of LID sites have been installed, maintenance agreements have been used if municipalities and/or developers are uncomfortable about on-site landscaping features (LIDC 2005).

As cited above, the key factor in the success of LID has been to ensure that landscape changes are perceived by the property owner as enhancing their property (Freeman and Associates 2008, LIDC 2005, Pudget Sound Action Team-Washington State University Pierce County 2005). If these LID practices are viewed as assets, property owners will maintain them (CWP 1998, LIDC 2005, Pudget Sound Action Team-Washington State University Pierce County 2005). In addition, the use of educational campaigns can successfully promote active public awareness for the need to protect water quality through the property owners maintaining their property (CWP 1998, LIDC 2005, Pudget Sound Action Team-Washington State University Pierce County 2005).

Table 8.3 provides further tools and resources to address concerns expressed by our watershed stakeholders.

Table 8.3 Watershed Stakeholders Concerns and Tools to Overcome Barriers

Watershed stakeholders concerns	*How CVC is addressing concerns*[1]
Contamination of groundwater and surface water	Urban watershed studies which involve fieldwork to identify pollution hotspots and work with stakeholders to implement pollution prevention plans. Pollution Prevention Website, Fact Sheet, and Workshops Pollution Prevention Pilot Study (in progress – see section on Pollution Prevention above) Mississauga Salt Study (to be released in 2011) GO Transit Salt Study (in progress) Identification of alternative winter snow removal operations (ploughing versus salting, optimized parking lot design to reduce salting requirements by implementing permeable pavement) CVC Spill Characterization Report which provides a summary of types, frequency of spills to help decision makers develop management priorities.
Lack of funding sources for retrofitting and maintaining stormwater infrastructure	CVC Sustainable Stormwater Funding Study provides an overview of various funding mechanisms available to municipalities, some examples include: Stormwater utility fees based on impervious cover, Developer funded stormwater maintenance trust Unlike centralized treatment facilities or large above and below ground storage systems that require large capital outlays all at once, LID can be implemented over time incrementally and incorporated into long term capitol budgets (US EPA, 2008).

Table 8.3 contd. ...

Table 8.3 contd. ...

Watershed stakeholders concerns	*How CVC is addressing concerns*[1]
Lack of LID knowledge amongst professionals	CVC/TRCA LID SWM and Planning Guide which provides guidance on design and landscaping of LID structural techniques Over the last 5 years CVC has hosted several LID workshops bringing experts in the field to share their experiences and lessons learned, target audience for these workshops has been Municipal and Provincial staff, Development community, and Consulting Engineers/Landscape Architects. In 2011, CVC hopes to target contractors and inspectors in future workshops. CVC Partnership with Canadian Standards Association (CSA) to develop training programs to expand audience beyond CVC watershed. LID Demonstration sites across the watershed to build consensus and confidence, lessons learned will be posted on CVC's website
Delayed approvals	Leadership from the top-down is required to provide staff with the structural support to work cross-functionally between departments and implement innovative solutions to overcoming current standards and barriers. CVC has hired LID specialists to assist designers and plan review staff with approvals. Some Municipalities in the US have tried to encourage developers to adopt LID by fast-tracking approvals and hiring a dedicated review team for LID projects. CVC Education and Outreach (see above) to educate municipal, CA and provincial approval staff In partnership with TRCA, in consultation with member municipalities and the development community new SWM Criteria are being developed for water quality, quantity, erosion and water balance to protect our watersheds. CVC/TRCA LID SWM and Planning Guide provides design and planting guidance for LID structural practices. In consultation with municipalities, development community, provincial agencies and CAs across Ontario, CVC is developing case study handbooks that provide options for overcoming the common barriers to implementing LID in ICI and multi-residential sector and in the public realm. Through urban watershed studies, identify and amend standards/bylaws in order to facilitate LID.

Table 8.3 contd. ...

Table 8.3 contd. ...

Watershed stakeholders concerns	*How CVC is addressing concerns*[1]
Perceived increased risk	Develop a marketing strategy based on demographics (refer to the Marketing Research section above) to promote LID, educate residents about West Nile virus. Organizing multi-stakeholder roundtable discussions to identify local solutions to perceived risks Identify willing stakeholders and implement LID demonstration sites to build confidence and consensus Monitor and publish data from demonstration sites to provide performance data, information for future sizing of infrastructure, etc. to watershed stakeholders CVC LID specialists to assist designers and plan review staff with approvals Case study handbooks being developed by CVC will provide watershed stakeholders with techniques to deal with or overcome perceived risks
Lack of knowledge among the construction professionals	In consultation with Contractors, Developers, and Municipalities, CVC is producing two LID Construction Guides, one tailored to Designers and the other to Inspectors and Contractors. These guides will be released in 2011 and will be applicable to practitioners in Ontario with the aim of developing training programs in 2012 to certify Contractors and Inspectors across Ontario.

[1] For further information and resources see:

Credit Valley Conservation (www.creditvalleyca.ca/low-impact-development/) This includes information on Pollution Prevention, LID Guidance, LID marketing and LID case studies on overcoming barriers.

US Environmental Protection Agency (www.epa.gov/owow/NPS/lid/),

Low Impact Development Center (lowimpactdevelopment.org),

Toronto Regional Conservation Authority (www.sustainabletechnologies.ca/),

and Canadian Standards Association CSA training website (learningcentre.csa.ca)

CONCLUSION

More than 33 million people live in the Great Lakes Basin, with 80% of the population residing in 17 metropolitan areas (Thorp et al. 1997). Thorp et al. (1997) found that physical growth (urban expansion) outpaced population growth within the basin. For example, the State of Michigan has predicted a population growth of 12% by 2020 while its physical area is projected to increase by 87% (Thorp et al. 1997). Likewise, Toronto's population is expected to grow by 2 million by 2020 with an additional 600 km^2 urbanized (Thorp et al. 1997).

The US EPA (1997) found that continued growth in major metropolitan areas and the sprawl of residential areas and other developments has severe social, environmental, and economic impacts that threaten the sustainability of the Great Lakes Basin. The US EPA (1997) also concluded that "Urban sprawl contributes to polluted runoff by replacing green open spaces and farmland with paved surfaces and requiring the building of additional roads and commuter highways; it contributes to air pollution by boosting commuter distances and vehicle miles traveled per person; and it results in the loss of viable habitat for animals and plants. Between 1981 and 1992, for example, basin farmland decreased by 9.6%" (US EPA 1997).

The amount of pollutant loading in urban stormwater reflects the rate of generation of pollutants (NAHB Research Center 2003, Guillette 2004a, US EPA 2000, Hsieh and Davis 2005, PGDER 1993). "Low Impact Development" (LID) is an alternate design strategy that has been implemented in Northeast USA and the Pacific Northwest to enhance groundwater recharge and reduce contaminant loading and erosion by maintaining pre-development hydrology for frequent storm events (6-mon to 2-yr storm) (Guillette 2004a, O'Connor et al. 2004, PGDER 1993). LID uses lot level controls and planning practices to maintain predevelopment hydrology (Guillette 2004a, Coffman 2004, PGDER 1993).

Given urban development pressure projected for the next 30-yr and the significant impact that urban non-point pollution and urban sprawl has on human health, it is important for policymakers to have adequate tools and knowledge to mitigate and protect the environment and human health from the impacts of urbanization.

The early experiences of LID stormwater programs have shown that LID guidance and policy, demonstration initiatives and targeted marketing are among the key tools for implementing LID on a wide scale.

LID guidance and training is required by planners, architects, engineers, and contractors on how to plan, design, and construct urban layouts and stormwater practices to mitigate the impact of urbanization on human health and the environment. In addition, provincial, state and municipal planning policies need to provide direction and structure in order to facilitate sustainable development for public and ecosystem health.

LID demonstrations serve several important functions. Despite volumes of research and case studies from other regions, most property managers and municipal engineers will not be comfortable with LID stormwater practices until they can see functional demonstration projects in local conditions. Demonstrations are also necessary to create the domino effect in which neighboring property owners incorporate LID projects. Implementing LID demonstrations like permeable pavement on important public properties or rain gardens in high profile residential areas will motivate others through beautified landscapes, peer association, and being "on trend".

LID targeted marketing approach overcomes perceived social barriers by drawing on emotion to effect behavioral changes. Before educating or informing property owners on why or how to incorporate LID into their property, it is imperative to give them an attractive vision of what their property could look like. The only means to overcome property owner negative perceptions of sustainable, naturalized, or water efficient landscapes is to provide them with visual images that counter this perception (Freeman & Associates 2008).

LID is an adaptive tool to solve a myriad of water resource issues for large and small municipalities alike. Simply forcing an engineered stormwater solution is not enough; successful LID programs must consider the right training, marketing, and guidance support tools.

ACKNOWLEDGEMENTS

We would like to acknowledge the efforts and contributions of City of Mississauga staff which include Michelle Charboneau, Jeff Smylie, Steve Barrett, Tom Wenzel, Lincoln Kan, Jeremy Blair, MOE and CVC Staff; funding from Region of Peel and support from Rae Horst, CAO John Kinkead, Director and CVC's Board.

REFERENCES

Bannernam, R. and R. Dodd. 1992. Sources of Pollutants in Wisconsin Stormwater. Wisconsin Department of Natural Resources, Madison, WI, USA.

Bitti, M.T. 2010. Facing Canada's Water Woes. Financial Post, Tuesday March 9, 2010.

Carson, J.S. 2002. Model Verification and Validation. Proceedings of the 2002 Winter Simulation Conference. E. Yucesan, C.H. Chen, J.L. Snowdon, J.M. Charmes [eds.]. http://www.informs-cs.org/wsc02papers/008.pdf

Center for Watershed Protection [CWP]. 1997. Stormwater BMP Design Supplement for Cold Climates. Prepared for: US EPA Office of Wetlands, Oceans and Watersheds. Washington, DC. USA.

Chicago Department of Transportation. 2007. The Chicago Green Alley Handbook. The City of Chicago, IL. http://www.cityofchicago.org/city/en/depts/cdot/provdrs/alley/svcs /greenalleys.html

City of Seattle. 2010. Seattle's Natural Drainage Systems. http://www.seattle.gov/util/groups/public/@spu/@usm/documents/webcontent/spu02_019984.pdf

Credit Valley Conservation and Toronto Regional Conservation Authority. 2010. Low Impact Development Stormwater Management Planning and Design Guide. Version 1.0. http://www.creditvalleyca.ca/sustainability/lid/stormwaterguidance/index.html

CWP. 1998. Better Site Design: A Handbook for Changing Development Rules in Your Community. Prepared for Site Planning Roundtable. Center for Watershed Protect. Ellicott City, MD.

CWP. 2000a. Better Site Design: Center for Watershed Protect. Center for Watershed Protect. Ellicott City, MD, USA.

CWP. 2000b. 2000 Maryland Stormwater Design Manual. vols. I and II. Maryland Department of the Environment Water Management Administration. Baltimore, MD, USA.

CWP. 2003. The Impacts of Impervious Cover on Aquatic Systems: Watershed Protection Research Monograph No. 1. Center for Watershed Protect. Ellicott City, MD, USA.

Coffman, L. 2001. Low Impact Development – A New Stormwater Management Paradigm: Micro-scale Source Management. Low Impact Development in Puget Sound: Innovative Stormwater Practices. June 5-6, 2001. Puget Sound Water Quality Action Team Office of the Governor. Olympia, WA, USA.

Coffman, L.S. 2004. Discussion of an Ecosystem Functional Basis for Protecting Receiving Waters. Proceedings of Protection and Restoration of Streams ASCE. pp. 383-391.

Environment Canada [EC]. 1992. Historical Stream flow Summary Ontario to 1990. Inland Waters Directorate, Water Resources Branch, Water Survey of Canada, Ottawa, Canada, and the 2002 HYDAT CD-ROM [Version 2002–2.04] published in July 2004.

EC. 1999. Summary and Update of the 1997 Science Assessment of the Impacts of Municipal Wastewater Effluents (MWWE) on Canadian Water and Human Health. Environment Canada, October 1999, Ottawa.

EC. 2005. Threats to Sources of Drinking Water and Aquatic Ecosystem Health in Canada. www.nwri.ca/threats/indiex-e.html. updated 2003-01-013

Freeman & Associates. 2008. Market Research and Marketing Strategy: Lot-level Stormwater Control in the Residential Sector. City of Mississauga. Ontario, Canada.

Gaffield, S.J. and R.L. Goo, L.A. Richards and R.J. Jackson. 2003. Public Health Effects of Inadequately Managed Stormwater Runoff. American Journal of Public Health. September 2003, vol. 93. No. 9. pp. 1527-1533.

Guillette, A. 2004a. Low Impact Development Technologies. Whole Building Design Guide (WBDG) © 2005. National Institute of Building Sciences Updated: 05-20-2004. http://www.wbdg.org/design/lidtech.php

Guillette, A. 2004b. Achieving Sustainable Design Through Low Impact Development Practices. Whole Building Design Guide (WBDG) © 2005, National Institute of Building Sciences Updated: 05-03-2004. http://www.wbdg.org/design/lidsitedesign.php

Hirschman and Woodworth. 2010. Design, Construction, and Maintenance of LID Practices: Results from a Field Assessment in Virginia's James River Watershed. Proceedings of the 2010 International Low Impact Development: Redefining Water in the City. April 11-14 2010, San Francisco.

Hsieh & Davis, (2005) Evaluation and Optimization of Bioretention Media for Treatment of Urban Storm Water Runoff. J. Environ. Eng. 131(11), 1521-1531.

IBC [Insurance Bureau of Canada]. 2010. Cities are Facing Increase Risk and Loss Due to Flooding. The Proceedings of New Challenge: A Climate for Change 2010 CSA/CMHC/CVC LID Conference. September 28, 2010, Mississauga, ON.

IJC [International Joint Commission]. 1978. Environmental Management Strategy for the Great Lakes System. Final Report to the International Joint Commission from the International Reference Group on Great Lakes Pollution from Land Use Activities Reference Group. Windsor. Ontario, Canada.

IJC [International Joint Commission]. 1987. 1987 Report on Great Lakes Water Quality. Great Lakes Water Quality Board Report to the International Joint Commission Windsor Ontario, Canada.

Kok, S. and J. Shaw. 2005. Wet Weather Flow Management in the Great Lakes Areas of Concern. Proceedings EWRI 2005. Copyright ASCE 2005.

Lee, G.F. and A. Jones-Lee. "Water Quality Aspects of Groundwater Recharge: Chemical Characteristics of Recharge Waters and Long-Term Liabilities of Recharge Projects," In: Proc. Of the Second International Symposium on Artificial Recharge, American Society of Civil Engineers, New York, NY, July (1994).

Low Impact Development Center [LIDC]. 2005. The Concept of Green Infrastructure: Using LID and Green Practices for Sustainable Infrastructure and Resource Management. The Proceedings of Lessons Learned from Implementing Low Impact Development in the Northern United States. October 7, 2008, Mississauga, ON.

Low Impact Development Center [LIDC]. 2010. Policies and Administrative Approaches for Overcoming Institutional Barriers to Low Impact Development. Report funded by Credit Valley Conservation.

Marsalek, J. 2002. Overview of Urban Stormwater Impacts on Receiving Waters. proceedings of the Urban Water Management: Science, Technology and Delivery. NATO Advanced Research Workshop. Borovetz, Bulgaria. pp. 3-14.

Marsalek, J. and H.Y.F. Ng. 1989. Evaluation of Pollution Loadings from Urban Non-Point Sources, Methodology and Application. J. Great Lakes Res. 15(3) 444-451.

Marsalek, J. and B.C. Anderson, and W.E. Watt. 2004. Suspended Particulate in Urban Stormwater Ponds: Physical, Chemical and Toxicological Characteristics. Proceedings Urban Drainage ASCE Copenhagen, Denmark 2004. pp. 1-12.

Maunder, D. 2010. Why LID for the Public Realm? What are the obstacles, barriers, challenges, opportunities and benefits? The Proceedings of New Challenge: A Climate for Change 2010 CSA/CMHC/CVC LID Conference. September 28, 2010, Mississauga, ON., Canada.

Ministry of the Environment [MOE]. 2003. Stormwater Management Planning and Design Manual. Queen's Printer Toronto Ontario. Canada.

Ministry of the Environment [MOE]. 2010. MOE Review of Stormwater Management Policy in Light of Climate Change. The Proceedings of New Challenge: A Climate for Change 2010 CSA/CMHC/CVC LID Conference. September 28, 2010, Mississauga, ON., Canada.

NAHB Research Center, Inc. 2003. The Practice of Low Impact Development. Prepared for the US Department of Housing and Urban Development. Upper Marlboro, MD. Contract No. H-21314CA.

O'Connor, T.P. and M. Clar, and B. Barfield. 2004. General Considerations in BMP Design. Proceedings of World Water Congress ASCE Critical Transitions in Water and Environmental Resources Management, Salt Lake City, Utah, June 27th - July 1 2004. pp. 1-11.

Paul, M.J. and J.L. Meyer. 2001. Steam in the Urban Landscape. Annual Review Ecological Systems. Vol. 32, pp. 333-365.

Pitt, R.E. and S. Chen, S. Clark and J. Lantrip. 2005. Soil structure effects associated with urbanization and the benefits of soil amendments." R. Pitt, S. Chen, S. Clark, and J. Lantrip. World Water and Environmental Resources Congress. ASCE/EWRI. Anchorage, Alaska. May 2005. pp. 1-12.

Prince George County Department of Environmental Resources [PGDER]. 1993. Design Manual for Use of Bio retention in Stormwater Management. Division of Environmental Management, Watershed Protection Branch. Landover, MD, USA.

Prince George County Department of Environmental Resources [PGDER]. 1999. Low Impact Development Strategies: An Integrated Approach. Division of Environmental Management, Watershed Protection Branch. Landover, MD, USA.

Puget Sound Action Team. 2005. Low Impact Development: Technical Guidance Manual for Puget Sound, January 2005. http://your.kingcounty.gov/solidwaste/greenbuilding/documents/Low_Impact_Development-manual.pdf

Thorp, S. and R. Rivers, and V. Peebles. 1997. Impacts of Changing Land Use. State of the Great Lakes Ecosystem Conference (SOLEC). Environment Canada and USEPA Ottawa and Washington.

Toronto Region Conservation Authority. 2007. Action Plan for Sustainable Practices: Implementation Strategies for the Residential and Business Sectors in the GTA. Oct 2007. pp. 28-30, Appendix 5 – verbatim responses to questions 80 through 84 and City of Mississauga. Market Research and Marketing Strategy: Lot-level Stormwater Control in the Residential Sector, July 2008. pp. 27-28, Appendix 2 – verbatim responses to question 66 through 82.

US EPA 1997. United States Great Lakes Program Report on the Great Lakes Water Quality Agreement. United States Environmental Protection Agency EPA 160 R 97005 Dec 1997.

US EPA. 2000. Low Impact Development (LID): A Literature Review. Office of Water Washington, DC. EPA 841-B-00-005 Oct. 2000

US EPA. 2001. Stream Corridor Restoration: Principles, Processes and Practices. Office of Water Washington, DC. EPA.

US EPA. 2005. Stormwater Management Fact Sheet: Dry Extended Detention Pond. Office of Wetlands Washington, DC. EPA. www.stormwatercenter.net/Assorted%20Fact%20Sheets/

US EPA. 2008. Managing Wet Weather with Green Infrastructure Municipal Handbook: Funding Options. EPA-833-F-08-007.

Watt, W.E. and J.D. Paine. 1992. Flood Risk Mapping in Canada: Uncertainty Conditions. Canadian Water Resources Journal, vol. 17, No. 2, pp. 129-138.

Watt, W.E. (ed.). 1989. Hydrology of Floods in Canada: A Guide to Planning and Design. National Research Council Canada, Associate Committee on Hydrology, Ottawa, Ontario, Canada.

Watt, W.E. and D. Waters, and R. McLean. 2003. Climate Change and Urban Stormwater Infrastructure in Canada: Context and Case Studies. Toronto Niagara Region Study and Working Paper Series, Report 2003-1. Meteorological Services of Canada, Waterloo, Ontario. Received April 2001.

CHAPTER 9

Conservation Ontario's Coordinating Mandate to Promote an Integrated Watershed Management Approach to Protection of the Great Lakes and St. Lawrence Ecosystem

Bonnie Fox[1] and Sonya Meek[2]

[1]Conservation Ontario, 120 Bayview Parkway, Box 11, Newmarket, Ontario, Canada, L3Y 4W3
E-mail: bfox@conservationontario.ca

[2]Toronto and Region Conservation Authority, 5 Shoreham Drive, Downsview, Ontario, Canada, M3N 1S4
E-mail: smeek@trca.on.ca

INTRODUCTION

Implementation actions for the protection and enhancement of the Great Lakes and St. Lawrence ecosystem happen at the local scale with local partners. Therefore, a strong framework is needed to link Great Lakes/St. Lawrence objectives to local priorities and to engage local partners and citizens. This framework is lacking in Ontario. Conservation Authorities' Integrated Watershed Management approach represents an existing, established model that is already contributing to the Great Lakes/St. Lawrence, but could be more fully utilized to advance objectives. The chapter describes Conservation Ontario, Conservation Authorities and their integrated watershed management approach. It discusses the benefits of integrated watershed management to the Great Lakes/St. Lawrence, the lessons learned from its application to date, and, illustrates leading edge efforts in the Rouge River watershed. The chapter describes strengthening integrated watershed management benefits across the Great Lakes and St. Lawrence Basins and makes recommendations for moving forward.

OVERVIEW OF CONSERVATION ONTARIO AND CONSERVATION AUTHORITIES

Conservation Authorities are community-level organizations that manage natural resources on a watershed basis across Ontario. They have a long history of partnering with municipal, provincial and federal governments, as well as landowners and other groups, to deliver community based practical solutions to a range of natural resource challenges. Conservation Authorities vary in size from smaller operations managing one or two watersheds to larger operations managing one larger or several adjoining watersheds. As per Fig. 9.1, 35 of the 36 Conservation Authorities drain into the Great Lakes/St. Lawrence Ecosystem (Mattagami Region CA drains into Hudson Bay).

Conservation Ontario is a non-profit corporation that represents the network of 36 Conservation Authorities and their common interests at provincial and federal levels. Conservation Ontario is governed by a Council comprised of representatives from each Conservation Authority and meets five times per year. The organization's operations are funded by the member Conservation Authorities and details regarding budget and staff resources can be found in the Annual Report and on Conservation Ontario's website (www.conservationontario.ca).

Provincial, federal and bi-national initiatives to address various Great Lakes and St. Lawrence ecosystem issues can have significant influence on how water and other natural resources are managed in Ontario and hence, Conservation Ontario places priority on providing the watershed level perspective with an emphasis on integrated watershed management. Conservation Ontario coordinates watershed level input to Great Lakes and St. Lawrence Basin initiatives through:

- partnered research initiatives on the State of the Science and State of the Practice for integrated watershed management;
- establishment of review teams of Conservation Authority technical experts;
- assigning representatives to participate in Great Lakes initiatives or consultations with responsibility for reporting back to Conservation Ontario Council; and/or,
- endorsement by Conservation Ontario Council of coordinated submissions.

Fig. 9.1 Map of Ontario's 36 Conservation Authorities

The above coordination provides an accountability framework between Conservation Authority representatives and the broader network of Conservation Ontario and provides an ability to develop shared positions. In the past 10 years, through Conservation Ontario, Conservation Authority technical and administrative experts provided verbal and written comments on such Great Lakes initiatives as, the Great Lakes Water Quality Agreement Review (Agreement Review Committee 2007), the Canada-Ontario Agreement Respecting the Great Lakes Basin Ecosystem (Environment Canada 2002 and 2007), the Great Lakes-St. Lawrence River Sustainable Water Resources Agreement (Great Lakes Governors and Premiers 2005) and its Water Conservation and Efficiency Objectives (Great Lakes-St. Lawrence River Water Resources Regional Body 2007). It is noted that for all of these Great Lakes/St. Lawrence reviews, some individual Conservation Authorities submitted their watershed-specific perspective as well. Additionally, Conservation Ontario and Conservation Authorities have participated in bi-national initiatives such as development of the Lake Ontario Biodiversity Conservation Strategy (Lake Ontario Biodiversity Conservation Strategy Working Group 2009) and the Lake Huron Biodiversity Conservation Strategy (Lake Huron Biodiversity Conservation Strategy Working Group 2010) to ensure that the watershed perspective is represented in the outcomes.

Overall, as watershed management agencies, Conservation Authorities share a common responsibility for the Great Lakes and St. Lawrence Ecosystem and recognize the need for collaboration. In 2006, with the assistance of acknowledged Great Lakes expert Dr. Gail Krantzberg (McMaster University), Conservation Ontario developed and endorsed a position statement on Great Lakes Sustainability. The collective mission for Great Lakes Sustainability directs that:

> *Conservation Ontario will work with all orders of government and basin residents as stewards to protect and improve the unique, shared Great Lakes and St. Lawrence ecosystem for present and future generations.*[1]

Additionally, the guiding principle for our efforts is to:

> *Apply an integrated watershed management approach to planning and implementation in order to protect and improve the Great Lakes and St. Lawrence Ecosystem.*[1]

CONSERVATION ONTARIO'S INTEGRATED WATERSHED MANAGEMENT (IWM) APPROACH

The Great Lakes Water Quality Agreement (1987) specifies that the governments of the United States and Canada will take an "ecosystem approach". The International Joint Commission's *Special Report to Governments* on the Great Lakes Water Quality Agreement with respect to utilization of an ecosystem

[1]The full Conservation Ontario Position Statement on Great Lakes Sustainability can be found at www.conservationontario.ca under "What We Do/Protect Water/Great Lakes".

approach states "The comprehensive watershed planning conducted by the Conservation Authorities in Ontario also constitutes an ecosystem approach" (International Joint Commission 2006). Conservation Ontario calls this ecosystem approach "integrated watershed management".

Integrated Watershed Management (IWM) is the process of managing human activities and natural resources on a watershed basis.[2] It requires partner and public involvement throughout the process. This approach focuses on the management of water and related resources, including aquatic and terrestrial natural heritage systems, and is increasingly addressing the inter-relationships between these systems and the broader socio-economic systems which they support. IWM considers the watershed's response to current and anticipated future management issues and opportunities, such as urban growth, climate change and associated management strategies. The holistic approach of integrated watershed management (also referred in the literature as integrated water resources management) is applied through a continuous and cyclical process involving the preparation of a plan that addresses identified issues and concerns that is then implemented, monitored, reported on and updated as required in order to adapt to changing or new emerging stressors or management approaches.[3] Effective partnership approaches and public involvement are key factors for success in integrated watershed management (see Fig. 9.2).

As summarized in "Watershed Management in Ontario: Lessons Learned and Best Practices" (Conservation Ontario 2003), public/stakeholder involvement in watershed planning:

> *starts early in the process and continues throughout; is focused at key milestones in the process; captures the public's preferences in terms of watershed goals, objectives and targets (helping to answer the question, "Where do we want to be?"); provides direction for identification, evaluation and selection of management options; creates local "ambassadors" who act as watchdogs, advocates, educators and catalysts for action; and encourages citizens to act as stewards - to get involved in protection, restoration, cleanup and monitoring projects in the watershed. In practice, the scale of a public involvement program and the techniques and tools used differ depending on the scale of a watershed planning study, the level of interest, and the issues involved.* (p.38).

CONSERVATION ONTARIO'S INTEGRATED WATERSHED MANAGEMENT APPROACH AND ITS BENEFITS TO THE GREAT LAKES

Given the basic premise that healthy watersheds contribute to healthy Great Lakes, Conservation Ontario's integrated watershed management approach offers

[2]To read more about integrated watershed management, see Conservation Ontario 2003, and Conservation Ontario 2010d. These can be found at www.conservationontario.ca under "Resources/Publications/Reports".

[3]This process in consistent with the principle of Adaptive Environmental Management.

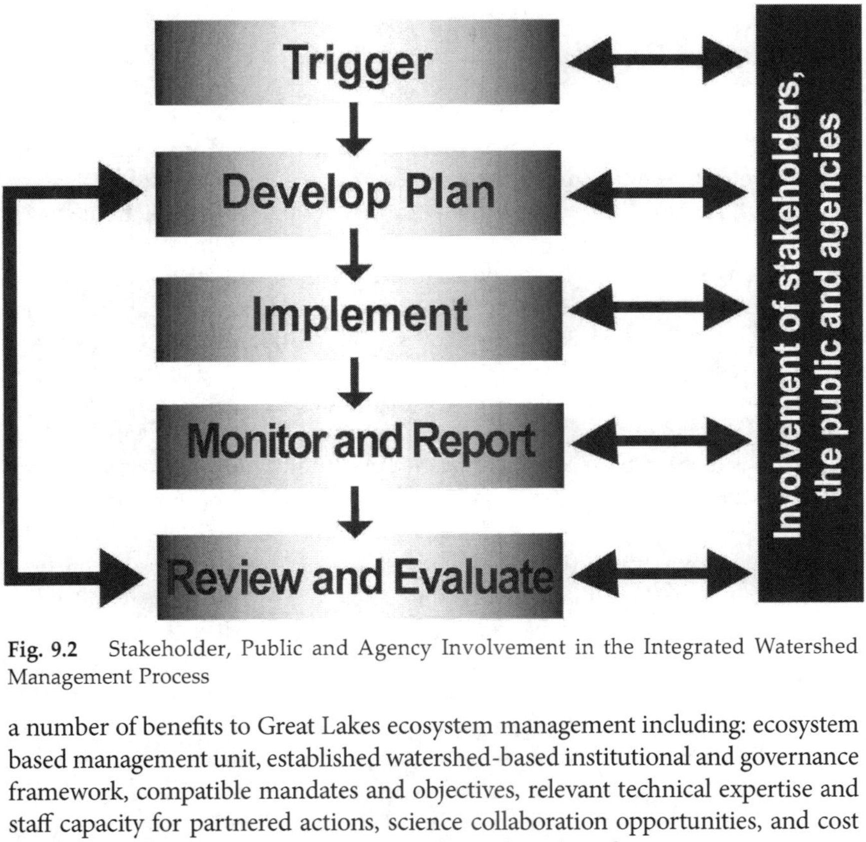

Fig. 9.2 Stakeholder, Public and Agency Involvement in the Integrated Watershed Management Process

a number of benefits to Great Lakes ecosystem management including: ecosystem based management unit, established watershed-based institutional and governance framework, compatible mandates and objectives, relevant technical expertise and staff capacity for partnered actions, science collaboration opportunities, and cost effectiveness. This section further expands on these benefits.

Ecosystem-based Management Unit

Conservation Ontario's integrated watershed management approach utilizes the watershed as the management unit rather then political boundaries. This provides the basis for understanding linkages at a relevant scale between local decision-making and the greater Great Lakes Basin from both a science and governance perspective. From a science perspective the watershed unit is at a relevant scale for the management of water for all uses and inputs to the Great Lakes and for modeling watershed responses to various land use and climate change scenarios. From a governance perspective the watershed is a relevant ecosystem boundary (International Joint Commission 2006) as it pertains to the language in the Great Lakes Water Quality Agreement which makes a commitment to an "ecosystem approach". Under the *Conservation Authorities Act*, the watershed boundary was chosen for Conservation Authority jurisdictions as an important ecosystem boundary to manage natural resources and particularly, water.

Established Watershed-based Institutional and Governance Framework

In Ontario, for the watershed community and implementation of the integrated watershed management approach, there is an established institutional and governance framework through the existence of Conservation Authorities. As per Fig. 9.1 (see map), Conservation Authority watershed jurisdictions cover the entire Canadian portion of the Lake Ontario and Lake Erie Basins and the southern Lake Huron Basin to the southern tip of Georgian Bay. Relatively small portions of the northern Lake Huron Basin and Lake Superior Basin and the St. Lawrence River Basin are covered by Conservation Authorities, however, it is important to note that approximately 90% of Ontario's population resides within Conservation Authority jurisdictional boundaries. Therefore taking an integrated watershed management approach to the Great Lakes within these jurisdictions alone represents a significant opportunity to address some major threats to the Great Lakes (e.g. urban and shoreline development, non-point source pollution).

Conservation Authorities are formed under the *Conservation Authorities Act* (enacted in 1946) at the request of municipalities in a watershed and most Conservation Authorities have been around for more than 40 years. Each Conservation Authority is governed by a Board of Directors whose members are municipally appointed. The Conservation Authority Board directs the watershed programs that are operated by experienced and trained staff and the Board approves the budget. As of a 2009 statistical survey, approximately, 67% of the Board members are Municipal Councillors/Mayors and the remaining 33% are citizen appointments. This watershed-based institutional and governance framework provides immediate and significant stability to commitment between the local and basin levels to address the Great Lake issues through an integrated watershed management approach.

Compatible Mandate and Objectives

The stability of the institutional and governance framework is further enhanced through the fact that a Conservation Authority's mandate and objectives match many of those of the Great Lakes programs. The mandate that all Conservation Authorities share is to ensure the conservation, restoration and responsible management of Ontario's water, land and natural habitats. This mandate is easily scaled up to address the protection and enhancement of the Great Lakes Basin ecosystem. Conservation Authorities work in partnership with watershed municipalities and others to provide a variety of natural resource programs and services. Emphasis may vary due to local priorities and needs but basically every Conservation Authority undertakes a watershed program based on these four key business objectives:

- Safeguard Ontario's rivers, lakes and streams;
- Protect and manage Ontario's woodlands, wetlands and natural habitats;
- Protect life and property from natural hazards; and
- Promote learning and respect for Ontario's environment.

Again, the mandate and these objectives[4] are compatible with the needs of the Great Lakes Basin ecosystem thus enabling Conservation Authorities to assist in addressing Great Lakes issues while addressing local priorities.

Relevant Technical Expertise and Additional Staff Capacity for Partnered Actions

Conservation Authority's mandate and objectives match many of those of the Great Lakes program which means that Conservation Authorities can have staff with relevant technical expertise in: stewardship, ecological restoration/rehabilitation, biology, engineering, monitoring, GIS, environmental planning, communications and even accounting. As of a 2009 statistical survey, Conservation Authorities employed approximately 2,000 full time professional staff. They provide science-based advice and services within their watersheds including: watershed/sub-watershed planning, water quality/quantity monitoring and modeling, natural heritage and forestry, source protection, watershed stewardship, technical input and review for municipal land use planning, as well as, outdoor education and recreation. These staff represent additional capacity for successfully managing and implementing with watershed stakeholders projects that are identified priorities in the Areas of Concern (AOC) Remedial Action Plans, Lakewide Area Management Plans (LaMPs) and various Great Lakes strategies (e.g. previously referenced Lakes Huron and Ontario Binational Biodiversity Conservation Strategies). Of course, this works most successfully where partnerships with watershed stakeholders have been nurtured as in Areas of Concern (see BOX 1). In general, Conservation Authorities have a tradition of partnering with watershed stakeholders and a demonstrated ability to build and sustain valuable partnerships with local municipalities, federal, and provincial and community agencies and landowners to plan and deliver programs and services.

A significant benefit of this relevant technical expertise and additional staff capacity for partnered actions is that it contributes to integration of Great Lakes issues into local decision-making. Conservation Authority staff provide science-based advice and services to key local decision-making processes that impact the Great Lakes, including: municipal planning and development;

[4]The full details can be found at www.conservationontario.ca under "About Us/Conservation Authorities/Mandate".

Box 1: Remedial Action Plans (RAPs) – Conservation Authority Link to Local Engagement and Actions

RAPs demonstrate how there can be better integration across scales of governance to benefit the Great Lakes. Of the total 15 Canadian Areas of Concern there are 11 within Conservation Authority boundaries and seven have a formalized Conservation Authority leadership/coordination role as of 2010. Ten AOCs have Conservation Authority involvement through representatives to the various committees and councils. The Committee structure varies across AOCs according to the social context of an area considering local strengths and needs to encourage participation and the Committee structures have evolved for these same reasons. For example, community engagement in the Toronto and Region AOC now occurs through the Conservation Authority's existing Watershed Councils; other AOCs have ongoing Public Advisory Councils created in direct response to the Great Lakes Program; and some have evolved from a combination of the two such as Bay of Quinte Remedial Action Plan's Restoration Council which is co-chaired by the General Managers of the two Conservation Authorities involved in that AOC. At least eight of the AOCs have had Conservation Authorities and their watershed partners implementing projects through Great Lakes funding programs.

Environmental Assessment projects for municipal infrastructure; and determining priority restoration projects to benefit local watersheds. With a strong network of municipal and community contacts, combined with an active involvement in municipal planning and development processes and restoration projects, Conservation Authorities are well positioned to assist in addressing Great Lakes issues. Conservation Authorities have a strong track record in delivery of watershed-based, on-the-ground partnered actions to protect and restore the waters and associated lands of the Great Lakes. For example, partnered actions in 2008[5] include water quality improvement projects (617 projects with 646 landowners), tree planting (2.6 million trees planted on 1,778 hectares of land; 1,357 projects with 1,496 landowners), habitat rehabilitation/restoration projects (346 hectares and 222 km of stream; 387 projects with 241 landowners), and, water supply and septic management projects (682 projects). Great Lakes water quality cannot be effectively maintained or enhanced without taking action to improve the water quality of the inland rivers and streams that drain into the Great Lakes.

Science Collaboration Opportunities

An integrated watershed management approach facilitates collaboration and data sharing and provides a mechanism for the existing science base at the

[5]Statistics taken from the 2009 Conservation Ontario Statistical Survey of Ontario's 36 Conservation Authorities for stewardship activities undertaken in 2008 (Conservation Ontario, Newmarket, Ontario)

watershed level to be used more fully. A large majority of Conservation Authorities have good and improving science as well as, monitoring and reporting (e.g. watershed report cards) capacity. The provincial and federal Great Lakes programs could fully integrate this capacity into the broader Great Lakes science and monitoring program. While there are good examples of partnerships, opportunities for watershed level agencies to influence and help achieve the science agenda should be maximized. Engaging the local and watershed levels in crafting the science agenda will ensure that the science program is responsive to regional stressors and directly informs decision-making and management actions in the watersheds. This can be achieved through the considerable effort that Conservation Authorities expend in communicating the science to stakeholders and decision-makers in a relevant and understandable manner.

Cost Effectiveness

Cost effectiveness is another feature and potential benefit in supporting an integrated watershed management approach to the Great Lakes. Conservation Authorities were originally established under a principle of cost sharing (originally 50/50 between municipal and provincial levels of government) and despite the fact that this financial relationship has shifted in the past dozen years, this founding principle of "cost sharing" still guides their current culture. Federal and provincial funds and other watershed partner funds do add weight at the decision-making table for a Conservation Authority's budget and watershed program priorities. Basically the municipally-appointed Boards of Directors are more willing to invest in a program that has matching or multiplying dollars from partner sources for a shared business interest. Leveraging partner sources of funding and delivering value for the dollar spent is a defining and surviving quality of Conservation Authorities. The Conservation Authorities are a valuable partner with a collective annual budget in 2009 resulting in $272 million CAD in expenditures and the revenue sources (see Fig. 9.3) are: 43% Municipal - levies and special projects, 36% self-generated revenue, 19% Provincial - transfer payments and grants, and 2% Federal grants.

The municipal funding consists of levies (which the *Conservation Authorities Act* enables) and municipal special project funding which when coupled with the self-generated revenue results in a total of 79% local investment. It is noted that this local investment at the watershed-level does not even take into account investments at the municipal level for infrastructure improvements (e.g. combined sewer overflows, stormwater management, etc.). The Provincial funding consists of provincial transfer payments for the flood reduction program and source water protection program as well as special project grants. Combined with Federal grants, senior government contributions is only 21% for delegated programs and special projects. Overall, 79% or $215 million

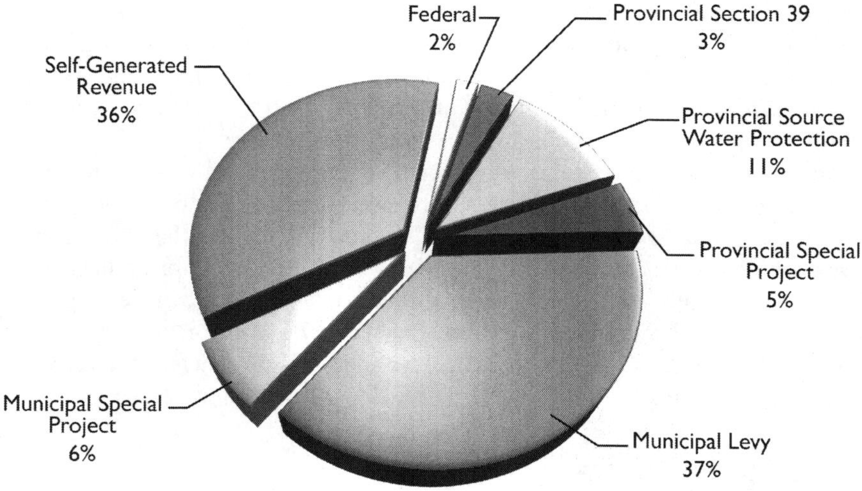

Fig. 9.3 Conservation Authorities Revenue—2009

CAD represents a significant local investment in watershed management that in general means healthier Great Lakes and is worthy of leveraging for greater Great Lakes and St. Lawrence benefits.

LESSONS LEARNED AROUND PARTICIPATORY GOVERNANCE

Conservation Ontario is committed to "continuous improvement" of integrated watershed management (IWM) approaches and is advised through surveys, research and analyses with its member Conservation Authorities and communication of the results. A 2008 survey of 33 Conservation Authorities with varying degrees of experience in integrated watershed management planning indicated a strong support for IWM which is hindered by such barriers as: staff capacity, data gaps, gaps in scientific knowledge, as well as, lack of public and political support. These barriers can be addressed through greater collaboration and resource sharing that could be supported by a strengthened participatory governance structure. Two specific initiatives that included examination of elements of participatory governance were published in 2003 and 2010. In the most recent example, Conservation Ontario entered into a partnership with the Ontario Ministry of Natural Resources and the Federal Department of Fisheries and Oceans to explore application of integrated watershed management in Ontario. This 2009 collaborative research initiative resulted in a series of four reports (Conservation Ontario 2010 a,b,c,d).

With regard to improvement of participatory governance, key lessons learned from these initiatives (2003 and 2010) are described in this section and include: partnership approaches and public involvement are key to success; governance shifts are necessary; and a broad sustainability context can encourage participation.

Partnership Approaches and Public Involvement are Key to Success

The integrated watershed management process is not only a technical process but has extensive inclusion of all stakeholders throughout the process as well. "Watershed Management in Ontario: Lessons Learned and Best Practices" concluded that watershed management "is built on the concept of shared responsibility for environmental protection and enhancement" (Conservation Ontario 2003). Key elements of success in watershed management as experienced by three leaders in watershed management (i.e. Grand River, Credit Valley, and, Toronto and Region Conservation Authorities) are summarized in the 2003 Conservation Ontario report and they all identified partnership approaches and public involvement. In their review of "lessons learned" they indicate that for partnership approaches, the factors for success are:

- *Involve key partners from the beginning of the process in Steering Committee and others (i.e., Technical Committees);*
- *Seek strong leadership at the political and staff level from key partners;*
- *Forge strong links to other programs and processes to maximize the use of information;*
- *Use consensus-based approaches to develop a shared vision; and*
- *Adopt a philosophy of collaboration in planning and implementation*

(Conservation Ontario 2003)

Additionally, they identified that for public involvement, the factors for success are:

- *Involve the public in determining the study objectives, goals and selection of the preferred plan;*
- *Be inclusive, open and unbiased;*
- *Aim to create "local ambassadors" – public participants in the process who can educate, motivate and serve as watchdogs in their own neighbourhoods; and*
- *Find a strong, enthusiastic and respected citizen to chair the project Steering Committee*

(Conservation Ontario 2003)

Through the 2009 collaborative research initiative this work was further expanded upon through a survey of 33 Conservation Authorities in 2008 with varying degrees of experience in integrated watershed management planning. "Lessons learned" from the survey of Conservation Authorities that are relevant to participatory governance include:

- One size does not fit all; need flexibility in approach as some communities have different needs;
- Importance of community champions;
- Communication is key – use an open process – see all perspectives, therefore more buy-in;
- Need a coordinator with a lot of time to spend on relationship building exercise;
- Implementers must be involved and own their piece of the plan;
- Studies have been well received but conflict exists where recommendations do not fit with individual desires;
- Conservation Authority's leadership on delivery of IWM (as opposed to a consultant fulfilling the role) is critical for Conservation Authority staff to build relationships with their watershed stakeholders and public and to be able to effectively respond to their needs and questions;
- IWM plan development and implementation is a significant opportunity to continue to build and/or strengthen local partnerships, to gain mutual trust and to influence key areas such as budgets and land use change;
- Local stakeholder involvement helps to streamline and set watershed priorities and to obtain implementation funding; and,
- Involving local stakeholders is critical.

(Conservation Ontario 2010a)

These lessons can guide continuous improvement in participatory governance at the watershed scale, however they can also inform effective governance at the Great Lakes scale.

Governance Shifts are Necessary

The 2009 collaborative research initiative revealed that participatory, ecosystem based management (i.e. integrated watershed management) is being called for by numerous groups around the world including United Nations organizations, and within numerous jurisdictions such as Australia, Brazil, Europe and Canada (Conservation Ontario 2010a). All highlight the need to get the citizens, citizen groups and stakeholders involved more closely in the management of watershed resources and to foster a greater involvement of the people most affected by decisions. As summarized in the Conservation Ontario 2010a report[6]:

> *From a national perspective, shifts include a focus on outcomes, recognizing the watershed as a unit of analysis, moving toward recognizing that everyone has a share in responsibility when it comes to the health of the watershed, and moving towards clarifying how decisions are made and by whom.*

The 2009 collaborative research initiative calls for an examination of governance and suggests that the challenges facing us are to develop coordination and decision-making frameworks that are resilient and allow for adaptation (Conservation Ontario 2010a). In particular, the 2009 collaborative research initiative suggests that adaptive co-management is a governance option particularly worthy of consideration by Ontario and indicates:

> *Adaptive co-management is an approach to environmental and resource management that brings together two aspects–adaptive management and co-management. The former refers to a capability to learn and adapt under certain conditions and in the face of changing circumstances, while the latter refers to the sharing of management power and responsibilities between governments, resource users, and resource-based communities.*
>
> *... the key elements of Adaptive Co-Management include learning by doing, information sharing, collaboration and shared decision-making, partnerships at regional and national levels and finally, flexibility in management approaches. This approach fits us best, given that the Ontario approach to IWM is already rooted in Adaptive Environmental Management."*
>
> (Conservation Ontario 2010a)

The existing watershed based institutional and governance framework of Conservation Authorities, coupled with their commitment to a partnership approach to IWM, provides a platform for greater application of adaptive co-management at the watershed scale. To improve the effectiveness at achieving both local watershed and Great Lakes objectives, the adaptive co-management approach to IWM needs to be applied at the Lake Basin scale and Great Lakes Basin scale.

A Broad Sustainability Context can Encourage Participation

The 2009 collaborative research initiative reports that considerations around IWM today (globally, nationally and locally) are concerned with the need for

[6]This summary is based in part on the review of work undertaken by a number of prominent non-government organizations that have waded into the issue of integrated watershed management including the Canadian Water Resources Association (2006), The Polis Project for Ecological Governance (2005), Pollution Probe (2008), and the Canadian Chamber of Commerce (2006). An updated 2010 reference is provided at the end of this paper for the Water for Sustainability 2006 Canadian Chamber of Commerce paper.

more effective collaborations to foster shared responsibility for action and ensure sustainable outcomes (Conservation Ontario 2010d). Conservation Ontario's Position on Great Lakes Sustainability recognizes that a sustainable and resilient Great Lakes and St. Lawrence ecosystem encompasses environmental, social, and economic health.[1] Integrated watershed management continues to be regarded as an appropriate framework to address environmental, social and economic concerns, particularly as they relate to water; ensuring adequate quantities of water of acceptable quality are available for all uses (Simms et al. 2010, Roy et al. 2009, Government of Canada 2009).

Although the traditional focus of IWM has been on actions (often led by governments) to achieve environmental outcomes, more expedient and effective delivery is necessary to achieve these and other objectives. The Sustainable Development Commission in the United Kingdom points to the need for integrated approaches to implementation which engage and enable a broader range of local partners and which can deliver a host of economic, environmental and social co-benefits more cost-effectively than previous single purpose oriented programs (Sustainable Development Commission - UK 2010). The 2009 collaborative research initiative suggests:

> *If we agree that ultimately the goal of IWM is to maintain and enhance watershed health which, in turn, links to human well being, then we need to shift towards greater economic, social and environmental integration if sustainability is to be achieved. We need to model how societies and economies function in the environment with each other and not independently. To achieve this, tools and methodologies need to be developed and key to the success is collaboration among all stakeholders.*
>
> (Conservation Ontario 2010d)

The lack of tools and methodologies for including social and economic considerations into watershed management plans was identified as a significant technical gap through the survey of 33 Conservation Authorities (Conservation Ontario 2010a). Such tools may assist in analysis of co-benefits and facilitate the meaningful participation of a broader range of partners in planning and implementation processes. Development of these tools/methodologies requires collaboration amongst all levels of government, academia, and social/environment/economic sectors of society; and ultimately will lead to greater support for implementation among a broader spectrum of stakeholders (CO 2010a, Swainson et al. 2010, Roy et al. 2009, Parkes et al. 2010).

A CASE STUDY: ROUGE RIVER WATERSHED PLAN – LEADING EDGE EFFORTS IN INTEGRATED WATERSHED MANAGEMENT SYNTHESIS AND ENGAGEMENT OF WATERSHED PARTNERS

The suggested benefits and the lessons learned can be illustrated through a Conservation Authority case study of leading edge efforts in integrated watershed management synthesis and engagement of watershed partners. The Toronto and Region Conservation Authority (TRCA) in partnership with the Rouge Park Alliance and a multi-stakeholder Task Force led the preparation of an integrated watershed management plan for the Rouge River watershed during 2004–2008. The watershed plan and its implementation, monitoring and reporting form the ongoing process of integrated watershed management (IWM). This case study describes the planning study approach and outcomes as an example of how the Great Lakes objectives can be effectively delivered through participatory approaches at the local watershed scale.

Local Watershed Context

The 336 km^2 Rouge River watershed lies in the central east portion of the Toronto region and drains from the Oak Ridges Moraine to Lake Ontario (see Fig. 9.4). It includes parts of the Regions of York and Durham, cities of Toronto and Pickering, and towns of Markham, Richmond Hill and Whitchurch-Stouffville. Land use in the watershed is approximately 40% rural, 35% urban, 24% natural and 1% open water. The lower watershed is dominated by Rouge Park. The middle and western parts of the watershed support urban and urbanizing communities, while the upper and eastern portions are primarily rural and agricultural with a few small villages. Urban growth pressures and the need to restore more natural hydrologic function and natural heritage connectivity, are key issues facing this watershed. Added to these current issues are the anticipated effects of climate change.

The Rouge River watershed is a component of many broader regional regeneration plans and strategies, including the Toronto and Region Remedial Action Plan (RAP). The Rouge is one of six RAP watersheds affecting the health of nearshore Lake Ontario. River water quality and its effects on nearby Lake Ontario swimming beaches and drinking water intakes are also concerns for watershed health, as is riverine habitat with respect to migratory salmonid populations.

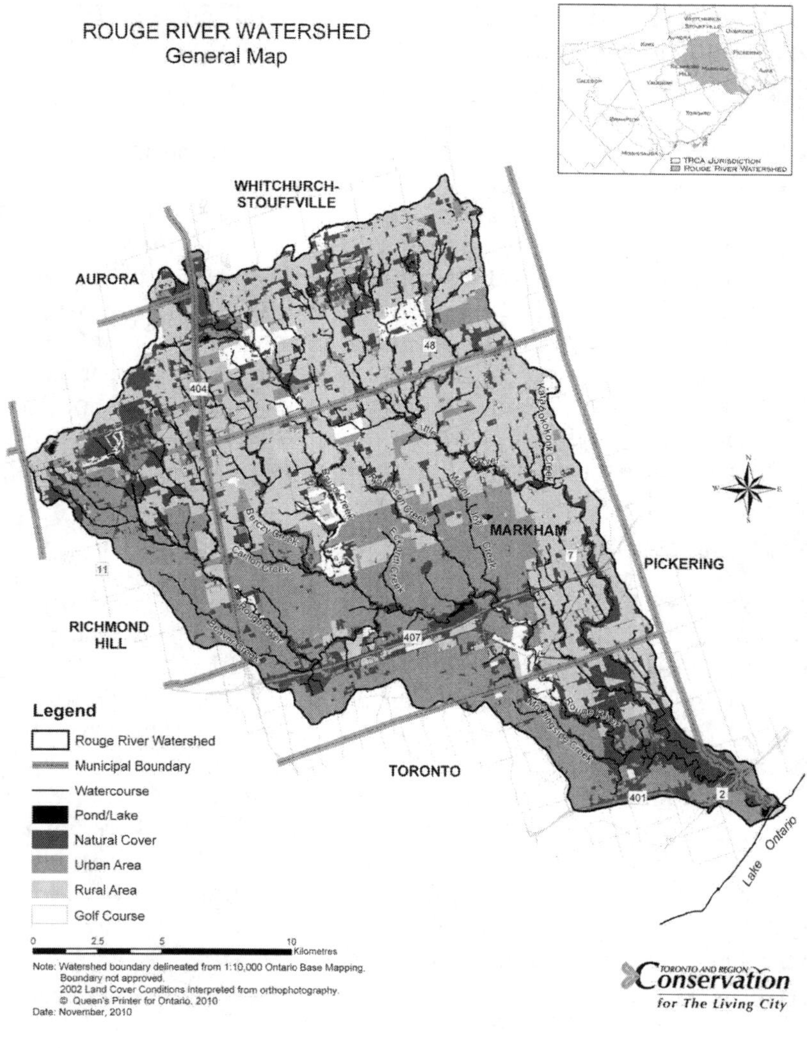

Fig. 9.4 Map of the Rouge River Watershed

One objective of the watershed planning study was to provide watershed scale direction for the implementation of RAP objectives and targets, however there were several other compelling and inter-related reasons for the study. This initiative was to assist York region municipalities in fulfilling the provincial Oak Ridges Moraine Conservation Plan (ORMCP) requirement to complete watershed plans in advance of major development approvals. The study was also designed to update the TRCA's original 1990 Comprehensive Basin Management Strategy for the Rouge River Watershed (MTRCA 1990); provide a watershed context for Rouge Park plans; augment the city of Toronto's Wet Weather Flow Management Master Plan (City of Toronto 2003) by providing direction for upstream municipalities; and, apply TRCA's jurisdiction-wide Living City vision (TRCA 2006) and its objectives for healthy rivers and shorelines, regional biodiversity and sustainable communities to the specific needs of the Rouge River watershed. Funding for the watershed study was provided by the Regional Municipality of York and the City of Toronto.

Key Components of the Rouge's Integrated Watershed Planning Approach

The goal of the watershed planning study was to make long term strategic recommendations for the protection and enhancement of watershed health. The plan is intended to inform and guide municipalities, provincial and federal governments, TRCA, Rouge Park, non-government organizations and private landowners as they implement policies and practices.

TRCA's integrated approach offers all stakeholders the opportunity for becoming engaged in the process of defining watershed goals, objectives and management strategies to achieve them, based on a sound understanding of watershed systems. Key components of this work are as follows:

Local Partnership Approach

TRCA established and supported the Rouge Watershed Task Force, a multi-stakeholder advisory committee whose membership consisted of elected representatives appointed by each watershed municipality, staff representatives from several provincial and federal government departments, watershed residents, and representatives of local environmental groups, industry, TRCA and the Rouge Park Alliance. A terms of reference set out the mandate, procedures and term of the Task Force, all of which focused on preparation of the watershed plan. A citizen member was elected as Chair of the Task Force.

The Rouge Watershed Task Force created a mechanism to bring diverse watershed stakeholders together to share their perspectives and prepare a common plan they could all support.

Establishment of a Common Management Framework

An overall goal, a set of principles, nine individual goals, 22 objectives and ultimately a series of management strategies make up the management framework for this watershed plan. For each objective, a set of indicators and

Box 2: Example Objective, Indicators and Targets for the Rouge River Watershed Aquatic System

Objective: Protect, restore and enhance the health and diversity of native aquatic habitats, communities and species.	
Indicator	**Example targets**
Fish	"Good" or better Index of Biotic Integrity Scores at monitoring stations upstream of urban development. Maintain or improve existing Index of Biotic Integrity Scores at monitoring stations within the urban boundary
Aquatic habitat features	Increase wetland cover to 10% of total watershed area
	50% of riparian zone treed 50% of riparian zone other natural cover
	No additional loss of stream length
Invasive and exotic species	Prevent the introduction of any invasive or exotic species
Fish passage	Only strategic barriers remain; barriers removed/mitigated in *priority* sequence

Note: For details on baseline conditions, data sources and rationale, see *Rouge River State of the Watershed Report* (TRCA 2007b).

associated targets were defined (see BOX 2). A rating of watershed health was assigned for each objective, based on an evaluation of current watershed conditions. The framework facilitates a consistent means for future reporting.

The overall goal for the Rouge watershed is:

To work towards a healthy and sustainable Rouge watershed by protecting, restoring and enhancing its ecological and cultural integrity within the context of a regional natural heritage system

(Toronto and Region Conservation Authority 2007a).

The goals, objectives, targets and management strategies address a broad scope of concerns, including:

- Groundwater
- Surface water
- Stream form
- Aquatic system
- Terrestrial system
- Air quality and climate change

- Cultural heritage
- Nature-based recreation
- Sustainable land and resource use.

When establishing the management framework, the partners recognized the broader basin and bio-region of which the Rouge watershed is a part, as well as any objectives that may have been previously set for those systems. For example, Toronto and Region RAP goals, objectives, and targets were embraced and refined as part of the Rouge watershed planning study, such that the plan is consistent with RAP goals yet provides more specific guidance based on local information. Similarly, individual watershed partners brought forward objectives as may have been established for their particular subject or geographic areas of interest and these were incorporated as appropriate into the watershed framework. The result was a common set of watershed goals and objectives that all partners could work toward achieving through their own areas of responsibility.

Science-based, Interdisciplinary Analysis

Multi-stakeholder involvement and a broad study scope enabled an interdisciplinary understanding of watershed systems under current conditions and possible futures. This information provided the basis for setting strategic management direction.

To determine effective management strategies, the study defined a series of potential future land use and management scenarios, representing progressively greater extents and different forms of urban growth coupled with other interventions, including expansion of natural cover and stormwater management infrastructure improvements. The analysis included two climate scenarios that bracketed the range of climate change predictions for this area. The choice of scenarios allowed the team to assess the relative effectiveness of different interventions and areas of relative sensitivity in the watershed.

A watershed response model[7] guided the overall evaluation of scenarios, by defining watershed ecosystem components and describing the order in which changes manifest themselves through these systems. The degree to which effects on each system and relationships between systems can be quantified varies, as does our ability to define thresholds of integrity, due to limitations in science. Subsequently, there was a range of sophistication in the tools available to predict the effects of changes in one system to another, from computerized models to professional judgement. To help overcome uncertainties, the study team drew upon multiple lines of evidence, such as observations of similar patterns in results from different disciplines or neighbouring watershed studies, a focus on relative differences among scenario outcomes, and expert involvement in interpretation during "integration workshops".

[7]A description of the watershed response model is provided in the Rouge River Watershed Scenario Modelling and Analysis Report (TRCA 2007).

The scenario modelling and analysis results showed that no single management intervention would be adequate to achieve watershed objectives, but rather the cumulative effects of a number of strategies (led by a number of implementers) would be needed throughout the watershed. The interdisciplinary analysis improved the science and understanding of watershed systems, their interdependencies and sensitivities. For example, this approach enabled the team to quantify the effects of expanded natural cover on peak flows, erosion potential and water quality in various stream reaches. It also permitted an assessment of the effects of impervious surfaces on runoff and infiltration to groundwater and the resultant effects on groundwater levels and discharge to streams. These predicted effects in turn allowed fisheries biologists to assess potential impacts on aquatic communities. This information served as a basis for setting strategic, cost effective management direction that would benefit multiple objectives. It also established clear expectations of results, and areas where limitations in science and technology leave us with challenges for continued improvement and adaptive management.

Consultation and Engagement of Target Audiences

Public consultation occurred through a variety of means, including public open houses, presentations to municipal councils and meetings with individual stakeholder groups, such as the urban development industry, agriculture and golf course associations and fishing clubs. The study solicited input on specialized issues from relevant agencies, groups and other experts during "management summits". Technical peer review of various aspects of the study occurred through formal and informal mechanisms.

In anticipation of significant recommendations for single family residential lots, a predominant land use in the Rouge watershed, extra effort was made to ensure representative input from homeowners. A community-based social marketing study recruited participants who represented the socio-economic demographic of local neighbourhoods (Freeman Associates 2006). Their input identified barriers and motivators to the effective implementation of sustainable practices on private lands, and helped guide management strategies. Barriers and motivators tended to be associated with personal values/perceptions and economic considerations, and therefore communication strategies and business case development form part of the resulting management strategies.

Implementation Guidance

An Implementation Guide was prepared to facilitate implementation of the recommendations from the Watershed Plan (TRCA 2008)[8]. Whereas the

[8]The "Rouge River Watershed Plan Implementation Guide" (TRCA 2008) is available at www.trca.on.ca.

Watershed Plan's recommendations address each of the watershed goals, the Implementation Guide organizes the recommendations according to how they are implemented and by whom and supplies additional reference tools and maps to assist in delivery. For example, all of the policy-related recommendations are summarized in one section, along with model policy concepts and reference maps. All of the recommended regeneration actions are presented on a series of subwatershed scale regeneration plans, and strategic priorities for implementation are identified based on an understanding of the urgency, scale and effectiveness of proposed regeneration actions and the multiplicity of benefits that would accrue from implementation. Environment Canada's Great Lakes Sustainability Office contributed funding to support the regeneration priorities planning work, as they were interested in using the outcome to guide effective investments in watershed implementation projects. The Guide also made an effort to link the watershed recommendations with ongoing programs and projects of implementing partners.

The preparation of this Implementation Guide involved many local implementing partners and was especially effective in stimulating involvement of "action-oriented" and front line individuals, who would be less likely to be involved in higher level planning initiatives. This tool helped to translate the scientific direction to practical, implementable action.

Outcomes

The *Rouge River Watershed Plan – Towards a Healthy and Sustainable Future* was approved by the TRCA in March 2008, and has been endorsed by all watershed municipalities, most by Council resolution. The plan has already been used to guide significant implementation activities.

Recommendations from the watershed plan have informed local municipal growth planning initiatives and policy updates, such as the York Region Official Plan (as adopted by Council December 2009) and Town of Markham environmental policies. York Region's Water and Wastewater Master Plan and Transportation Master Plan guide Regional programs and operations. Updated scientific information on watershed systems has been used to inform environmental assessment planning studies associated with Regional sanitary sewer and transportation construction projects. These public infrastructure projects have incorporated environmental regeneration work, including reforestation, wetland creation, stormwater management improvements and trail creation all of which contribute to more sustainable and resilient communities.

Leaders within the urban development and building industry have partnered with TRCA to pilot new technologies including green building design and lot level innovations, such as low impact development type stormwater management, which helps to advance watershed objectives through monitoring, education and market transformation. TRCA's Sustainable Technology Evaluation Program (STEP)[9], supported in part by Environment Canada's Great Lakes Sustainability Office, serves as a monitoring and evaluation forum for industry, academia, and government.

Involvement of local partners in the watershed planning process enables all actions to be viewed through the watershed lens. This often opens up innovative opportunities for implementation through initiatives which happen to be the focus of local attention, as well as in day-to-day operations.

The ongoing multi-stakeholder Rouge Park Alliance, combined with an annual festival of Rouge watershed public events and meetings and publication of watershed reports, continue to serve as mechanisms for regular collaboration among partners and tracking of implementation progress now that the plan is complete.

Summary and Lessons Learned in Rouge River Watershed Plan

Multi-stakeholder leadership and strong community engagement in the development of the Rouge watershed plan, combined with sound science, has ensured support for implementation that is founded on a shared vision and common objectives. This shared vision is essential for coordinated action among the many partners. Several specific aspects of the integrated watershed planning process contributed to its success.

Meaningful engagement of implementing partners, by empowering them with the responsibility of developing the plan, led to its wide adoption and quick initiation of implementation activities.

Development of a common framework of watershed management goals and objectives provided a mechanism for partners to identify their shared interests in the planning study. It facilitated the application and technical refinement of broader, Great Lakes Basin and RAP objectives, at the watershed scale. Similarly, it helped partners understand how their local geographic or systems focused objectives fit within the overall watershed ecosystem and contribute to the nearshore health of Lake Ontario.

The interdisciplinary analysis used in this study drew upon new data and modelling techniques to provide a sound understanding of watershed systems, their sensitivities and interdependencies, and their predicted responses to potential future scenarios. The study was enriched by sharing of local data and expertise. Interdisciplinary studies take more time, but result in more defensible science and improved watershed understanding among team members.

---

[9]http://www.sustainabletechnologies.ca/

Community-based social marketing studies provided insights into effective implementation strategies associated with sustainability practices on private lands. This provided a strong basis for selecting strategic, cost-effective strategies and predicted outcomes. These studies illustrate the compelling need to integrate environmental actions within a broader context of social and economic considerations in order to increase the rate of implementation.

Development of more specific implementation tools, such as model policy directions and priority locations for regeneration projects, engaged a much greater interest among implementing partners. This led to more meaningful input and greater uptake of the direction provided immediately following completion of the study.

The integrated watershed planning process used in the Rouge watershed can be and has been applied in other watersheds, with appropriate tailoring to address local issues and interests. The scope and sophistication of each approach is determined based on local watershed needs and roles and by the capacities and interests of the various players. This approach illustrates an example of adaptive co-management at the watershed scale. Conservation Authorities, with their well established local partnerships and integrated watershed management approach, represent an avenue to more fully deliver Great Lakes objectives.

STRENGTHENING INTEGRATED WATERSHED MANAGEMENT BENEFITS ACROSS THE GREAT LAKES BASIN

As highlighted in the preceding sections, there is strong support locally, nationally and globally for integrated watershed management (Conservation Ontario 2010a) and through application of integrated watershed management in Ontario there are lessons learned that illuminate the way forward for integrated watershed management practitioners. However when Conservation Authorities were asked whether or not their watershed/subwatershed studies considered Great Lakes Objectives, only 30% responded that they had and these seemed to be linked to the presence of a Great Lakes Water Quality Agreement "Area of Concern" and Remedial Action Plans in the watershed (Conservation Ontario 2010a). While it is widely accepted that the positive actions taken in watersheds to improve water quality benefit the Great Lakes' nearshore, this is an indication that the majority of Conservation Authorities are disconnected from Great Lakes issues. Thus it is likely that local investments are not being maximized for Great Lakes benefits. This represents a lost opportunity especially since the majority (i.e. 36 of 38 respondents or 95%) of agency respondents felt that IWM was considered important to Great Lakes management (Conservation Ontario 2010a).

Overall, a solid framework is needed to focus the efforts of the many partners and interests on producing positive results for the Great Lakes across the entire Basin. This section describes two key improvements that could strengthen the benefits of integrated watershed management for the Great Lakes, including: engagement of watershed stakeholders at Great Lakes tables; and, making the watershed/nearshore connection.

Engagement of Watershed Stakeholders at Great Lakes Tables

While watershed planning by Conservation Authorities in the Great Lakes Basin has always been cognizant of the impacts of watersheds on the lakes, it has not been truly well integrated with binational basin-wide planning, such as the Lakewide Management Plans (LaMPs). The overall outcome across the Basin is that watershed plans are not linking to Great Lakes objectives (Conservation Ontario 2010a) and thus efforts and resources at all levels are not maximized for Great Lakes benefits. There is a lack of a solid governance framework to focus the efforts of the many partners and interests on producing positive results for the Great Lakes; in other words there are many agencies and organizations, both government and non-government, that are doing vital work but the pieces are not being fitted together to produce a more positive result.

Through various Great Lakes reviews/initiatives, Conservation Ontario continues to submit that municipal and watershed representatives should be engaged at Great Lakes Tables that traditionally only include provincial/state and federal representatives. This engagement would better enable integration of the continuum of planning efforts, in particular Lakewide Management Plans (LaMPs) to watershed plans to municipal (land use) plans. As well, work plan priorities and investments on actions can be better coordinated because significant planning has already been done at the local watershed scale and needs to be implemented. Municipal representatives and watershed management organization representatives at binational decision-making tables (e.g. IJC Boards, Bi-national Executive Committee, LaMPs) would ensure more effective coordination and integration of Great Lakes objectives and LaMPs with watershed plans/actions and municipal plans/actions. For example, Conservation Ontario Council would endorse the Conservation Ontario representatives that would be responsible for liaison between the Conservation Authorities and the various Great Lake tables to which they would be invited/appointed. This liaison would be further facilitated via one window contacts for each Conservation Authority whose responsibility would be to connect Great Lakes issues/questions to the appropriate staff expertise within their Conservation Authority. Overall, this approach requires major commitment of time and resources from the network of Conservation Authorities however, it is probably more feasible than having provincial and federal Great Lakes experts at every watershed planning table.

An integrated approach to governance as described above is consistent with the co-management concept and facilitates collaboration on policy development and implementation, research and monitoring, capital assistance programs and education and communication/outreach programs. Such collaboration is necessary for maximizing benefits and producing positive results for the Great Lakes and the St. Lawrence ecosystem.

Making the Watershed/Nearshore Connection

The nearshore of the Great Lakes is a vital resource that contributes to the social, economic and environmental health of the Great Lakes Basin; it is the zone of greatest primary productivity, a location of diverse habitats for fish and wildlife, and the place where humans interact most closely with the lake, through fishing, swimming, boating and other activities (Kidd 2009). The contributing watersheds (as one of the major threats to the nearshore) must be recognized as part of the geographic scope for nearshore science and assessment activities. With regard to the watershed/nearshore relationship, an integrated watershed management approach would enable the assessment and subsequent adjustment of watershed actions to benefit the nearshore. Watershed-based water balance analyses and modelling linking quantity to quality will be critical in assessing priority watershed actions to benefit the nearshore. It will be important to examine population growth projections and land use scenarios that are watershed-based and modelled for climate change predictions that demonstrate the range of variability for which we need to manage. Best management actions for adapting to climate change can then be targeted to specific watershed characteristics in watershed plans.

Conservation Ontario, with Federal funding, hosted a multi-agency technical workshop "Managing Watersheds for Great Lakes Benefits – Nutrients in the Nearshore" in March 2009, to discuss the watershed and nearshore relationship. The workshop identified current threats as being population growth, changing land use practices, invasive species, and climate change and further identified that without action, eutrophication and the growth of nuisance algae will increase, and will affect fisheries, recreational resources, waterfronts and drinking water. Workshop recommendations (http://www.conservation-ontario.on.ca/great_lakes_workshop/index.html) detail an action plan for implementing a Great Lakes Nutrient Strategy that: promotes collaboration and integration, is science-based and adaptive, and encourages stewardship actions in the watersheds; an executive summary can be found at the above link.

From the above referenced 2009 workshop, priority must be placed on development of a Great Lakes Watershed Action Agenda to:

- Incent development of watershed management plans by funding watershed-based actions that are identified in a watershed plan/study;

- Support "best bet" actions to benefit the Great Lakes; and
- Re-engage the public with the Great Lakes through its watershed communities.

Also from the above referenced workshop, priority must be placed on development of a Great Lakes Nearshore Research Agenda to:

- Understand the relationships between watersheds and the nearshore;
- Transfer knowledge to practitioners;
- Confirm threats to Great Lakes (e.g. individual tributaries, point sources, non-point sources); and
- Enable targeting of watershed restoration actions with greatest nearshore benefit.

To address the significant nearshore issues of the Great Lakes through implementation of the above workshop recommendations regarding collaborative action and research agendas, it is noted that the watershed/nearshore relationship needs to be pursued within a recognized adaptive co-management approach to governance at both the Great Lakes scale and local watershed scale. Additionally, this model can be applied to addressing issues in the St. Lawrence.

MOVING FORWARD

Conservation Ontario's Position on Great Lakes Sustainability (2006) identifies as its mission to "work with all orders of government and basin residents as stewards to protect and improve the unique, shared Great Lakes and St. Lawrence ecosystem for present and future generations". The Guiding Principle is to "apply an integrated watershed management approach to planning and implementation in order to protect and improve the Great Lakes and St. Lawrence Ecosystem." While the commitment and foundation are in place to advance Great Lakes objectives through local IWM delivery, a stronger framework of collaboration is needed among Conservation Ontario, Conservation Authorities, and the Municipal, Federal and Provincial governments to enable this to happen more effectively.

Integrated watershed management provides a mechanism for protecting the Great Lakes and St. Lawrence ecosystem and reconnecting Ontarians with the Great Lakes and St. Lawrence through planning and restoration at the local level. It is a potentially powerful stimulant for re-initiation of grassroots action because IWM plans provide a framework for the watershed community to identify priorities, responsibilities, actions, monitoring and reporting. As local, provincial and federal players in protection and restoration of the Great Lakes/St. Lawrence Basin Ecosystem move forward, integration between local watershed level and Great Lakes/St. Lawrence Basin scale initiatives would be facilitated by an adaptive co-management approach at the Great Lakes scale that:

- engages watershed stakeholders at Great Lakes/St. Lawrence tables, through representative organizations such as Conservation Authorities;
- recognizes the need for a holistic integrated watershed management approach through adaptive co-management implementation;
- ensures collaborative development of a funded, watershed-based Action Agenda for the Great Lakes/St. Lawrence Basin that re-engages watershed communities through taking immediate "best actions" and that incents development of watershed management plans and studies (e.g. only fund actions identified in a watershed plan/study);
- ensures collaborative development of a concurrent Research/Science Agenda to address the watershed/nearshore relationship and key watershed science issues;
- recognizes the broader sustainability context for water management and ensures collaborative development of tools and methodologies for including social and economic considerations into watershed management plans; and
- transfers knowledge, tools, and techniques learned in priority areas (e.g. Areas of Concern; degraded watersheds contributing significantly to nearshore issues) to areas of "outstanding resource value" so as to address the need for a preventative approach across the Basin.

ACKNOWLEDGEMENTS

The co-authors gratefully acknowledge the insights and improvements provided to this chapter through review by Hazel Breton (Hamilton Region Conservation Authority; also the Project Manager for the 2009 collaborative research initiative) and Conservation Ontario staff, Don Pearson (General Manager) and Jane Lewington (Marketing & Communications Specialist).

REFERENCES

Agreement Review Committee. 2007. Final Report to the Great Lakes Binational Executive Committee, vols I and II – Review of The Canada – U.S. Great Lakes Water Quality Agreement. USEPA-Great Lakes National Program Office and Great Lakes Environment Office Environment Canada – Ontario. (http://binational.net/glwqa_2007_e.html)

Canadian Chamber of Commerce. 2010. Water for Sustainability – A Canada-wide National Water Strategy. The Canadian Chamber of Commerce. Ottawa, ON. (www.chamber.ca)

City of Toronto. 2003. Wet Weather Flow Management Master Plan. City of Toronto. Toronto, Ontario. http://www.toronto.ca/water/protecting_quality/wwfmmp/index.htm

Conservation Ontario. 2003. Watershed Management in Ontario–Lessons Learned and Best Practices. Conservation Ontario. Newmarket, ON (http://www.conservationontario.ca/projects/pdf/reports/PHASE%20I/lessons_learned_best_practices.pdf)

Conservation Ontario. 2010a. Overview of Integrated Watershed Management in Ontario in series: Integrated Watershed Management – Navigating Ontario's Future. Conservation Ontario. Newmarket, ON (http://www.conservation-ontario.on.ca/watershed_management/reports/IWM_OverviewIWM_Final_Jun2.pdf)

Conservation Ontario. 2010b. A Water Management Framework for Ontario in series: Integrated Watershed Management – Navigating Ontario's Future. Newmarket, ON: Conservation Ontario. Newmarket, ON: (http://www.conservation-ontario.on.ca/watershed_management/reports/IWM_WaterMgmtFramework_Final_Jun2.pdf)

Conservation Ontario. 2010c. A Water Budget Overview for Ontario in series: Integrated Watershed Management – Navigating Ontario's Future. Conservation Ontario. Newmarket, ON

Conservation Ontario. 2010d. Summary Report in series: Integrated Watershed Management – Navigating Ontario's Future. Conservation Ontario Newmarket, ON.

Environment Canada. 2002 & 2007. Canada-Ontario Agreement Respecting the Great Lakes Basin Ecosystem. (www.ec.gc.ca)

Freeman Associates. 2006. Action Plan for Sustainable Practices – Implementation Strategies for the Residential and Business Sectors in the Greater Toronto Area. Toronto and Region Conservation Authority. Downsview, ON, Canada.

Government of Canada. 2009. Government Response to the Fourth Report of the Standing Senate Committee on Agriculture and Forestry Entitled Beyond Freefall: Halting Rural Poverty. Her Majesty the Queen in Right of Canada, 2009-11-19.

Great Lakes Governors and Premiers. 2005. Great Lakes – St. Lawrence River Basin Sustainable Water Resources Agreement. (www.mnr.gov.on.ca)

Great Lakes – St.Lawrence River Water Resources Regional Body. 2007. Great Lakes – St. Lawrence River Basin Water Conservation and Efficiency Objectives (http://www.glslregionalbody.org/Docs/Misc/ConservationEfficiency_Objectives.pdf)

International Joint Commission. 2006. Special Report: Advice to Governments of Canada and the United States on their Review of the Great Lakes Water Quality Agreement. Windsor, ON, International Joint Commission (IJC), Canada and Detroit, Michigan, IJC United States. (www.ijc.org)

Kidd, J. 2009. Managing Watersheds for Great Lakes Benefits: Technical Workshop on Nutrients in the Nearshore. Conservation Ontario. Newmarket, ON (http://www.conservationontario.ca/great_lakes_workshop/PDF/Final_Full_Rprt_Mar_34_Techl_wrkshp.pdf)

Lake Huron Biodiversity Conservation Strategy Working Group. 2010. The Sweetwater Sea – Strategies for Conserving Lake Huron Biodiversity. Prepared in Collaboration with the Lake Huron Binational Partnership; U.S. Environmental Protection Agency and Environment Canada. (http://lakehuron.ca/uploads/pdf/Lake-Huron-Biodiversity-Strategy-2010.pdf)

Lake Ontario Biodiversity Conservation Strategy Working Group. 2009. The Beautiful Lake – A Binational Biodiversity Conservation Strategy for Lake Ontario. Prepared in Cooperation with the U.S.–Canada Lake Ontario Lakewide Management Plan. (http://www.epa.gov/greatlakes/lakeont/reports/lo_biodiversity.pdf)

Metropolitan Toronto and Region Conservation Authority. 1990. A Comprehensive Basin Management Strategy for the Rouge River Watershed. Metropolitan Toronto and Region Conservation Authority. Downsview, ON, Canada.

Parkes, M. and K.E. Morrison, M.J. Bunch, L.K. Hallström, R.C. Neudoerffer, H.D. Venema, and D. Waltner-Toews. 2010. Towards integrated governance for water, health and social-ecological systems: The watershed governance prism in Global Environmental Change 20 (2010) 693–704. 2010 Elsevier Ltd. (www.elsevier.com/locate/gloenvcha)

Roy, D. and B. Oborne, and H.D. Venema. 2009. Integrated Water Resources Management (IWRM) in Canada – Recommendations for Agricultural Sector Participation. Prepared for Agriculture and Agri-Food Canada. International Institute for Sustainable Development. Winnipeg, Manitoba (www.iisd.org)

Simms, G. and D. Lightman, and R. de Loë.. 2010. Tools and Approaches for Source Water Protection in Canada. Governance for Source Water Protection in Canada, Report No. 1. Water Policy and Governance Group. Waterloo, ON. (http://www.governanceforwater.ca/)

Sustainable Development Commission – UK. 2010. The Future is Local – Empowering Communities to Improve their Neighbourhoods. Sustainable Development Commission – UK. London, United Kingdom.

Swainson, B. and G. Simms, and R. deLoe. 2010. Exploring the Changing Roles of the Natural Resource Sectors in Canadian Water Governance – Implications and Opportunities Workshop Report. Water Policy Governance Group and the National Roundtable on Environment and the Economy. Waterloo, ON. (http://www.governanceforwater.ca/)

Toronto and Region Conservation Authority. 2006. Moving Toward the Living City – Strategic Plan. Toronto and Region Conservation Authority. Downsview, ON, Canada.

Toronto and Region Conservation Authority. 2007a. Rouge River Watershed Plan – Towards a Healthy and Sustainable Future. Report of the Rouge Watershed Task Force Downsview, ON. (http://www.trca.on.ca/protect/watersheds/rouge-river/resources.dot)

Toronto and Region Conservation Authority. 2007b. Rouge River State of the Watershed Report. Toronto and Region Conservation Authority. Downsview, ON, Canada.

Toronto and Region Conservation Authority. 2007c. Rouge River Scenario Modelling and Analysis Report. Toronto and Region Conservation Authority. Downsview, ON, Canada.

Toronto and Region Conservation Authority. 2008. Rouge River Watershed Plan Implementation Guide. Toronto and Region Conservation Authority. Downsview, ON, Canada.

Section 3

Case Studies

CHAPTER 10

The Remedial Action Plan Program, Historical and Contemporary Overview

Gail Krantzberg

Centre for Engineering and Public Policy, School of Engineering Practice
McMaster University, ETB 510, 1280 Main St. W Hamilton,
ON,L8S 4L7, 905 525 9140 x 22153
E-mail: krantz@mcmaster.ca

INTRODUCTION

Lakes Superior, Michigan, Huron, Erie, and Ontario, which are joined by five connecting channels, form the largest surface freshwater system in the world. The Province of Ontario and the states of Minnesota, Wisconsin, Illinois, Indiana, Michigan, Ohio, Pennsylvania, and New York, all lie along the shores of the Great Lakes. Dependence on the Great Lakes has led to cooperation by the Canadian and American governments on improving and maintaining the quality of their boundary waters.

More than 33 million people inhabit the Great Lakes Basin, including about a third of Canada's population and 15% of United States' population. The Great Lakes and other lakes and rivers in the Basin provide drinking water to millions. On both sides of the border, the Basin supports multibillion dollar manufacturing, service, tourism and outdoor recreation industries as well as strong maritime transportation systems and diversified agricultural sectors. It provides the foundation for trade between Canada and the United States, equaling 50% of Canada's annual trade with the United States. Degradation of environmental quality directly impairs the viability and vitality of the region. The reliance of the economy on a healthy Great Lakes Basin Ecosystem is unequivocal and the imperative to restore ecosystem health is clear. To achieve sustainability, ecology and economics must be integrated. Sustainability can be defined as a balanced relationship between the dynamic human economic systems and the dynamic, but generally slower-changing ecological systems in which: 1) human life can continue indefinitely; 2) people can flourish, 3) cultures can develop, but within bounds such that human activities do not destroy the diversity, complexity, and function of the ecological life-support system (Costanza 1992).

BINATIONAL ACCORDS

The 1909 Boundary Waters Treaty, signed between Canada and the United States, stated that "boundary waters and water flowing across the boundary shall not be polluted on either side to the injury of health or property on the other". The Treaty created the International Joint Commission to prevent disputes over the use of boundary waters and to deal with boundary problems between the two countries.

Commencing in 1912, the International Joint Commission conducted several studies on Great Lakes' pollution. The Commission recommended the treatment of all water supplies and the installation of proper sewage treatment.

By 1930, 70 Ontario municipalities had sewage treatment plants and 237 municipalities had water works. In the next 15 years with urban and industrial growth, the systems became taxed. A 1966 detailed investigation of pollution problems in Lakes Erie and Ontario and the St. Lawrence River resulted in an in-depth report on their water quality and the recommendation for an international lower lakes clean-up effort. The report resulted in the signing of the Great Lakes Water Quality Agreement in 1972.

The Agreement coordinated an international clean-up effort to enhance the water quality of the Great Lakes. The International Joint Commission became actively involved in analyzing and disseminating information. The Commission advised both governments on effectiveness of programs and provided water quality updates.

In 1978, the Canadian and U.S. governments reviewed the Agreement of 1972 and revised it to re-affirm the commitment of each country to restore and maintain the chemical, physical and biological integrity of the Great Lakes Basin Ecosystem. Significant problems such as toxic substances found in fish and wildlife, urban and agricultural runoff, and airborne contaminants had been documented and were now addressed. Even more comprehensive that the original agreement, the 1978 Great Lakes Water Quality Agreement placed greater emphasis on toxic substance management, dredging and shipping regulations, and continuation of the phosphorus control program started in 1972.

In 1987, a Protocol was signed amending the 1978 Agreement. The amendments were aimed at strengthening the programs, practices and technology described in the 1978 Agreement and to increase accountability for their implementation. Timetables were set for implementation of specific programs. New annexes addressed atmospheric deposition of toxic pollutants, contaminated sediment, groundwater, and non-point sources of pollution. Annexes were also added to incorporate the development and implementation of Remedial Action Plans for Areas of Concern and Lakewide Management Plants to control critical pollutants.

ANNEX 2

Remedial Action Plans (RAPs) for Great Lakes Areas of Concern (AOCs) are perhaps the best example of community-based environmental protection in existence (EPA n.d.) Through the collaboration between public and private institutions, the RAPs apply a watershed approach to ecosystem regeneration and protection, as they progress towards the re-establishment of beneficial uses. The location of the geographic Areas of Concern is present in Fig. 10.1.

Annex 2 in the 1987 Protocol identifies 14 beneficial use impairments and initiated programs to restore these uses to the Great Lakes. These are:

(i) restrictions on fish and wildlife consumption;
(ii) tainting of fish and wildlife flavour;
(iii) degradation of fish wildlife populations;
(iv) fish tumours or other deformities;
(v) bird or animal deformities or reproduction problems;
(vi) degradation of benthos;
(vii) restrictions on dredging activities;
(viii) eutrophication or undesirable algae;
(ix) restrictions on drinking water consumption, or taste and odour problems;
(x) beach closings;
(xi) degradation of aesthetics;
(xii) added costs to agriculture or industry;
(xiii) degradation of phytoplankton and zooplankton populations; and
(xiv) loss of fish and wildlife habitat.

(GLWQA, Canada and United States 1987)

Two scales of programs were initiated: Remedial Action Plans for Areas of Concern and Lakewide Management Plans for each Great Lake.

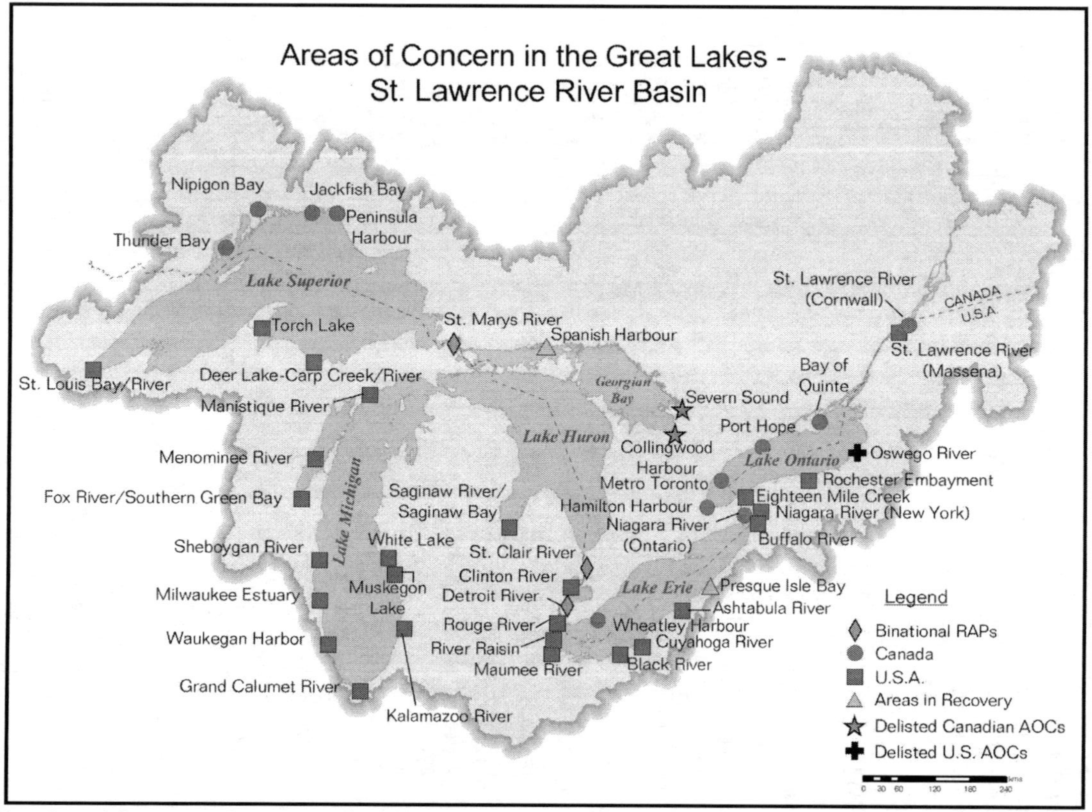

Fig. 10.1 The location of the geographic Areas of Concern for which Remedial Action Plans are underway or completed
Color image of this figure appears in the color plate section at the end of the book.

The restoration experiments, as suggested by Sproule-Jones (2002), promise a way in which resource users, regulators, and those with an interest in restoring the local ecosystem can collaborate towards a common purpose. They promise to empower local stakeholders to determine their own solutions to ecological degradation, and open new venues for collaboration.

With the assistance of governments, residents in most AOCs formed an advisory council/committee to work with federal/state/provincial technical and scientific experts. These committees typically have or had representatives from diverse community sectors, including, agriculture, business and industry, citizens-at-large, community groups, conservation and environment, education, fisheries, health, labour, municipal governments, native peoples, shipping, tourism and recreation. Engaging stakeholder groups in the plan design minimizes the risk of future polarization. Advisory Committee participants possess unique knowledge and represent the interests of their particular stakeholder groups. A key premise is that community residents pose important knowledge, and can provide an informed perspective of the social impacts of the decisions (Harris et al. 2003). The importance of involving communities in the management of water resources was one of the strongest and most consistent messages coming forward from a recent conference (Managing Shared Waters 2002). It is a matter of recognizing the value of traditional knowledge and the public's anecdotal and experiential expertise. Good public processes use plain language to communicate clearly, are supported by commitments in institutional workplans, demonstrate clearly how public input will be used, include mechanisms to resolve disputes, provide the community with access to technical experts, celebrate successes and train community leaders.

Stakeholders have been instrumental in helping governments be more responsive to and responsible for restoring uses in AOCs. Further, stakeholders have been the primary catalyst for implementing actions which have resulted in ecosystem improvements. Such broad-based partnerships among diverse stakeholders can best be described as a step towards grassroots ecological democracy in the Great Lakes Basin (Hartig and Zarull 1992). The collective objective is to work with governments and develop a plan to revitalize ecosystem health and implement the plan to achieve agreed-upon targets that indicate when beneficial uses are restored.

A key concept in the RAP process is accountability for action. This is established through open sharing of information, clear definition of problems (including identification of indicators to be used in measuring when the desired state is reached), identification of causes, agreement on actions needed, and identification of who is responsible for taking action. From this foundation, the responsible institutions and individuals can be held accountable for progress (Hartig and Zarull 1992).

Having been involved in RAPs since their inception, I can point to notable differences in the progress across the 43 (now 38) AOCs. The first stage for each Remedial Action Plan is to identify environmental problems, impaired beneficial uses, and their probable causes. This stage is for the most part complete. The second stage is to develop a recommended set of remedial actions and preventative initiatives to improve environmental quality in support of the beneficial uses. Targets are set by which RAP practitioners can recognize that they have met their goals surrounding beneficial uses. In some AOCs, the targets set science-based and quantitative, where ever possible. In other cases, general statements guide the practitioners, making it difficult to recognize when success has been achieved.

There are significantly divergent approaches across the AOCs in the preparation of Stage 2 Plans and the degree to which their implementation will actually restore beneficial uses. In fact, there is an emerging dichotomy between those who perceive that completing the implementation of the actions is synonymous with the restoration of beneficial uses, and those who acknowledge that the ecosystem will take time to respond to human intervention, and that a period of recovery will be required for beneficial uses to be restored. The interpretation of the Annex varies among and within jurisdictions, and the final decision to delist an AOC, that is, declare all beneficial uses restored, carries with it significant implications depending upon the direction one is placed within this dichotomy.

Stakeholders in various AOCs in the United States and Canada have made considerable investments of time and money, and several well-documented successes are highly visible (IJC 2003). Gurtner-Zimmermann (1995) notes that the commitment of individuals who participate in the RAP process, local support for the RAP goals, and the scientific basis and sound analysis of environmental issues contribute to the positive outcomes.

Major successes include Collingwood Harbour, Severn Sound, and Wheatley Harbour in Ontario, and Oswego River in New York, where conditions have improved to the point that these locations are no longer considered to be Areas of Concern. Spanish Harbour in Ontario and Presque Isle Bay in Pennsylvania are now recognized as in a stage of recovery due to completion of all selected remedial actions, while monitoring continues to measure recovery of beneficial uses.

Other examples of successes include the removal of over 1.3 million cubic yards (1 million cubic meters) of sediment contaminated with polychlorinated biphenyls (PCBs) from the Kalamazoo River, Manistique River, Maumee River, Rouge River, Saginaw River, Saginaw Bay, and St. Lawrence River. Approximately Can $270 million and at least US $3 billion has been invested over the last 10 years to improve the condition of wastewater infrastructure in various AOCs (IJC 2003).

Since 1987, incremental progress has been made to restore beneficial uses in the Areas of Concern. Approximately 25 years since the inception of the RAP program, hundreds of kilometers of riparian vegetation and thousands of hectares of wetlands have been rehabilitated (Canada-Ontario 1999). More than 9,700 hectares of waterfowl habitat were protected and restored in Ontario (Canada-Ontario 2007). Over Can $500 million was invested through Canadian provincial and municipal government partnerships to reduce pollution from municipal sewage and wastewater (Canada-Ontario 2007). Sediment quality is improving in some locations because of pollution control and some sediment clean-up (IJC 2003). In the US alone, there have been at least two dozen sediment remediation projects undertaken in 14 different Areas of Concern (EPA nd). More fish are edible in more places and swimming is again possible in parts of our urban centres for the first time in decades (Krantzberg et al. 1999). Tens of thousands of volunteers are giving their energy to revitalize their homes. Scores of funding partners have collaborated. Research is being advanced basin-wide on the insidious nature of toxic chemicals. Technologies are emerging to better manage stormwater and wastewater, and contaminated sediment. There have been declines in chemical concentrations in Great Lakes fish (SOLEC 2005).

Notwithstanding these strides forward, human health is still being compromised by toxic chemicals, particularly for those consuming fish that are contaminated at unsafe levels, and particularly for children exposed to contaminant in utero (Schwartz et al. 1983, Davidson et al. 1995, Jacobson and Jacobson 1996, Lonky et al. 1996). More aggressive action to revitalize the Lakes is essential to protect the health of all their residents (IJC 2003). The chemical, physical, and biological integrity of the Great Lakes Basin ecosystem remains threatened. Lack of resources and lack of inter-program coordination continue to impede progress (Gurtner-Zimmermann 1995). In some AOCs, environmental problems remain ill-defined both in terms of the magnitude of degradation and the societal costs to either maintain the status quo or undertake adequate remedial actions (IJC 2003).

Differences in the local context of the plans have resulted in a diversity of individual planning and implementation experiences. Notwithstanding this diversity, the motivation and political clout of RAP participants are strongly intervening factors. Resource input from upper levels of government, in particular financial commitment for plan implementation, is also a necessary ingredient for progress due to the RAPs' weak regulatory and institutional framework (Gurtner-Zimmermann 1996).

THE ECOSYSTEM APPROACH

The Parties developed the concept of Lakewide Management Plans (LaMP). There has been much debate about what constitutes a LaMP. One of the more contentious issues surrounding the development of the LaMPs is the boundary of the undertaking. According to the 1987 Agreement: "Such Plans shall be designated to reduce loadings of Critical Pollutants in order to restore beneficial uses." Yet the Agreement also calls for the embodiment of "a systematic and comprehensive ecosystem approach". Progress on LaMPs has been slow in part due to confusion as to whether the breadth of LaMPs includes such ecosystem issues as critical fish and wildlife habitat, biodiversity, and land-use, or should be limited to the identification of critical pollutants and their loading reduction plans.

The current language in the Articles and Annex 2 is somewhat equivocal on the issue of ecosystem approach compared to virtual elimination. However, the Parties' actions over the last 25 years is not. Habitat, biodiversity, and land-use have been and are expected to be a component of the RAP/LaMP programs.

Efforts to separate "ecosystem issues" and "toxics issues" have proven to be artificial. The risk to humans and wildlife posed by various toxic chemical substances (i.e., the chemical integrity of the Great Lakes) cannot be understood or addressed without understanding the physical transport of contaminated water and sediment (i.e., physical integrity) and its biological food webs and pathways for bioaccumulation (i.e., biological integrity). Similarly, the integrity of binational fisheries cannot be sustained simply by addressing contaminants without providing adequate physical habitat for spawning and shelter, limiting predation and competition of invasive exotic species, and ensuring that existing genetic diversity is not lost. Ecosystem and toxic chemical issues must be addressed together to properly restore and protect the "chemical, physical, and biological integrity of the Great Lakes Basin".

Documenting progress

Almost everyone who has been involved in the RAP process has learned a lot over the past two decades. There is a school of thought developing that, under some conditions, following the full implementation of all practical remedial measures, nature may be the best source of recovery and restoration. The Parties should consider recognizing "Area of Concern in Recovery" as an interim step to delisting at sites where remedial measure have been implemented, yet the ecosystem is still recovering. Since Annex 2 stipulates that the final step in RAPs prior to delisting is the achievement of the restoration of beneficial uses, recognizing Areas that are in Recovery would signal a huge success in the advancement to the stage of delisting. Ongoing monitoring of the recovery would be a necessary component of this designation. Because it is an interim designation, it takes into account the difficulty in determining the limits to restoration, because there is no way of knowing the unforeseeable advances in technology, availability of resources, or public will.

The 2006/2007 review of the GLWQA

As noted in the IJC's Advice To Governments On Their Review Of The Great Lakes Water Quality Agreement (IJC 2006), "Article VII, a permanent reference under Article IX of the Boundary Waters Treaty, requires that the International Joint Commission (the Commission), among other things, issue a biennial report concerning progress by the Parties and the state and provincial governments toward achieving the Agreement's general and specific purposes. Article X requires that the Parties conduct a comprehensive review of the Agreement's operation and effectiveness following every third such biennial report. The Commission's *12th Biennial Report*, issued in September 2004, triggered the requirement for the review that took place in 2006 and concluded in 2007.

The review found ambiguity about the Annex's stated purpose. There is a lack of clarity in several areas. According to the Agreement Review Committee (ARC 2006) these include:

- "There is ambiguity regarding whether the Annex takes an ecosystem approach or simply a water quality approach.
- There is ambiguity regarding whether the Annex focus is on the open waters only or on nearshore, inland, tributaries, and watersheds.
- Beneficial Use Impairments are poorly defined, particularly with regard to human health.
- There is a general question about the purpose of the Annex regarding whether it uses an ecosystem approach or a water quality approach.
- There is a question related to whether the Remedial Action Plans and Lakewide Management Plans are to be prepared and implemented in relation to Critical Pollutants using an ecosystem approach to the multi-media sources, pathways and distribution of this narrow group of contaminants or are they for general ecosystem management and stewardship within the Great Lakes basin?"

When the IJC (nd) engaged the binational citizenry in the basin to ask questions of the implementation successes and deficits associated with the GLWQA, Remedial Action Plans attracted the most attention and were often used as examples of shortcomings in Agreement implementation:

"They were probably the source of greatest hope for visible, tangible improvement on an AOC-by-AOC level," said one retired government official who is still active in environmental issues. "Many questions were raised in connection with Remedial Action Plans (RAPs). "Is the concept of RAPs fundamentally flawed?" asked one participant. "Did we not invest enough money? Were they not high priority enough? Did they not fit with other programs? Did we not manage them effectively enough? Were the local government people not involved enough?" Overall, insufficient funding, bureaucratization, inadequate or ineffective public participation, and a lack of accountability provisions were the factors most often cited."

CONCLUSION

Annex 2 is perhaps the most public of the GLWQA's Annexes, because the activities required therein involved interest groups and Great Lakes stakeholders extensively.

The RAP process, where successful, clearly embraces the ecosystem approach. Here, the ecosystem approach is based on the man-in-system concept rather than a system-external-to-man concept (IJC 1978), where the ecosystem is composed of the interacting elements of water, air, land and living organisms including man. While Lee et al. (1982) discuss several variants of the ecosystem approach, most share a focus on the responsiveness of ecological systems to natural and human activities, and a readiness to strike a programmatic compromise between detailed understanding and more comprehensive holistic meaning. This flexible pragmatism is perhaps the most productive feature for addressing Great Lakes environmental problems. As the Parties conclude their renegotiation of a revised GLWQA (anticipated for 2012) the ability to learn from the past on what has worked, what has not worked and why, should inform a renovation to Annex 2, and enhance place-based engagement and implementation for Great Lakes regeneration. By learning from doing, evaluating limitations and opportunities, we can collectively make the Lakes Great.

REFERENCES

ARC. 2006. Agreement Review Committee Report to the Great Lakes Binational Executive Committee, vol. 2. Review of the Canada–U.S. Great Lakes Water Quality Agreement: Final Review Working Group Reports to ARC http://binational.net/glwqa/v2_glwqareview_en.pdf

Canada-Ontario. 1999. Third Report of Progress Under the Canada-Ontario Agreement Respecting the Great Lakes Basin Ecosystem 1997-1999. http://www.on.ec.gc.ca/coa/third-progress-report

Canada-Ontario. 2007. 2002-2007 Progress Under the Canada-Ontario Agreement Respecting the Great Lakes Basin Ecosystem. http://www.ene.gov.on.ca/en/water/greatlakes/coa/2002-2007progress-report_e.pdf

Costanza, G. 1992. Ecological economics of sustainability: investing in natural capital. pp. 106-118 *In:* R. Goodland, H.E. Daly and S.E. Serafy [eds.] Population, Technology and Lifestyle. Washington DC Island Press. Washington DC, USA.

Davidson, P.W. and G.J. Myers, C. Cox, C.F. Shamlaye, D.O. Marsh, M.A. Tanners, M. Berlin, J. Sloane-Reeves, E. Cernichiari and O. Chloisy. 1995. Longitudinal neurodevelopmental study of Seychellois children following in utero exposure to methymercury from maternal fish ingestion: outcomes at 19 and 29 months. Neurotoxicology 16: 677-688.

Environment Canada. 1997. Great Lakes 2000 Cleanup Fund Project Summaries Report. http://www.on.ec.gc.ca/glimr/data/cleanup-project-summaries/intro.html ISBN 0B662-26232-8

EPA. Great Lakes Areas of Concern. http://www.epa.gov/greatlakes/aoc/rap.html (accessed 11/5/10)

Gurtner-Zimmermann, A. 1995. A mid-term review of Remedial Action Plans: Difficulties with translating comprehensive planning into comprehensive actions. J. Great Lakes Res. 21: 234-247.

Harris, C.C. and E.A. Nielsen, W.J. McLaughlin and D.R. Becker. 2003. Community-based social impact assessment: the case of salmon-recovery on the lower Snake River. Impact Assessment and Project Appraisal 21: 109-118.

Hartig, J.H. and M.A. Zarull [eds.]. 1992. Under RAPs. Toward Grassroots Ecological Democracty in the Great Lakes Basin. University of Michigan Press. Ann Arbor, Michigan, USA.

International Joint Commission. 1978. Great Lakes Research Advisory Board. The Ecosystem Approach. Special Report to the International Joint Commission. Windsor, Ontario, Canada.

International Joint Commission. 1997. Overcoming Obstacles to Sediment Remediation. Report of the Sediment Priority Advisory Committee (SedPAC) to Water Quality Board.

International Joint Commission. 2003. The Great Lakes Areas of Concern Report. Ottawa, Washington, Windsor.

International Joint Commission. 2006. Advice To Governments On Their Review Of The Great Lakes Water Quality Agreement: A Special Report to the Governments of Canada and the United States. ISBN 1-894280-67-9

International Joint Commission. Nd. Consolidation of consultations. http://www.ijc.org/en/activities/consultations/glwqa/synth_5.php

Jacobson J.L. and S.W. Jacobson 1996. Intellectual impairment in children exposed to polychlorinated biphenyls in utero. New England Journal of Medicine 335: 783-789.

Krantzberg G. and H. Ali, and J. Barnes. 1999. What progress has been made in the RAP program after ten years of effort? Aquatic Restoration in Canada. pp. 1-13. *In:* T. Murphy and M. Munawar [eds.]. Ecovision World Monograph Series, Backhuys Publ., The Netherlands.

Lee, B.J. and H.A. Regier, and D.J. Rapport. 1982. Ten ecosystem approaches to the planning and management of the Great Lakes. J. Great Lakes Res. 8: 505-519.

Lonky, E. and J. Reihman, T. Darvill, J. Mather, Sr. and H. Daly. 1996. Neonatal behavioral assessment scale performance in humans influenced by maternal consumption of environmentally contaminated Lake Ontario fish. J. Great Lakes Res. 22: 198-212.

Managing Shared Waters. 2002. http://www.pollutionprobe.org/managing.shared.waters/

Ontario Ministry of Environment. 2001. Guide to Eating Sport Fish 2001-2002. Queen= Printer. ISSN 0826-9653

Samy, M. and H. Snow, and H. Bryan. 2011. Integrating social impact assessment with research: the case of methylmercury in fish in the Moblie-Alabama River Basin. Impact Assessment and Project Appraisal 21: 133-140.

Schwartz, P.M. and S.W. Jacobson, G. Fein, J.L. Jacobson and H.A. Price. 1983. Lake Michigan fish consumption as a source of polychlorinated biphenyls in human cord serum, maternal serum, and milk. Amer. J Public Health 73: 293-296.

SOLEC 2005. State of the Lakes Ecosystem Conference, Toronto, Canada. http://binational.net/solec/English/SOLEC%202004/Indicator%20Summary%20Series/Can%20we%20Eat%20the%20Fish%20Draft%20-%20tagged.pdf

Sproule-Jones. 2002. The Restoration of the Great Lakes. University of British Columbia Press. Vancouver BC, Canada.

CHAPTER 11

First off the List: The Collingwood Harbour Story

Gail Krantzberg

ArcelorMittal Dofasco Centre for Engineering and Public Policy,
McMaster University, Hamilton Ontario, Canada
E-mail: krantz@mcmaster.ca

INTRODUCTION

Collingwood Harbour is located on the south shore of Nottawasaga Bay in Lake Huron's Georgian Bay (Fig. 11.1). Collingwood's location on the southern shores of Georgian Bay and at the base of the Niagara Escarpment has featured prominently in the growth of tourism. Collingwood has focused on establishing itself as a four-season tourist destination area, with its web site advertising "a superior quality of life and tourism amenities".

Yet historical abuse resulted in degraded environmental quality in Collingwood Harbour. Nuisance algal growth plagued the harbour waters up until the mid 1980s as a result of excessive phosphorus inputs to the harbour from the Collingwood sewage treatment plant (STP), which at the time was a primary treatment facility (Collingwood Harbour RAP 1992). The harbour, as an industrial port for ship construction for over a century, suffered from habitat and wetland loss, shoreline hardening, and contaminated sediment.

In 1987 the Ontario Ministry of Environment (MOE) assigned the author to coordinate the Collingwood Harbour RAP. This assignment began vaguely defined, as the RAP process was completely new and no one had any experience to inform the specific role of the RAP Coordinator.

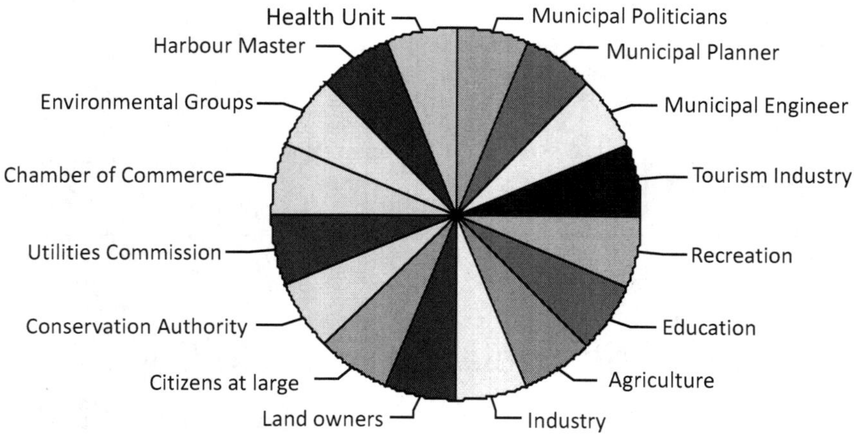

Fig. 11.1 Composition of the Public Advisory Committee

Beginning with the principle that unique to the process was extensive community engagement, we began by consulting with various sectors in the community that had activities that either would affect or be affected by ecosystem health of the harbour. To that end we identified municipal staff, tourism, industry, commerce, agriculture, land owners, recreation, education, environmental groups, public health, water utilities and others (Fig. 11.1) and invited them to be members on a Public Advisory Committee (PAC). Our approach was to attract each viewpoint to the table, with the philosophy of "one view, one voice" such that the member represented the sector, rather than having multiple members from any given sector. Individuals in the position to make decisions were recruited, as the PAC was viewed as a decision making body. The RAP team was composed of government scientists and engineers possessing knowledge relevant to the problems plaguing Collingwood Harbour, and spanned the Ontario Ministry of Environment, the Ontario Ministry of Natural Resources, and Environment Canada.

The first task of the PAC was to identify its goals and desired future uses for the harbour. This would enable the entire group to understand the degree to which ecosystem quality needed to be improved. Reasonably, the PAC wanted information on funding available for remediation and protection, however, the purpose of setting goals and targets was finally understood as a collaborative visioning exercise, rather than a work plan with associated budgets and time lines. As such, at its very first gathering, the PAC recognized that Collingwood Harbour was and could continue to be a site for a blend of industrial and recreational uses. In keeping with that recognition, the PAC decided upon balanced goals and uses, such as recreation, nature observation, fishing, industry, and the retention of the basin as receiving water for the sewage effluent, with each element conducted in a manner to ensure others could be supported. These views met with inclusive public favour early in the RAP process (Collingwood Harbour RAP 1992).

SUSTAINABLE COMMUNITIES IN RELATION TO COLLINGWOOD

The Collingwood Harbour RAP practitioners[1] acknowledged and deployed an ecosystem approach using the man-in-system definition (IJC 1978). Practitioners conceived of the ecosystem as an interacting lattice of water, air, land and living organisms including humans.

With human inclusion in the ecosystem approach, societal values were overtly expressed. From the start, town participants recognized that the local economy and quality of life is inextricably bound to environmental excellence. This unequivocal reality effectively defused what can become a divisive economics versus environment debate, which is inconsistent with the ecosystem approach to resource revitalization.

Could Collingwood be considered a sustainable community? If sustainable cities are vibrant, harmonious and lasting (Sustainable Cities 2004), then the fact that Collingwood is a pleasure to live in supports a positive resolution to the question. Florida (2002) maintains that environmental excellence exceeds other factors including housing costs, climate, government services and public safety in the selection of places to live. If such is the case, the strategic goal of delisting the harbour serves the purpose of demonstrating environmental excellence and attracting talented trades to town.

Sustainable communities have also been defined as places that have made it their business to remain robust over the long term. "Sustainable communities have a strong sense of place. They have a vision that is embraced and actively promoted by all of the key sectors of society, including businesses, environmentalists, civic associations, government agencies, and more. They are places that build on their assets. These communities value healthy ecosystems and actively seek to retain and enhance a locally based economy... Public debate in these communities is engaging, inclusive, and constructive" (Institute for Sustainable Communities 2004).

The RAP was effective and efficient in resolving the consequences of historic misuse (Krantzberg and Houghton 1996). By explicitly seeking the inclusive and active participation of a broad range of stakeholders during the vision and goal setting process, the RAP achieved inclusivity which nurtures legitimacy,

[1]Provincial and federal government experts, community representatives from a cross section of sectors including local government, education, environment, agriculture, industry, tourism and recreation, land owners, conservation authority, human health, citizens at large.

accountability, and can galvanize distributed decision making. With the aim of leaving a sustainable legacy, in its final publication to Town Council, the RAP's "Sustainability Subcommittee", chaired by the Town Planner, wrote, among other things, that "development (at the shipyards) is acceptable so long as certain conditions are met to ensure water quality and the promotion of fish habitat." (Collingwood Harbour RAP 1994).

Ten years later, in 2004, plans were being finalized for the development of a mixed residential commercial project occupying the shipyard's property, approximately 16.6 hectares along the Harbour Waterfront. This presented an excellent opportunity to determine whether the principles of the RAP were effective in enabling sustainable growth along the waterfront and within the Town a decade after RAP implementation was completed and the harbour was delisted as an Area of Concern.

SUSTAINABLE COMMUNITIES, A BROADER PERSPECTIVE

Boyd (2004) refers to the emergence of sustainable communities that convey ambitious new environmental, economic, and social agendas. In order to advance a prosperous, just and sustainable future, Boyd lists nine critical challenges. I emphasize three here:

- PROTECTING AND CONSERVING WATER: Recognizing and respecting the value of water in our laws, policies, and actions
- CONSERVING, PROTECTING AND RESTORING CANADIAN NATURE: Taking effective steps to stop the decline of biodiversity and revive the health of ecosystems, and
- BUILDING SUSTAINABLE CITIES: Avoiding urban sprawl in order to protect agricultural land and wild places, and improve our quality of life.

Civic governance works well when there is a civic community marked by an active, public-spirited citizenry, by egalitarian political relations, by a social fabric of trust and cooperation (Putnam et al. 1993). Such qualities were cultivated during the development and implementation of the Collingwood Harbour RAP. The social fabric of the community encouraged cooperation.

A strong sense of place-based consciousness sways individuals to act cooperatively. It provides the context for cooperative action according to ethics despite economic and immediate, need-meeting motivations which encourage one to do otherwise (Moore 1994). Civic consciousness supports sustainability.

When the local governments choose to be inclusive, involvement of citizens in decision making requires transparent and expansive dialogue. Such engagement is a catalyst of local ownership of a community's programs and public policies. Collaborative decision-making of this nature creates a positive feedback loop for ongoing community involvement. These are characteristics demonstrated within Collingwood and among the RAP practitioners (Krantzberg 2003).

As reported by Krantzberg (2006) the 10 Aalborg Commitments help local governments set qualitative and quantitative targets to implement the urban sustainability principles of the Aalborg Charter (Aalborg 2004). The Aalborg Commitments address 10 themes that are instructive to the North American context at the local level for ensuring sustainability. Some of these are:

- Governance – Local governments increase citizens' participation and cooperation with all spheres of governance in their efforts to become more sustainable.
- Urban management – Local governments formulate, implement and evaluate management schemes aimed at improving urban sustainability.
- Natural common goods – Local governments preserve natural common goods.
- Planning and design – Urban planning is vital in addressing environmental, social, economic, and health issues.
- Local action for health – Local governments have a duty to protect the health of their citizens.
- Sustainable local economy – Local governments are committed to creating a vibrant local economy that promotes employment without damaging the environment.

RESULTS AND DISCUSSION

"Vision 2020 Report No. 6 – The Waterfront" (Town of Collingwood 2002) was compiled by a community committee. The report recommends to Town Council:

> "*that a proactive* ***Ecosystem approach*** *be taken when considerations are made. Within an ecosystem approach everything is connected to everything else. These links are air, soil, water, wildlife, land uses, communities, economic activities and much much more. If we do this we will understand how the parts affect one another and we will understand the complexities of the whole.*
>
> *As a result of our ecosystem approach we can identify ways in which human activities can be reintegrated into the ecological process to ensure efficient use of resources, reduce waste and pollution, etc. It is also incumbent upon an ecosystem approach to hold that the economy, social issues, and the environment are interrelated. An ecosystem approach would make the most of the quality of Collingwood's waterfront area and ensure our responsibilities to future generations. A vibrant,*

> *healthy waterfront is what we need to accomplish the above. The waterfront is our* ***crowning glory.*** *It represents best the future of our community. It is what the citizenry aspires for."*

Earlier it was noted that the shipyards featured prominently in a report from the RAP committee to Town Council at the time of delisting in 1994. The final development design of this critical piece of waterfront was revealed in 2003/04 (Fig. 11.3). The lands affected by the proposed Official Plan Amendment (the former C.S.L. shipyard property) are on the southern shore of the Collingwood Harbour, at the northern end of Hurontario Street, which is the main street connecting the downtown core to the harbour and Georgian Bay. The site is approximately 16.6 hectares (41.0 acres) in area.

Citing from the Development Plan

"Something monumental is happening to Collingwood's downtown. The entire shoreline is about to be transformed.... The pleasure of downtown conveniences, the beauty of Georgian Bay's breathtaking water and mountain views, and the chance to discover four-season living at your doorstep makes this community the only one of its kind. Be drawn to the water's edge. There is so much to experience at the Shipyards. Embrace the best of all-inclusive waterfront living at the Shipyards" (Fram Building Group and Slokker Canada Corp. 2004).

Included in the development plans is an open space situated on the west side of the property and will be accessible to the public, a Natural Common Good (Fig. 11.4). This feature was recommended by the RAP's "Sustainability Subcommittee" in 1994. Urban planning and design reflected good governance in adapting to the desires of the citizenry. Local action for health is being ensured, in that before any of the construction takes place, soil remediation of the brownfield will be completed. Importantly, a wetland feature, providing fish and wildlife habitats is included in the proposal, and is in keeping with the "Sustainability Subcommittee" and 1994 RAP document (Collingwood Harbour RAP 1994). Waterfront trails, a public plaza and a recreational facility for public use, are also included in the plans, and are consistent with the purpose of sustainable towns, to support the environmental, economic, and social fabric of the community.

Years after the delisting of the Harbour, bringing people back to the revitalized waterfront has been a Town priority. Harbourlands Park was created in 2000 and is one of the most beautiful areas in the community. Residents and visitors alike are enjoying the rugged beauty of a once active shipping/grain storage area. The backdrop of the Collingwood Terminals with its huge white columns rises up from the once wasteland "spit area", now a series of beautifully landscaped walkways and gardens with a history of the area on massive granite plinths. Harbourlands Park offers the ever-changing grandeur and scenic beauty of Georgian Bay for the many people who drive or walk to the Park. There are benches for reflective moments or to watch the quiet

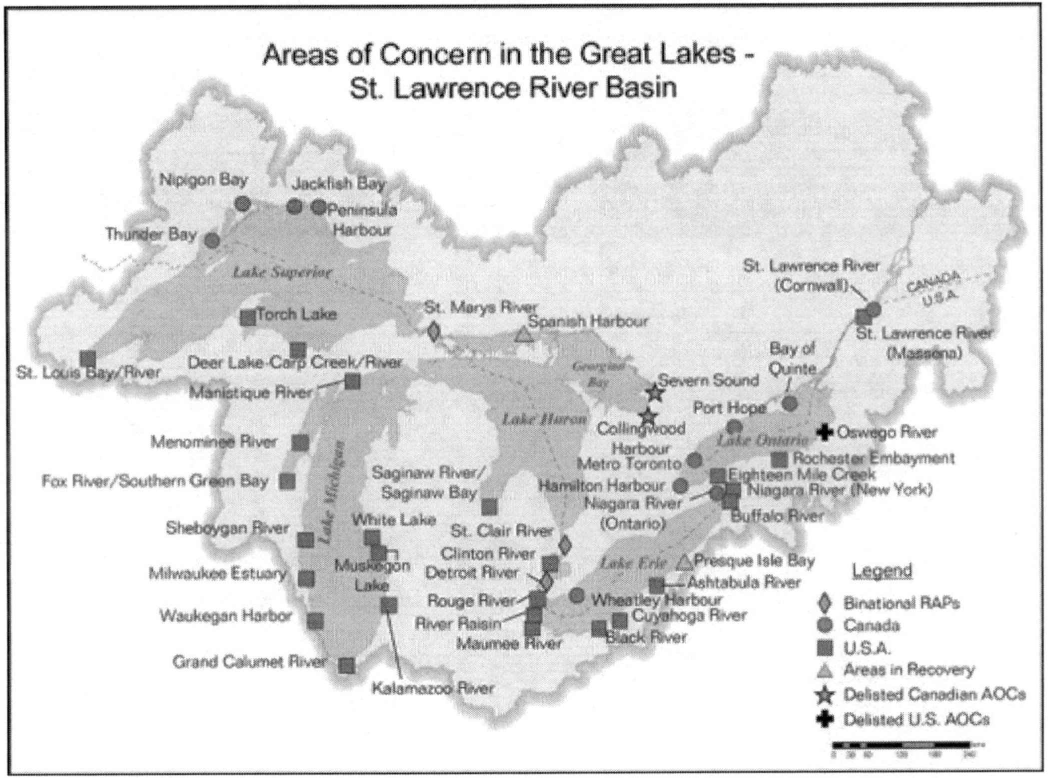

Fig. 11.2 Location of the Great Lakes Areas of Concern
source: http://www.epa.gov/glnpo/aoc/
Color image of this figure appears in the color plate section at the end of the book.

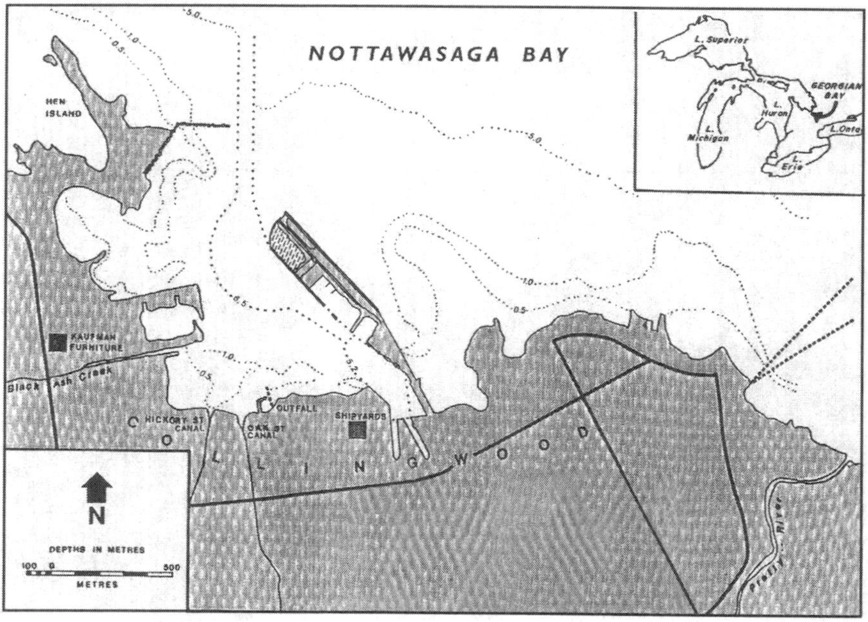

Fig. 11.3 Location of Collingwood Harbour, Ontario

beauty of sailboats filling their sails as they make their way out of the historic Collingwood Harbour (Krantzberg 2006, Town of Collingwood, 2004).

CONCLUSION

According to the IJC (1996), the Remedial Action Plan process is breaking ground in community-based and ecosystem based management processes. In 2012, some might argue that the ground breaking in many locations has stalled. Where this is the case, from the Collingwood experience, it is apparent that governments need to view themselves as a mechanism that nurtures community capacity. Participation of the appropriate actors, development of mutually agreed upon decision making processes, common objectives, political support, public participation and funding are all central prerequisites to achieving a sustainable community, and are central to the philosophy that is behind successful RAP programs (Krantzberg 2003).

The Collingwood Harbour RAP employed a combination of human, scientific, technological, organizational, institutional, and resource capabilities to generate and sustain the capacity for the changes required to solve the Habour's environmental problems. As defined by Hartig et al. (1995) capacity-building enhances the ability of a community to identify and achieve

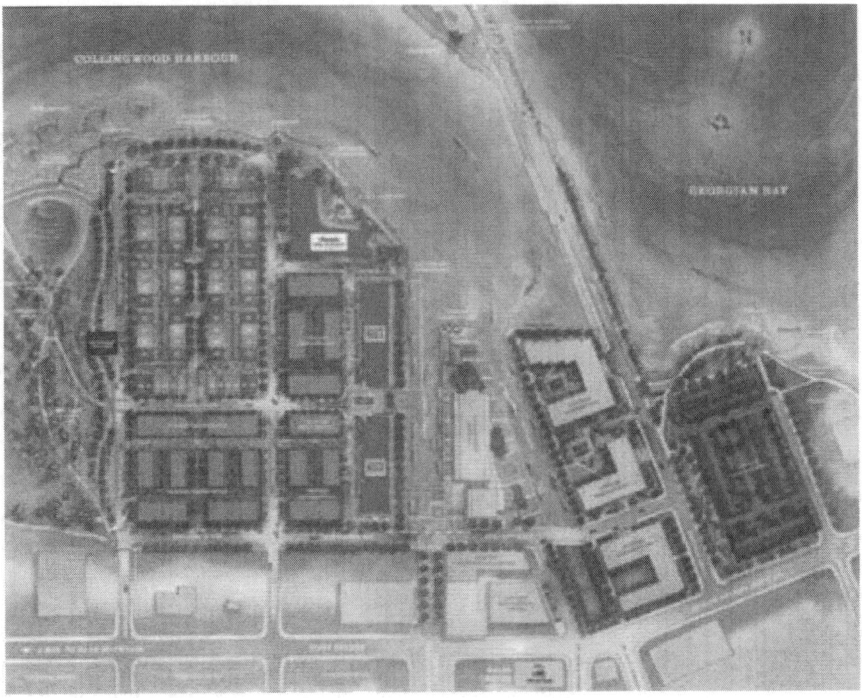

Fig. 11.4 The Shipyards proposed development plan.
Color image of this figure appears in the color plate section at the end of the book.

consensus on problems, develop policies and programs to address them, and marshal appropriate resources to carry out the policies and programs. Further, democratic dialogue and participatory decision making enabled consensus and ownership of the RAP and its legacy. This is evidenced beyond the RAP by the commitment of the politicians and citizens to ensure the sustainability of the Town's economy, environment, and social fabric. Cumulatively, the consequence encompassed a community fiercely protective of its excellence that embedded sustainable design in the approved redevelopment plan of the brownfield property at the Shipyards.

ACKNOWLEDGEMENTS

I thank my most esteemed associate, Douglas Markoff for providing the stimulus to document this remarkable experience. My continued thanks go to Ed Houghton, the Collingwood Harbour Public Advisory Committee Chairman, unwavering leadership of the mayors of the Town of Collingwood, and the proud people of this extraordinary Town.

REFERENCES

Aalborg 2004. http://www.iclei.org/home/documents/press_release_11june2004.doc

Boyd, D.R. 2004. Sustainability within a Generation. A New Vision For Canada. ISBN 0-9689731-6-7. David Suzuki Foundation, Canada.

Collingwood Harbour RAP. 1992. Stage 2 Document. Prepared by the Ontario Ministry of Environment and Energy, Toronto, Ontario; Environment Canada, Ministry of Natural Resources, and the Collingwood Harbour Public Advisory Committee. ISBN # 0-7778-0162-0

Collingwood, 2004. http://www.town.collingwood.on.ca/

Collingwood Harbour RAP. 1994. Planning for the Future, A report to Collingwood Town Council. Coastline Development and Sustainability Subcommittee, Collingwood, Ontario, Canada.

Costanza, G. 1992. Ecological economics of sustainability: investing in natural capital. pp. 106-118. *In:* R. Goodland, H.E. Daly and S.E. Serafy [eds.]. Population, Technology and Lifestyle. Washington DC Island Press. Washington DC., USA.

Florida, R. 2002. Competing in the age of talent: quality of place and the new economy. http://www.heinz.cmu.edu/~florida.

Gurtner-Zimmermann, A. 1995. A mid-term review of Remedial Action Plans: Difficulties with translating comprehensive planning into comprehensive actions. J. Great Lakes Res. 21: 234-247.

Hartig, J.H. and N.L. Law, D. Epstein, K. Fuller, J. Letterhous and G. Krantzberg. 1995. Capacity-building for restoring degraded areas in the Great Lakes. Int. J. Sustain. Dev. World Ecol. 2: 1-10.

International Joint Commission. 1978. Great Lakes Research Advisory Board. The Ecosystem Approach. Special Report to the International Joint Commission. Windsor, Ontario, Canada.

International Joint Commission. 1996. Position statement on the future of Great Lakes Remedial Action Plans. Report of the Water Quality Board. Windsor, Ontario, Canada.

International Joint Commission. 2003. The Great Lakes Areas of Concern Report. Ottawa, Washington, Windsor.

Krantzberg, G. 1997. International Association for Great Lakes Research Position Statement on Remedial Action Plans. J. Great Lakes Research 23: 221-224.

Krantzberg, G. 2003. Keeping Remedial Action Plans on target: lessons learned from Collingwood Harbour. J. Great Lakes Research (29): 641-651.

Krantzberg, G. 2006. Sustaining the Gains Made in Ecological Restoration: Case Study Collingwood Harbour, Ontario. Environment, Development and Sustainability (2006) 8: 413-424.

Krantzberg, G. and E. Houghton. 1996. The Remedial Action Plan that lead to the cleanup and delisting of Collingwood Harbour as an Area of Concern. J. Great Lakes. Res. 22: 469-483.

Moore, J.L. 1994. What's Stopping Sustainability? http://www.newcity.ca/Pages/mooreindex.html

Putnam, R. and R. Leonardi, and R.Y. Nanetti. 1993. Making Democracy Work: Civic Traditions in Modern Italy. Princeton University Press, Princeton, N.J., USA.

Sustainable Cities. 2004. http://www.sustainable- cities.org.uk/institute/index.html

Town of Collingwood. 2002. Vision 2020 Report No. 6 – The Waterfront.http://www.town.collingwood.on.ca/living_news.cfm?category=25

Town of Collingwood. 2004. http://www.collingwood.ca/visiting_tourism.cfm?action=list&type=1&topic=1

United States and Canada. 1972. Great Lakes Water Quality Agreement.

United States and Canada. 1987. Revised Great Lakes Water Quality Agreement of 1978 as Amended by Protocol Signed November 18, 1987. Consolidated by the International Joint Commission.

CHAPTER 12

Remedial Action Plan Case Study: Participatory Governance Used in Hamilton Harbour

John D. Hall[1,a] and Kristin M. O'Connor[1,b]

[1]Hamilton Harbour Remedial Action Plan, 867 Lakeshore Road, Burlington, Ontario, L7R 4A6
[a]E-mail: john.hall@ec.gc.ca
[b]E-mail: kristin.oconnor@ec.gc.ca

The story of the Hamilton Harbour Remedial Action Plan (HHRAP) begins 150 years ago when the City of Hamilton established its waterworks system. It was a time when cholera and other deadly diseases broke out in the community and a safe and secure source of fresh drinking water was critical to the community's health and prosperity. Thomas Keefer, the engineer engaged to develop a water supply system, recommended taking water from Lake Ontario and pumping it up to Escarpment reservoirs thereby providing a reliable clean water service for drinking and fire protection. Wastes would find their way to the Harbour and in this manner remain separated from Hamilton's clean water supply (City of Hamilton, Public Works 2009).

This concept of separating the water supply from the waste stream by using Hamilton Harbour as the repository for waste set the pattern which underlies today's Remedial Action Plan. In 1992 and again in 2002 HHRAP stakeholders endorsed a recommendation, "That diversion of [wastewater treatment plant] effluent to Lake Ontario be considered only after all other technically feasible and practical options have been implemented" (HHRAP 1992b). This approach reflects an environmental ethic held by the Stakeholders to not solve the Harbour's problems by exporting them to Lake Ontario (G.K. Rodgers, personal communication).

Incrementally over the years, the Harbour was filled to provide for the port and lands for the steel industry, eventually losing 25% of its surface area along the south and east shoreline (HHRAP 1992b). Industries discharged effluents directly into the Harbour and into the many small streams feeding the Harbour from the base of the Niagara Escarpment. By 1953 the condition of the Harbour was described as "…ringed by an oil slick "from canal to canal and shore to shore" for most of the summer" (Hughes 2002, NIE5) and in 1969 in the House of Commons, Hamilton Harbour was called "a stinking, rotten quagmire of filth and poisonous waste" (Hughes 2002, NIE5). Hamilton Harbour had become the busiest Port on the Canadian side of the Great Lakes. The cost of prosperity was born by the Harbour and this appears to have been a trade off that many community leaders reluctantly accepted (Bouchier and Cruikshank 2004, Cruikshank and Bouchier 2004).

But times were changing and with prosperity people wanted their Harbour back. The 1960s and 1970s saw the beginnings of environmental monitoring, research, and programs to curb pollution. A community prosperous from its good industrial jobs started to look at the condition of the Harbour. Community activation sprung up with groups such as CHOP (Clear Hamilton of Pollution) in 1969 and SOB (Save Our Bay) in 1971 (Bouchier and Cruikshank 2008). At the same time the Federal Government constructed the Canada Centre for Inland Waters (CCIW), a freshwater research institute located at the east end of the Harbour. The Province of Ontario began to monitor the condition of the Harbour and published several reports in the 1960s, 1970s and 1980s with the most notable being the 1985 report which gave a technical summary of environmental conditions and general management options for remediation (OMOE 1974, 1975, 1977, 1978, 1981, 1985, OWRC 1964, 1967).

CREATING A REMEDIAL ACTION PLAN

By 1985 Hamilton Harbour's time for environmental action had come and a broad based stakeholder group was formed to begin the task of determining the state of the Harbour and defining a course of action to clean it up. Hamilton Harbour was one of the first "Area of Concern" (AOC) on the Great Lakes to initiate remedial action planning. It was not until 1987 that the Great Lakes Water Quality Agreement designated AOCs; Hamilton Harbour began its process in 1985.

Preparing the Remedial Action Plan

As has been described in an earlier chapter in this book, the International Joint Commission (IJC) established a format for developing Remedial Action Plans (RAPs) based on identifying the condition of the AOC with respect to Beneficial Use Impairments (BUI). The strength of the HHRAP was the process established to develop the plan. The HHRAP utilized a scientific writing team in conjunction with an inclusive Stakeholder Group.

The writing team was led by Dr. G. Keith Rodgers, the former Director of the National Water Research Institute (NWRI). Dr. Rodgers moved from his position as head of the NWRI to lead the HHRAP writing team. This was a fortunate opportunity for those involved in remedial action planning, because Dr. Rodgers had a well established network of scientists in the community that could be prevailed upon to join this effort. The HHRAP with the CCIW literally on the shore of the Harbour was uniquely positioned to easily draw upon skilled scientific experts. The Ontario government was also committed to remedial action planning and similarly invested staff time and expertise to the process. Local institutions such as the Royal Botanical Gardens filled out the writing team. Unlike some other AOCs, the research community had already been involved in programs monitoring the Harbour and the adjoining Cootes Paradise Marsh. Perhaps just as importantly, many of the writing team members lived in the community and shared a personal commitment to restoring the Harbour.

Possibly the most important component of the RAP process in Hamilton Harbour was the establishment of a Stakeholder Group to guide the development of the Stage 1 and 2 Reports. A broad range of stakeholders were invited to participate. The Stakeholder Group included agencies, organizations, institutions, government departments, industries, and private citizens who made use of, who wished to make use of, or in some manner had jurisdiction over the Harbour. Approximately 60 stakeholders were interviewed with 49 initially joining the group and 40 groups maintaining participation from 1985 to 1992 (HHRAP 1992a). The HHRAP reports were updated between 1998 and 2002 and over 40 stakeholder groups continued to be represented along with past RAP Coordinators (HHRAP 2003).

As mentioned previously, there was pent up energy within the community to move forward with some form of planning for the future of Hamilton Harbour (A. Redish, personal communication). This energy was focused and organized by a very capable facilitator, Sally Leppard, and Chair of the Stakeholder Group, Anne Redish. Leppard designed a stakeholder process specifically intended to be broad based and inclusive. Every effort was made to ensure no group having an interest in the Harbour was left out. The Writing Team and science community had seats at the Stakeholder Group table and meetings were advertised and arranged to allow members of the public to attend. Simply described, all of the stakeholders including the scientists, civil servants, industry, and citizens were at the same table. This holistic approach required the joint sharing of knowledge and engagement in issues. While records of meeting show that controversial issues from time to time created conflict among the members, the strength of the process was the ability of the members to stay at the table and work through the conflicts. Now a common practice in community planning, this may well have been one of the first stakeholder engagement exercises used as a planning approach in the Great Lakes.

To their credit the Stakeholder Group and Writing Team allowed themselves to be influenced by the then modern concept of an ecosystem approach. The direction provided by the IJC was to identify BUIs within AOCs. The HHRAP contains the ecosystem approach as one of its principles (Hartig and Vallentyne 1989). As stated in the 1992 Stage 2 Report: "Three propositions underpin the ecosystem approach, 1. knowledge that our species and its associated technology originated in The Biosphere and hence is part of Nature, 2. ecological behaviour that takes account of feedback at diverse levels of behaviour from personal to planetary, and 3. ethical behaviour based on an ethic of respect for other systems of Nature, comparable to an ethic of respect for other persons (Vallentyne 1982)" (HHRAP 2003).

The ecosystem approach, in its holistic connection of all parts of the ecosystem, was a dramatic variation on the simple application of efforts to restore BUIs. The ecosystem approach, as applied in Hamilton Harbour, utilized key environmental conditions required to support natural communities of animals rather than simply specifying desired physical/chemical conditions appropriate for the Harbour waters. For example the dissolved oxygen target for the Harbour may be updated to specifically reflect the needs of a sensitive desired fish type, which should cover the needs of the larger community of fish and wildlife. As a result, the HHRAP is broader in its undertakings, for example including such matters as public access and education, clearly not matters intended by the IJC but absolutely important to the local community. In fact the 2002 Stage 2 Update Report (HHRAP 2003) organized the plan within the following seven headings rather than under the BUIs: Water Quality and Bacterial Contamination; Urbanization and Land Management; Toxic Substances and Sediment Remediation; Fish and Wildlife; Public Access and Aesthetics; Education and Public Information; and Research and Monitoring. Additional principles underpinning the plan are "zero discharge of inputs of persistent toxic substances [and] sustainable communities" (HHRAP 2003).

It is important to note that the HHRAP does contain specific BUIs that form the basis for delisting, but the RAP and its associated work plan follow the categories listed above with one component's success very much tied to the success of other complementary actions.

A public process complemented the work of the Writing Team and Stakeholder Group. Two public open houses were held, and celebrity Martin Short, a Hamilton native was used in radio advertisements to publicize the HHRAP. At the conclusion of the process in 1992 all the members of the Stakeholder Groups were asked to sign on to the Stage 2 Report (HHRAP 1992b, Appendix F).

In 1999, the Stakeholder Group was reconvened and began the process of updating the 1992 Stage 2 Report. This involved having technical task groups update the original 1992 RAP Recommendations and Delisting Objectives. For each recommendation the Task Groups: determined the current status of the recommendation; reconfirmed, deleted or revised the recommendation; and finally laid out specific targets, timelines and implementing agencies involved in meeting the updated recommendation. The Task Groups also added new recommendations to cover any perceived gaps in the original list. The Stage 2 Update went through a public comment process in the fall of 2002, and was approved as a draft document in December 2002 by the Stakeholder Group and forwarded to the IJC.

IMPLEMENTING THE RAP

With the publication of the Remedial Action Plan Stage 1 and Stage 2 Reports in 1992, the Stakeholder Group was decommissioned and replaced by a Bay Area Implementation Team (BAIT) and a Bay Area Restoration Council (BARC). This arrangement was a strategic decision of the Stakeholder Group and the terms of reference for the BARC were included in the Stage 2 RAP Report 1992 (HHRAP 1992b). Anne Redish, Chair of the Stakeholders Group and President of the newly formed Bay Area Restoration Council, in her letter of transmittal for the Stage 2 Remedial Action Plan Report of 1992 sets out the intention of creating the BAIT and the BARC as follows. "The Stakeholders recommended that during the implementation stage of the Remedial Action Plan they be re-organized into two groups with distinctly different responsibilities. One group, the Bay Area Implementation Team (BAIT), would be responsible for the implementation of the Remedial Action Plan while the other group, the Bay Area Restoration Council (BARC), would be charged with monitoring the progress of the Remedial Action Plan implementation and in keeping the public informed" (HHRAP 1992b).

BAY AREA IMPLEMENTATION TEAM (BAIT)

BAIT is made up of the following members (as of January 2012): ArcelorMittal Dofasco, Bay Area Restoration Council, City of Burlington, City of Hamilton,

Conservation Halton, Environment Canada, Fish & Wildlife Habitat Restoration Project, Fisheries and Oceans Canada, Hamilton Conservation Authority, Hamilton Halton Home Builders' Association, Hamilton Harbour RAP Office, Hamilton Harbour Technical Team, Hamilton Port Authority, Hamilton Waterfront Trust, McMaster University, Ontario Ministry of the Environment, Ontario Ministry of Natural Resources, Regional Municipality of Halton, Royal Botanical Gardens, and U. S. Steel Canada.

BAIT contains agencies, government departments and several technical teams that contain a broad range of scientific experts and project managers. Charged with the implementation of remedial actions the BAIT follows a well defined work plan and is provided with secretariat support by the Remedial Action Plan Office. Chairmanship is not elected among the members but rotates between Environment Canada and the Ontario Ministry of the Environment. This form of senior government chairmanship was purposefully determined by the BAIT partners when BAIT was first convened in order to secure the active involvement and leadership of the senior levels of government. Normally the chairmanship will rotate from the federal representative to provincial representative every two to four years depending on the best arrangement for maintaining continuity of actions.

BAIT performs a number of roles including: to develop and coordinate local arrangements for implementation of the RAP; to provide consolidated reporting of actions and progress; to consult with BARC, the general public and other Stakeholders on issues related to implementation of the RAP; to create and direct technical committees to support the RAP; to evaluate the effectiveness of remedial actions on an ongoing basis; and to assist the various implementation projects locate sources of funding.

BAIT is not a table for debate or for providing overall direction for the HHRAP. Its function is to implement the remedial actions established as direction from the Stakeholder Group/Forum as set out in the Stage 2 Reports. In the early days of the HHRAP, BAIT did not have a well defined work plan but followed the recommendations contained in the 1992 Stage 2 Report on an ad hoc basis. In some cases, individual comprehensive work plans were created, such as the Fish & Wildlife Habitat Restoration Project. In other cases, BAIT attempted to collaborate on actions as opportunities and priorities best presented themselves. With the updating of the Stage 2 Report between 1998 and 2002 specific work plan components were listed and in effect a work plan was defined through to 2005. This work plan was then updated by the BAIT in 2006 to provide not only specific tasks, but identify lead agencies responsible for implementing the tasks out to 2011. Bringing this rigour to the work plan resulted in a strong commitment and a substantial increase in the level of effort and funding allocated to remedial actions.

One indication of this increase in effort is the expenditure on the HHRAP. Prior to 1990 it is estimated that Can $600 million was invested by industry and the municipalities on pollution abatement directly related to Hamilton Harbour (HHRAP 2003), from 1990 to 2000 Can $205 million was invested by BAIT partners (HHRAP 2003), and from 2001 to 2005 inclusive Can $210 million was invested (HHRAP 2007). The pace from 1990 to 2000 was doubled in the following five years. The 2006 to 2010 investments have not yet been assessed; however, it is expected that the pace remained close to the 2001 to 2005 level. It is anticipated to reach its maximum investment as upgrades to the two largest wastewater treatment plants and containment of toxic sediments at Randle Reef are undertaken at a cumulative cost of close to Can $900 million (J.D. Hall, personal communication).

BAY AREA RESTORATION COUNCIL (BARC)

BARC was established in 1991 (HHRAP 1992b). As an incorporated citizen group BARC was organized with a Board of Directors elected from the membership at large. A President and other officers are elected by the Board of Directors. An Executive Director is hired by the Board of Directors and reports directly to them. Additional program staff members are hired as funding is available to complement their various programs. Describing itself as "a confederation of community stakeholders devoted to revitalizing Hamilton Harbour and its watershed" (BARC 2002). "The mandate of [BARC] is to: promote, monitor and assess implementation of the Hamilton Harbour RAP; promote protection and appreciation of the Hamilton Harbour watershed ecosystem; assess and enhance public awareness and understanding of the environmental status of Hamilton Harbour and its watershed; promote and develop public information and education programs; provide advice and information to relevant government bodies in support of environmental protection and restoration plans; and collaborate with other organizations and individuals with complementary objectives" (BARC 2010).

During the course of each year BARC will run a series of programs and pubic workshops. Programs have included: Adopt a Creek, presented in cooperation with the Royal Botanical Gardens (RBG); Mini Marsh, a joint program between BARC and RBG to educate school children through the growing of marsh plants in their classrooms; Don't Feed the Birds, an education program focused on preventing the feeding of waterfowl at Harbour beaches and parks; Earth Day Tree Planting; Marsh Volunteer Planting; Stream of Dreams™, where students receive environmental education and paint their own dream fish to be hung together on a nearby fence as a large mural; and the Yellow Fish Road™, a program that educates students and the public about the impacts of pollution entering urban storm drains. In addition to the ongoing programs, BARC holds three to four community workshops during the course of the year. Often the workshops have acted as forums to bring the public and politicians together for education and discussions. A newsletter is published four times annually and bulletins advertising upcoming events are circulated to the BARC members.

With a membership at the present time of approximately 700, BARC is able to exercise political influence to secure commitment to Harbour remedial actions. This from time to time is a role that BARC's President and Executive Director engage in on behalf of the local community. It should be pointed out that Hamilton has several well organized and well respected local environmental groups. Depending on the issue these groups effectively inform the local community and mobilize support to secure political discussion and involvement as they deem appropriate.

One of the most important functions of BARC is the production of an annual report monitoring the progress of the RAP. Typically these reports select a specific topic for investigation and will contain recommendations to BAIT for particular actions to improve the effectiveness of the HHRAP process. BARC has begun producing a Harbour Report Card every 5 years (BARC 2002, 2007).

Over the years BARC has positioned itself as an objective advocate for HHRAP and worked hard to be recognized as a voice of reason to move HHRAP projects forward. It has refrained from confrontational advocacy in all but a few circumstances. It is often sought out for its advice and positions on issues by the media especially for any new or controversial issues that arise related to the Harbour.

BARC in cooperation with BAIT has hosted an annual breakfast meeting of Hamilton Harbour watershed Mayors, Members of Federal Parliament (MPs) and Members of Provincial Parliament (MPPs) since 2000. At these meetings local issues of concern and major HHRAP projects are brought to the attention of this group of politicians with a request for their support.

COLLABORATIVE GOVERNANCE

As with all remedial action plans implementation is a voluntary commitment on behalf of the partners (Hartig and Vallentyne 1989). While incentives are built into the remedial action planning process to encourage and secure support for actions, in the end, it is the individual and corporate commitment of the partners that results in actions.

BARC has no legal ability to force BAIT to take on any particular actions. It does have the moral authority it gains from its membership and interaction with politicians and the public. Similarly, BAIT has no control over BARC. Both organizations work at arms length from each other, but are connected by

BARC having a seat at the BAIT table and the RAP Coordinator representing BAIT attending all BARC board meetings as a resource person. BARC also is provided with a grant for a portion of its operating budget from Environment Canada and may receive funding directly or indirectly for some of its public projects through BAIT member partners. In the opinion of the authors, both BAIT and BARC have been careful over the years to maintain their independence of actions and positions as separate groups.

Hamilton Harbour Remedial Action Plan Office

A key to the functioning of the HHRAP is the secretariat known as the HHRAP Office. It consists of two full time and one part time positions and is responsible for coordinating and providing secretariat support to the BAIT. Figure 12.1 shows the range of activities coordinated by the HHRAP secretariat.

Since 2000 the HHRAP Office has provided the secretariat for between 25 and 30 committees and groups. This normally accounts for 75 to 100 meetings per year scheduled and arranged directly by the HHRAP Office. Meeting summaries are provided for all meetings. While some committees have existed for years, individual committees and groups are formed and eliminated on an ongoing basis. In addition, on an annual basis a meeting is organized to brief watershed Mayors, MPs and MPPs. This meeting while organized by the HHRAP Office is hosted by the BARC. In order to facilitate the sharing of research and monitoring among the BAIT partners, three research and monitoring workshops are held annually, one focused on the Harbour, a second on Cootes Paradise marsh, and a third on the Harbour watershed. From time to time the HHRAP Office will make presentations to BAIT partners and convene special workshops to develop work plans or visioning the Harbour's future.

Developing 2006 to 2011 BAIT Work Plan

The development of the 2006 to 2011 BAIT work plan is a good example of how BAIT and the HHRAP Office functions. There were five steps in this process.

Step 1: BAIT Workshop. A workshop was held for BAIT members and their colleagues. The workshop's theme was "Share the Experience... Plan for our Future!" Open Space Technology (OST), a specialized facilitation technique that allows those in attendance to develop and lead their own agenda, was used to facilitate input. Discussion and reports ranged through 27 different agenda topics posted by participants. The workshop report was provided to the participants.

Step 2: Preliminary Work Plan. The RAP Office prepared a preliminary work plan utilizing the input from the OST workshop, existing BAIT partner updated information, and input from the BARC Toward Safe Harbours reports on Monitoring for Delisting (BARC 2003, 2004, 2005, 2006).

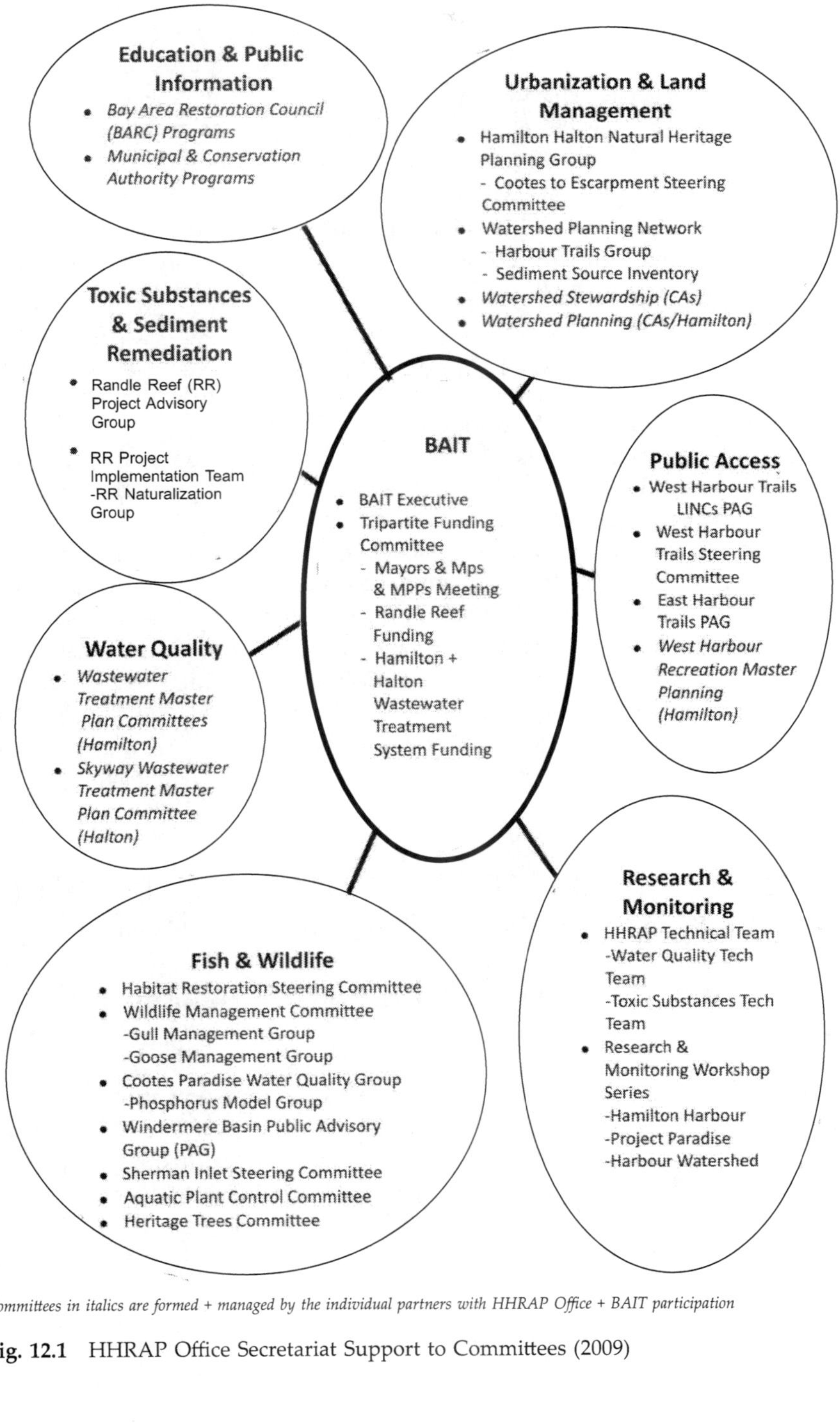

Committees in italics are formed + managed by the individual partners with HHRAP Office + BAIT participation

Fig. 12.1 HHRAP Office Secretariat Support to Committees (2009)

Step 3: Task Group Input and Review. The RAP Office organized meetings with three task groups comprised of resource people suggested by the RAP Office, BAIT members, and the pre-selected Task Group Chairs. Groups reviewed and proposed changes to the preliminary work plan sections relevant to the expertise of that particular group. For example one group reviewed all the items proposed under the Water Quality section and the Toxic Substances and Sediment Remediation section.

Step 4: Approval of BAIT 2006–2011 Work Plan. A draft work plan was prepared and reviewed by the BAIT. Changes were made and the final work plan was sent out with the agenda and approved at the next BAIT meeting.

While overall progress on the HHRAP is monitored by BARC, yearly progress on the work plan is peer reviewed by BAIT at three meetings held per year. Since 2000 the meetings of BAIT have been organized to focus on the specific tasks identified in the work plans. The first meeting of the year reports on new and ongoing projects and tasks related to Harbour water quality and bacterial contamination, and toxic substances and sediment remediation; the second meeting reports on urbanization and land management, fish and wildlife, and public access and aesthetics; the third meeting reports on education and public information and research and monitoring. At the BAIT meetings presentations are provided on ongoing major projects and a formal reporting procedure has been used to track the numerous actions contained in the current work plans. If a member of BAIT has not initiated or completed a work plan action they are required to report this to BAIT and explain what steps they propose to take to ensure a particular action they are designated to lead does in fact move forward.

CASE STUDIES OF PARTICIPATORY GOVERNANCE IN HAMILTON HARBOUR PROJECT IMPLEMENTATION

Case Study Number 1: Fish & Wildlife Habitat Restoration Project

One of the first initiatives of the HHRAP was the development of a Fish & Wildlife Habitat Restoration Project (F&WHRP). The Federal Department of Fisheries and Oceans (DFO) acted as the catalyst for this project pulling together local, provincial and federal agencies. Two major steps were involved in the process. First was the creation of an overall fish and wildlife habitat restoration plan, followed by a second stage of defining the specific site details for individual projects.

Defining the F&WHRP

In September of 1990, DFO convened a two day workshop on the plan to restore fish and wildlife habitat in Cootes Paradise. Nineteen local and international experts "provided an objective analysis of a variety of management options..." (CWS and GLLFFAS 1990).

In November of 1990, 18 professionals attended a meeting on the enhancement of fish and wildlife habitat in Hamilton Harbour. The proceedings of the workshop and the meeting formed the basis of the F&WHRP (CWS and GLLFFAS 1990).

The first action of the F&WHRP was to hire a Project Manager. The DFO stepped forward as the lead partner securing funding from Environment Canada's (EC) Great Lakes Clean-up Fund. An agreement was reached with a local Conservation Authority, the Halton Region Conservation Authority (later know as Conservation Halton) to secure the services of John Hall, a professional planner, as the full time Project Manager. The Hamilton Harbour Commissioners (HHC), later becoming the Hamilton Port Authority (HPA), provided space for a project office. Since most of the project sites involved the shoreline and bottom of the Harbour, this early relationship with the HHC proved to be critical to successful implementation of the site specific projects.

A formal steering committee reporting to BAIT was established to direct the project to which the Project Manager reported. This committee developed a restoration plan which was taken to the public under the terms of the federal Environmental Assessment and Review Process. A detailed process of stakeholder involvement and public meetings was part of the environmental assessment (EA) and exceeded the normal EA process. The provincial government of Ontario determined that a provincial EA was not required due to the extensive collaborative approach used to develop the overall fish and wildlife habitat restoration plan and extensive stakeholder consultation used to develop the individual projects as part of the federal EA.

In order to raise the profile of the F&WHRP and secure public input a major consultation was undertaken that included: speaking to over 50 special interest groups; a number of neighbourhood/community meetings; a public launch at the Royal Botanical Gardens covered extensively by the print and television media; and a boat cruise of Hamilton Harbour attended by most local, provincial and federal elected politicians from the Hamilton and Burlington area (FWHR 1992).

F&WHRP Site Specific Project Design and Implementation

The F&WHRP included eight specific project sites. Each site established a project advisory group (PAG) to work with the consultant hired to develop the details for the design of the particular site. The PAGs were made up of scientists, citizens, industries, and civil servants responsible for approvals or implementation of the projects. The PAGs often contained between 12 and 24 persons, but no limit on size was established as it was intended to have at the same table all those stakeholders/groups that had an interest in or could contribute to the design or development of the project. The PAGs provided advice only and decisions in the end were left with the governing structures of the particular implementing agency.

The Project Manager for the F&WHRP acted as the facilitator for the individual projects. Several steps were followed to bring together the PAG and reach consensus on final design and implementation.

Step 1: Determining PAG membership and securing individual participation. Using the F&WHRP Steering Committee the Project Manager identified a preliminary list of potential PAG members. Normally these were individuals representing a stakeholder group. A letter was then sent to the individuals asking them to consider being a PAG member, explaining the project, the commitment required and advising them that they would receive a call from the Project Manager to discuss their potential participation. From this step an initial PAG was formed.

Step 2: First meeting of PAG. The first meeting of the PAG was used to give a very brief overview of the potential project, introduce PAG members and the consulting team, explain the process and time commitment, define the responsibilities and commitment required of PAG members and to determine if any group or individual should be added to the PAG. The PAG members then worked through defining their own code of conduct and identified site visits or tours they felt would assist them with their work.

Step 3: Ongoing meetings of the PAG. The meeting schedule and timeline for working through the design was identified and confirmed at the second meeting of the PAG. It was normal that the entire process would attempt to be completed in six months to one year but in several cases more time was required, often due to unforeseen delays.

The intention of the process was to reach consensus on a design. The first step in the process was to develop the general purpose or objective of the project. Time was then spent defining the components of both the natural environment and social/cultural environment existing at the site that are of high value to the PAG. Once it was established what shouldn't be lost from the site, time was spent defining how to enhance the site for the particular purpose or objective for which the project was initially defined. For example, a site heavily armoured to protect it from wave erosion would require erosion protection as an objective, but the form of protection may be changed to provide a habitat function. Figure 12.2 shows an example of the complex relationships that often had to be accounted for and resolved in this process (Trotter et. al. 1998).

The fish and wildlife projects are good examples of expert opinion driving the project design. By its very nature fish and wildlife habitat construction is more an art than a science even though substantial science drives the general basis for design.

Step 4: General public consultation. A process for public consultation was always established that provided for a public open house and presentation at the outset of the PAG process. The first public consultation was used to introduce the project and canvas the individual's interests and concerns for the project. This information then fed into the PAG deliberations. A second public open house was held once the PAG reach consensus on design and implementation. This second meeting allowed the public to get a good understanding of the project and provided an opportunity to make changes if warranted.

Step 5: Environmental screening and permitting. Once consensus was reached by the PAG and public input received on the project design, it was taken through a federal EA screening. The Project Manager would prepare

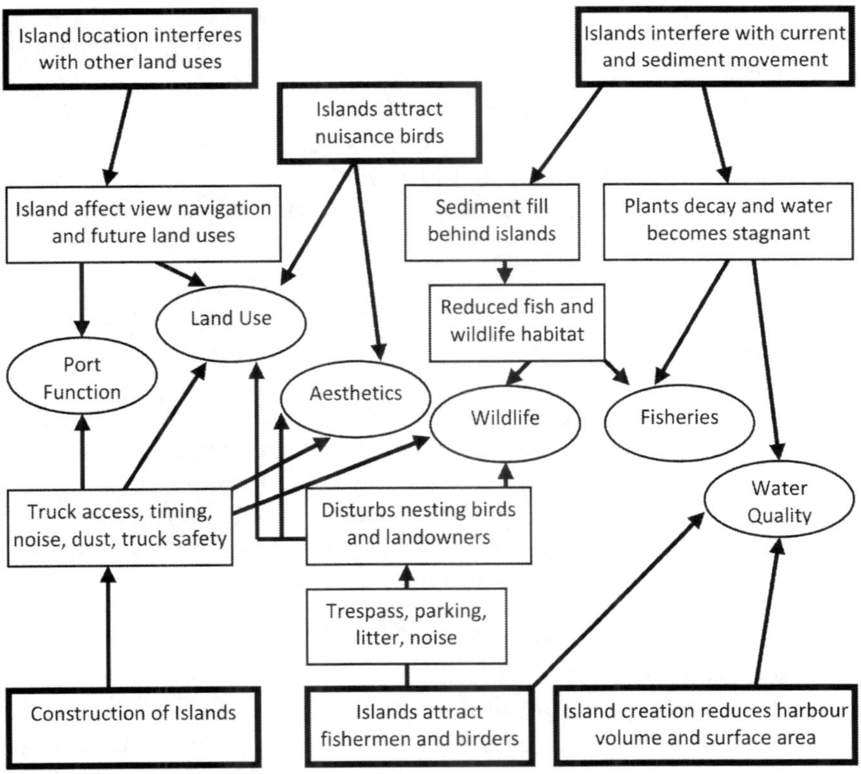

Fig. 12.2 Relationship of Potential Generic Impacts (Trotter et al. 1998)

an overview EA report that was utilized by the various federal departments responsible for the EA screening document. The PAG process included representatives of the federal departments involved in funding, lands and regulations. This often allowed for an expedited approval process because of the individual staff members familiarity and confidence in the project. The Project Manager would in addition file for all external permits required from the municipalities, conservation authorities, and province.

Construction and Timing of Project Implementation

Most of the major excavation and rock work in constructing shoreline modifications, islands, a major fishway, and trails was carried out during the winter months. This minimized disruption to neighbourhoods as people are not outside as much and noise is attenuated within homes. While the winter construction season can make for miserable working conditions it is often a time of year when better contracting prices can be obtained. Environmental impacts were also often reduced in the winter as animals have migrated or hibernated and those that remain are easier to identify for protection. Sites were prepared for spring landscaping. This proved in most cases to be a significant environmental and social benefit.

When the projects were within a neighbourhood, attempts were made to contact the community and let them know when construction would take place. This in part occurred as a result of the PAG process, as local community PAG members provided information back directly to their community groups and were able to advise the PAG on local issues that may be of concern.

Case Study Number 2: Wastewater Treatment Plant Upgrades

Hamilton Harbour has three wastewater treatment plants (WWTPs) that outlet within its watershed. The two largest, Woodward and Skyway, account for almost 50% of the flow to the Harbour and outlet directly to the Harbour. The remaining 50% of flow comes from the watershed. These flows are mixed with the exchange of water through the Burlington Ship Canal with Lake Ontario. Wastewater effluent is therefore a significant influence on the water quality of Hamilton Harbour. The Stage 2 Reports (HHRAP 1992b, 2003) defined water quality targets for the Harbour and associated effluent loading limits for the wastewater treatment plants.

Both the City of Hamilton and Region of Halton initiated the process of determining required upgrades to their wastewater treatment systems by first completing integrated water and wastewater master plans. These plans projected treatment capacities out 25 years and took into account both the

HHRAP effluent loading targets and the recommendation "that diversion of WWTP effluent to Lake Ontario be considered only after all other practical and technically feasible options have been implemented" (HHRAP 2003). These master plans followed a provincial EA process for water and wastewater treatment systems.

The next step in the planning was to define the specific effluent targets and identify the technology and upgrades required to meet the targets. In the case of the Woodward WWTP, the City of Hamilton established two advisory groups to provide input into their decisions. The HHRAP was asked to provide technical representatives to a Technical Advisory Committee (TAC) and representatives to a Community Liaison Committee (CLC). The TAC members representing the HHRAP included: the chair of the HHRAP Technical Team, experts from Environment Canada (EC) and the Ontario Ministry of Environment (OMOE), and the HHRAP coordinator. Through this process the HHRAP Technical Team worked closely with the City to define loading limits to meet HHRAP objectives for the Harbour (KMK 2007). In parallel to the City's exercise the HHRAP Technical Team prepared a review of water quality conditions in the Harbour and evaluated the effluent loading targets proposed by the City (HHTT 2007a, 2007b). In this manner the Technical Team felt confident in recommending approval of the City's loading targets. The City of Hamilton and the Region of Halton along with EC, OMOE and other experts sit on the HHRAP Technical Team.

The CLC had representation from the HHRAP Coordinator and BARC. This allowed the HHRAP Coordinator to speak directly to HHRAP loading limits and the integrated work of the City's TAC and the HHRAP Technical Team. This cross representation by the HHRAP Coordinator provided an improved level of understanding and acceptance of the City's process as the CLC could directly question and receive information from the HHRAP Coordinator rather than accept information second hand through City of Hamilton staff or their consultants. While the normal HHRAP process is to have all stakeholders at the same table, the process used by the City of Hamilton was effective and did provide for good communication in the opinion of the author (J.D. Hall, personal communication).

The Region of Halton's process, while following the same provincial EA requirements, was not as thoroughly integrated with the HHRAP as the City of Hamilton's process. The major difference was the lack of representation on the CLC by the HHRAP Coordinator or BARC. This in part resulted from establishing the CLC through soliciting membership from newspaper advertisements and not making a direct request for HHRAP or BARC representation. Nevertheless, the process followed by the Region of Halton did result in a project to upgrade the Skyway WWTP in accordance with provincial EA requirements and which is supported by the HHRAP Technical Team.

Case Study Number 3: Randle Reef Contaminant Remediation Project

The Randle Reef Remediation Project is perhaps the best known project of the HHRAP. It involves an underwater area of the Harbour contaminated by extremely high levels of polycyclic aromatic hydrocarbons (PAH). The site covers an area in excess of 25 ha and the plan is proposing to contain 630,000 m^3 of contaminated sediments. This is the largest and most severely contaminated site on the Canadian side of the Great Lakes and is second only to the Sydney Tar Ponds for PAH contamination in Canada (EC 2010).

Several attempts to define a remediation project have been tried without resolution since the discovery of the contamination in 1985. In 2002 a process was established to define a remediation project that could move forward to implementation. Similar to the F&WHRP, a PAG was formed and this group worked through a process attempting to reach consensus on a final design and implementation. The same steps as set out previously for the F&WHRP case study were followed and are therefore not repeated here; however, it should be noted that because of the more complex nature of the Randle Reef Project, additional expertise was brought to bear and additional time was required by this PAG. Emphasis was also placed on securing representation from local neighbourhood groups surrounding and downwind of the site along with the union at Stelco (currently U. S. Steel Canada). Both of these groups had been left out of the previous exercises and took issue with those proposals. Previously a Steering Committee of government agencies had directed all input into the project; albeit with a public consultation process at the end.

Two additional components were added to the normal PAG process. In order to assist with defining objectives and valued ecosystem components, a matrix was used to identify objectives, priorities and issues. As the project began to define particular options it was necessary for PAG members to chose a specific option to recommend for final design. At this point in the deliberations a local respected member of the community, Dr. Brian McCarry, Chair of Chemical Engineering Department of McMaster University was asked to facilitate an exercise to reach consensus on the preferred option. Meetings were held over two days, one week apart, and a significant majority of the PAG members were able to agree upon a preferred option.

The Randle Reef PAG has been kept informed of the progress of the project design and environmental assessment and subsequent meetings were held in 2007 and 2008 to inform the PAG of final design details and environmental screening. Public consultation followed a similar format to that of the F&WHRP with the difference that larger halls were utilized for the public open houses. At the time of writing this chapter, the environmental screening and permitting process is still underway.

LESSONS LEARNED .... SECRETS OF SUCCESS

There are a number of "lessons learned" to take away from the Hamilton Harbour RAP experience: contextual lessons learned (don't lose the original, learn from the past, creating a Remedial Action Plan); process lessons learned (get all the stakeholders at the table, good facilitation and well defined roles, science based writing team and task groups, value of senior level representation, good communication, and motivated people); implementation lessons learned (implementing structure, well defined work plan, accountability, technical teams and advisory committees, and project advisory groups); and communication lessons learned (communication with the public and engaging politicians).

Contextual Lessons Learned

Don't Lose The Original!

It needs to be said before learning anything from our HHRAP, if you have a healthy diverse ecosystem don't lose it because remediation is a poor substitute for the original. In Hamilton Harbour's case we lost 65% of the wetlands, mostly along the south shore of the Harbour. These wetlands were lost to the steel industry and the steel industry drove the economy of Hamilton delivering prosperity for over 100 years. Could Hamilton have had an industrial heart land and port while retaining the south shore wetlands? It is impossible to know. Times have changed and the idea of simply trading one community attribute for another, in this case the environment for steel and prosperity, are no longer acceptable and must be rigorously scrutinized through environmental assessments and all forms of legislative and political review.

Learn from the Past

Perhaps in the larger context of the area our Hamilton Harbour community has learned from the past. One recent project, The Cootes to Escarpment Park System (Wong 2009), approved as a concept in 2010, envisions protection for 3,500 ha of important natural lands between the Niagara Escarpment and Cootes Paradise/Hamilton Harbour. This concept planning exercise used the HHRAP as a catalyst to pull together eight partner agencies to set out a plan to secure and enhance the natural integrity and rich biodiversity of the area, while recommending appropriate educational and recreational land uses. In the future it is highly likely this area will be ringed by urban development, but rather than simply react to that urban form after it is set in place, the park system plan allows future planning to fully embrace the importance of this park system (Steelman and Hess 2009). In many respects the future urban form will be in part fashioned in synchrony with a park system.

Creating a Remedial Action Plan

The HHRAP was based on a stakeholder process to reach the best consensus of opinion and recommendations in formulating remedial actions. While no survey of stakeholders was taken that would formally gauge results, it is fair to suggest that those involved in the Hamilton Harbour process by and large found the process successful (A. Redish, personal communication). If this process was successful, what made it successful? In simple terms one can point to a good process with competent and motivated people involved. Let's consider the process.

Process Lessons Learned

Get All The Stakeholders at The Table

In the case of Hamilton Harbour considerable effort was made to ensure all stakeholders that could be imagined to have involvement with Hamilton Harbour were included (Hartig and Vallentyne 1989, Hartig et al 1996, Law and Hartig 1993). Over 40 stakeholders were eventually recruited representing industry, businesses, recreational users, naturalists, rate payers, special interest groups, academia, and all levels of governments and associated agencies. It can't be over stressed that good stakeholder processes requires all stakeholders to be at the table, even though some may historically be adversarial towards each other (Dalton 2005, Mehan III 2011, Orr et al 2007, Steelman and Hess 2009, Stringer et al. 2006).

Good Facilitation and Well Defined Roles

The Stakeholder Chair, chosen by the committee, worked in conjunction with a consultant facilitator and the RAP Coordinator to prepare meeting agendas and facilitate meeting processes. Facilitators need to be strong, impartial, and adept at securing input from all stakeholders (Dalton 2005). This type of professional assistance was a critical part of the HHRAP's success in securing ongoing and motivated interest by the stakeholders. The Chairperson needs to be a strong advocate, a good spokesperson for the process, and able to competently run orderly meetings.

Science-based Writing Team and Task Groups

A core writing team representing a cross section of interests from fish and wildlife to water quality and toxics in turn connected with the larger scientific community to prepare documentation on the state of the Harbour and options for its remediation (Hartig and Vallentyne 1989, Hartig et al. 1996, Marmorek and Peters 2001, Stringer et al. 2006). To update the Stage 2 Report, a system of technical task groups were asked to review and update recommendations to provide advice to the larger Stakeholder committee.

Value of Senior Level Representation

The HHRAP did have times of conflict, but resolved these at the table in large part because one requirement for stakeholders was that they be as senior a position in their organization as possible (Marmorek and Peters 2001). This allowed stakeholders to speak with authority for their constituent group, thereby allowing fulsome debate and the securing of agreement on issues.

Good Communication

The HHRAP was reported frequently in local news media. In addition, in its early years a series of radio commercials publicizing the HHRAP were broadcast featuring Martin Short a well know actor and local Hamiltonian.

Motivated People

The single most important ingredient in the success of creating the HHRAP is the motivation of the people involved (Dalton 2005). For years there was a pent up demand to improve the quality of Hamilton Harbour. The Stakeholder group contained many representatives that were keen to take advantage of the opportunity the HHRAP process offered. The Facilitator and Stakeholder Chair were extremely motivated in part because this was an opportunity to work with a process that truly valued stakeholders by asking them to create their own plan. The writing team contained researchers and project managers that lived locally and wanted to make a difference to their community (Hartig and Vallentyne 1989). Interestingly, this motivation extended into the 2002 update of the Remedial Action Plan some 10 years after it was first published in 1992. It is likely that this ongoing motivation came from seeing a measure of success in implementing the HHRAP.

Implementation Lessons Learned

Implementing Structure

The Stakeholders were determined that their plan would not sit on a shelf. They devised a system of implementation using a Bay Area Implementation Team (BAIT), agencies identified to carry out projects, and a Bay Area Restoration Council (BARC), a citizens' group to report on implementation and to communicate with and on behalf of the public. This structure put pressure on the BAIT to implement projects and programs while BARC scrutinized progress. It also allowed BARC to engage the public in workshops, debate over additional actions, and gave licence for the public to make the HHRAP their own.

Well-defined Work Plan

The 1992 Stage 2 Report contained a list of 50 recommendations in order of priority. Fortunately the priority ordering was not followed; instead, projects and programs were implemented as partners took the initiative. In the 2002 Update the Stakeholders recognized that a well defined work plan for 2002 to 2005 was required and should identify the participating partners (Hartig et al. 1996, Steelman and Hess 2009). For 2006 to 2011, BAIT developed an even more detailed work plan, this time identifying lead partners for each task. Subsequent work plans will be developed as required.

Accountability

The BAIT process was reformed in 2000 to require standardized meetings with a peer accountability process of reporting on work plan progress (Hartig et al. 1996). This discipline brought to the process resulted in a doubling of output as represented by dollars spent following 2000 from the period prior to 2000.

Technical Teams and Advisory Committees

The BAIT utilizes a Technical Team with two sub-committees and numerous advisory groups at any one time. These expert groups give advice on programs and projects (Hartig et al. 1996, Marmorek and Peters 2001). It is important to note that these groups do not function in a public forum and do not pass formal resolutions. Their meetings are summarized and their advice assists and allows for coordination of implementing actions.

Project Advisory Groups

Project Advisory Groups (PAGs) are formed to allow scientists, citizens, and civil servants to meet around one table, learn about a project and provide their best advice on how the project should proceed (Dalton 2005, Orr et al. 2007, Stringer et al. 2006). This has been a very successful tool in securing fulsome input into projects and has complemented public forums. PAGs have played an important role in securing input and providing feedback on a project to community interest groups. In many cases projects that may have seemed controversial have moved forward without opposition simply through better communication and meaningful community input as part of the PAG.

Communication Lessons Learned

Communication with the Public

BARC is the primary group charged with communicating the HHRAP. They are a citizen group which allows citizens to speak to citizens about HHRAP progress and issues (Hartig et al. 1996, Law and Hartig 1993, Mehan III 2011, Stringer et al. 2006). BARC has enjoyed good leadership both at the staff and board levels over the years and as a result it has been a meaningful group to mobilize public and political interest. Part of BARC's success is it takes responsibility for implementing some educational and stewardship programs in addition to communication. It has become the place where private citizens can get involved with the HHRAP.

Engaging Politicians

The HHRAP Office and BARC often work together to inform political representatives at all levels of government (Marmorek and Peters 2001, Steelman and Hess 2009). New municipal councils are provided presentations on the HHRAP; MPs and MPPs similarly are visited after they have been elected. Perhaps one of the most important opportunities for engagement comes from an annual breakfast meeting hosted by BARC for MPs, MPPs and Mayors in the HHRAP watershed. This event has always been well attended with highlights of upcoming projects and programs requiring political endorsement outlined.

It is important to note that the BARC takes the lead role in political contact with municipal and parliamentary staff. The RAP Coordinator is called in to some meetings for information purposes. This separation of responsibilities allows the BAIT agency staff and RAP Coordinator to give and be perceived as giving science based advice untainted by particular political or community interests.

CONCLUSION

Participatory governance may seem like a lot more work, but consensus may be easier to achieve when there is open communication and direct involvement with all parties. Stakeholders in Hamilton Harbour own this RAP, because they created it and are still involved with implementation.

ACKNOWLEDGEMENTS

We would like to thank Ken Cruikshank, Vic Cairns, and Anne Redish for their inputs.

REFERENCES

[BARC] Bay Area Restoration Council. 2002. Toward Safe Harbours 2002 Report Card.

[BARC] Bay Area Restoration Council. 2003. Toward Safe Harbours. Progress Toward Delisting - Work Plan.

[BARC] Bay Area Restoration Council. 2004. Toward Safe Harbours. Progress Toward Delisting - Fish and Wildlife.

[BARC] Bay Area Restoration Council. 2005. Toward Safe Harbours. Progress Toward Delisting - Water Quality.

[BARC] Bay Area Restoration Council. 2006. Toward Safe Harbours. Progress Toward Delisting - Toxic Substances and Sediment Remediation.

[BARC] Bay Area Restoration Council. 2007. Toward Safe Harbours 2002 Report Card.

[BARC] Bay Area Restoration Council. 2010. Who We Are. http://www.hamiltonharbour.ca/about-who.htm Website retrieved October 25, 2010.

Bouchier, N.B. and K. Cruikshank. Nature's playground: environmental change and planning for pleasure on the Burlington Bay, 1860-2000. pp. 329-339. *In*: J-M. Delaplace, S. Villaret, and W. Chameyrat. [eds.] 2004. Sport et nature dans l'histoire. Sankt-Augustin, Academia Verlag, Germany.

Bouchier, N.B. and K. Cruikshank. 2008. Remembering the struggle for the environment: Hamilton's Lax Lands/Bayfront Park, 1950s-2008. Left Hist. 13(1): 106-128.

City of Hamilton, Public Works. 2009. 2010 Water, Wastewater & Storm Rate Book & Services Overview.

Cruikshank, K. and N.B. Bouchier. 2004. Blighted communities and obnoxious industries: constructing environmental inequality on an industrial waterfront, Hamilton, Ontario, 1890-1960. Environ. Hist. 9: 464-496.

[CWS and GLLFFAS] Canadian Wildlife Service, Environment Canada and Great Lakes Laboratory for Fisheries & Aquatic Sciences, Fisheries and Oceans Canada. 1990. A Plan for Restoration of Fish and Wildlife Habitat in Hamilton Harbour and Cootes Paradise.

Dalton, T.M. 2005. Beyond biogeography: a framework for involving the public in planning of U.S. Marine Protected Areas. Conserv. Biol. 19(5): 1392-1401.

[EC] Environment Canada. 2010. Cleaning Up Randle Reef in Hamilton Harbour - Part of the Government of Canada's Action Plan for Clean Water. http://www.ec.gc.ca/default.asp?lang=En&xml=17131147-75BB-4BFB-AE8C-12B5BEB13D1E Website retrieved November 2, 2010.

[FWHR] 1992. Fish & Wildlife Habitat Restoration in Hamilton Harbour and Cootes Paradise. Preliminary Scoping and Consultation.

[FWHRP] Fish & Wildlife Habitat Restoration Project. 1998. General Information Fact Sheet.

Hartig, J.H. and J.R. Vallentyne. 1989. Use of an ecosystem approach to restore degraded areas of the Great Lakes. Ambio. 18(8): 423-428.

Hartig, J.H. and R.L. Thomas, and R. Iwachewski. 1996. Lessons from practical application of an ecosystem approach in management of the Laurentian Great Lakes. Lakes Reserv. Res. Manag. 2: 137-145.

[HHRAP] Hamilton Harbour Remedial Action Plan. 1992a. Environmental Conditions and Problem Definition. Second Edition of the Stage 1 Report.

[HHRAP] Hamilton Harbour Remedial Action Plan. 1992b. Goals, Options and Recommendations, vol. 2 – Main Report. RAP Stage 2.

[HHRAP] Hamilton Harbour Remedial Action Plan. 2003. Stage 2 Update 2002. .

[HHRAP] Hamilton Harbour Remedial Action Plan. 2007. 2001–2005 Stakeholder Investments.

[HHTT] Hamilton Harbour Technical Team – Water Quality. 2007a. Hamilton Harbour RAP Water Quality Goals and Targets Review. Part 1: Response to the City of Hamilton's Proposed Wastewater System Upgrades. Summary Report.

[HHTT] Hamilton Harbour Technical Team – Water Quality. 2007b. Hamilton Harbour RAP Water Quality Goals and Targets Review. Part 1: Response to the City of Hamilton's Proposed Wastewater System Upgrades. Technical Appendix.

Hughes, R. Hamilton hungers for its harbour. p. NIE4-6. *In:* J. Poling [ed.]. 2002. Newspaper in Education: A revealing look at Hamilton Harbour. The Hamilton Spectator. Hamilton, Canada.

KMK Consultants Limited. 2007. City of Hamilton: Summary Report on Woodward Avenue WWTP Wastewater Treatment and Combined Sewer Overflow (CSO) Management Strategy. Water and Wastewater Master Plan.

Law, N. and J.H. Hartig. 1993. Public participation in Great Lakes Remedial Action Plans. Plan Can. March. 31-35.

Marmorek, D. and C. Peters. 2001. Finding a PATH toward scientific collaboration: insights from the Columbia River Basin. Conserv. Ecol. 5(2): Art. 8.

Mehan III, G.T. 2011. Water management in the collaborative mode. WE&T (Water Environment & Technology) Magazine. 23(2): Viewpoint.

Nagy, S. 2010. Conserver Society of Hamilton & District. http://www.conserversociety.ca/ Website retrieved November 8, 2010.

[OMOE] Ontario Ministry of Environment. 1974. Hamilton Harbour Study.

[OMOE] Ontario Ministry of Environment. 1975. Hamilton Harbour Study 1974.

[OMOE] Ontario Ministry of Environment. 1977. Hamilton Harbour Study 1975.

[OMOE] Ontario Ministry of Environment. 1978. Hamilton Harbour Study 1976.

[OMOE] Ontario Ministry of Environment. 1981. Hamilton Harbour Study 1977, vol. 1.

[OMOE] Ontario Ministry of Environment. 1985. Hamilton Harbour Technical Summary and General Management Options.

[OWRC] Ontario Water Resources Commission. 1964. Summary report of the industrial waste loadings discharged to Hamilton Harbour by the Bayfront Industries.

[OWRC] Ontario Water Resources Commission. District Engineers Branch. Division of Sanitary Engineering. 1967. Water pollution survey of sewer outfalls and tributary streams to Hamilton Bay.

Orr, P. and J. Colvin, and D. King. 2007. Involving stakeholders in integrated river basin planning in England and Wales. Water Resource Management. 21: 331-349.

Steelman, T.A. and G.R. Hess. 2009. Effective protection of open space: does planning matter? Environ. Manag. 44: 93-104.

Stringer, L.C. and A.J. Dougill, E. Fraser, K. Hubacek, C. Prell, and M.S. Reed. 2006. Unpacking "participation" in the adaptive management of social-ecological systems: a critical review. Ecol. Soc. 11(2): Art. 39.

Trotter, K. and J.D. Hall and V. Cairns. 1998. Fish & Wildlife Habitat Restoration Project. General Information Fact Sheet.

Vallentyne, J.R. Making Ecology Personal. pp. 9-17. *In:* A.D. Misener and G. Daniel [eds.] 1982. Decisions for the Great Lakes. A project of Great Lakes Tomorrow and the Purdue Foundation. Hiram, Ohio, USA.

Wong, J. 2009. Cootes to Escarpment Park System: Conservation and Land Management Strategy. Protecting Ontario's Greenbelt. Phase II Report.

CHAPTER **13**

As the Water Flows: Community Based Decision-Making and Participatory Planning for the Maumee Area of Concern, Ohio

Patrick L. Lawrence

Department of Geography and Planning, University of Toledo
Toledo, Ohio 43606
E-mail: patrick.lawrence@utoledo.edu

INTRODUCTION

Within many Great Lakes communities, the flow of water over the landscape in the form of watershed processes, land drainage, and the discharge of rivers and streams, represents an important aspect of the natural environment and defining characteristic of the interaction of humans with the land and water they are so dependent upon. Since 1987, local communities, along with state and federal agencies, have been engaged in efforts to address the complex mired of water quality concerns and related environmental issues within the Great Lakes Area of Concern for the Lower Maumee River and associated watersheds in northwest Ohio. Following from provisions of the Great Lakes Water Quality Agreement, this effort for the lower Maumee has been undertaken within the framework of the establishment of a Remedial Action Plan (RAP) approach as has been the common undertaking at the Great Lakes Areas of Concern (AOC) sites within Canada and the United States. However, at each AOC within the Great Lakes Basin, the RAP planning process and resulting approaches and outcomes have been site specific and unique, often reflecting local interests, the political jurisdictions, issues to be addressed, and perhaps most importantly, the presence and roles of the local communities and citizens.

Over the last 20 years a number of studies have reviewed the process and progress by which community based planning efforts have been undertaken to complete the necessary steps towards completion of remedial action plans (RAPs) in various Great Lakes Areas of Concern (AOCs) in both the United States and Canada. The most noteworthy of such reviews include Grima et al. (1983), Gurtner-Zimmermann (1995, 1996), Hartig et al. (1996), Krantzberg (2003), Krantzberg and Houghton (1996), and Hall et al. (2006) – all of which do an excellent job of assessing and highlighting the key issues and challenges all too common in attempts to create and maintain successful grass root structures for RAPs – and to which readers are referred to for additional background and discussion. In addition, several writers have examined the specific aspects of community organization and engagement within Great Lakes Areas of Concern, most notably Beeker et al. (1991), Beierle et al. (1999), and Kellogg (1998), in which each paper considers how the need for public participation was identified as a key element in addressing Great Lakes water quality issues in the context and structure of Areas of Concern and the problematic difficulties that, along with opportunities, were presented during the Remedial Action Plan process within local communities.

This chapter will examine the history, planning framework, successes and challenges of the RAP planning process within the Maumee AOC and reflect upon the fundamental barriers and opportunities present in the community, importance of leadership and innovation, and the driving forces needed to achieve the improvements necessary to return the rivers and streams to fishable, drinkable, and swimmable waters for our community. The key evolutionary steps or approaches to community engagement and involvement via a local RAP Advisory Committee will be discussed and conclusions drawn as to the elements of each step, changes that occurred as progress continued and also waned over the last 20 years, and the current organizational arrangements developed as a response to varied challenges and opportunities that emerged during the history of the Maumee RAP and for its continued future success.

MAUMEE AREA OF CONCERN

The difficulties in managing the entire Maumee Basin (a collective area of over 20,000 square kilometers covering the portions of three states: Ohio, Michigan, and Indiana) has long been recognized as a serious barrier to watershed management efforts within the basin, resulting in a long history of stalemate and struggles (Nelson and Weschler 1998). The lower Maumee River and local streams in NW Ohio were designated as a Great Lakes Concern in 1987 by the

International Joint Commission (IJC). The Maumee AOC (Fig. 13.1) comprised a land area of 2,000 square kilometers that consists of several watersheds within northwest Ohio that have streams directly discharging into the western basin of Lake Erie. The population of the Maumee AOC is approximately 450,000 with several major urban areas, but over 70% of the land area is used for rural agriculture (MRAC 2006). Over the past 20 years, a local partnership of citizens, government, business and non-governmental environmental interest groups have worked together as the Maumee RAP Advisory Committee to address a wide range of water quality issues facing the Maumee AOC. Several of the problems facing the rivers and streams of this area include: contaminated sediments; nutrient loadings; wetland loss; aquatic habitat degradation; bacteria; combined sewer overflows (CSOs); eutrophication; and non-source pollutants from rural runoff.

The Maumee AOC was originally identified as the area extending from the Bowling Green water intake near Waterville along the Maumee River downstream to Maumee Bay (MRAC 1990). The area includes direct drainage into the waters that are within Lucas, Ottawa and Wood counties. This includes Swan Creek, Ottawa River (Ten Mile Creek), Duck Creek, Otter Creek, Cedar Creek, Grassy Creek, and Crane Creek. In 1992, this area was extended to the east to include Turtle Creek, Packer Creek, and the Toussaint River. Heavy metals and organic chemical sediment contamination are what led to the lower Maumee River being classified as an Area of Concern (MRAC 1997). Also a noted concern was that the Maumee River contributes the largest tributary load of suspended sediments and phosphorus to Lake Erie, which would become an issue of focus for water quality planning within the AOC (MRAC 1997). In 2010, USEPA Great Lakes National Program Office (GLNPO) and Ohio EPA approved the further extension of the Maumee AOC boundaries so as to incorporate the full watershed units with the AOC in which water quality improvements and related restoration projects had been targeted and in some cases implemented. In this way efforts to address the beneficial use impairments assessed on the basis of watershed management units could be addressed within the Maumee AOC by the Maumee RAP and local watershed partners and agencies.

The Maumee RAP as an organization was created after the first public meeting in October 1987. It has grown and changed over the years, but has always been a public-private partnership working to restore the health of our area's waterways to fishable and swimmable conditions.The Maumee RAP involves a diverse cross-section of environmentally concerned businesses, industries, government agencies, non-profit organizations, educators, and citizens. The Maumee RAP Advisory Committee (MRAC) makes the official decisions for the organization and provides general program oversights. The MRAC as formed includes action groups (or sub-committees) that are integral

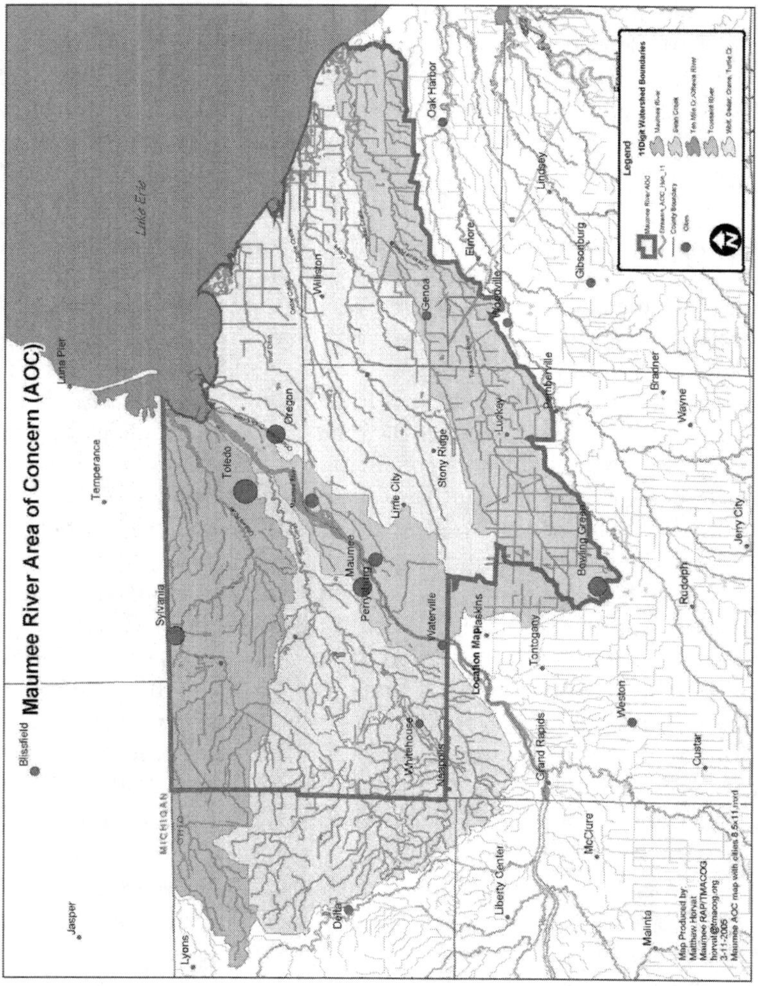

Fig. 13.1 Maumee Area of Concern, NW Ohio

Color image of this figure appears in the color plate section at the end of the book.

to the progress of the Maumee RAP. These action groups address specific issues that affect the Maumee AOC, such as open spaces, wetlands, agriculture, rural, and urban concerns. There have been two action groups utilizing a comprehensive watershed approach to improving Swan Creek and the Ottawa River. The Maumee RAP has also historically focused on public outreach and education.

The Maumee RAP planning process began on October 1, 1987, when the first public meeting was jointly held by the Ohio Environmental Protection Agency (EPA) – represented by Ohio statewide RAP coordinator – and the Toledo Metropolitan Area Council of Governments (TMACOG), who served as the local coordinator for the Maumee RAP. In 1988 the Maumee RAP Advisory Committee (MRAC) was officially formed. The 74-member MRAC included representatives from all levels of government, business and industry, universities and other interested individuals. Problem definition was completed in 1990 with the submission of the Maumee RAP Stage 1 Report (MRAC 1990) to Ohio EPA. Stage 1 of the RAP process was officially concluded in March 1992 when the Maumee RAP Stage 1 Report was reviewed and accepted by the IJC.

After Stage 1 was finished and the Advisory Board was abolished, the Maumee RAP Implementation Committee (MRIC) was formed to oversee all the restoration activities of Stage 2 that occurred throughout the Maumee AOC. In 1997 MRIC released a new Strategic Plan for the organization that utilized much of the same structure with a few minor differences including a few new action groups, a new logo, and a new simplified name for the organization, Maumee RAP Committee (MRIC 1997). The Strategic Plan classified action groups into three categories: issue, support or watershed. The MRIC was still the formal decision-maker for the organization that resided under the umbrella of TMACOG. The 1997 Strategic Plan also recommended the reformation of the Maumee RAP Advisory Committee (MRAC) to oversee several smaller action groups.

A 10-year Activities and Accomplishments Report was completed in 2002 setting the stage for identifying next steps toward restoration (MRAC 2002). The next planning step was when the Maumee RAP undertook an intensive and ambitious effort to create the Maumee AOC Stage 2 Watershed Restoration Plan. This plan combines the IJC requirements of a Stage 2, USEPA and Ohio requirements for a watershed action plan, with attention to the non point source management measures of the Ohio Coastal Management Plan, and consideration of Total Maximum, Daily Load (TMDL) and Natural Resource Damage investigations in the AOC. The plan underwent public review in November 2005 and a draft was submitted to Ohio EPA for review in early 2006 (MRAC 2006).

Over the two decades a great deal of information has been compiled and developed concerning the Maumee AOC. Some of the reports prepared by the Maumee RAP, TMACOG, and the Ohio EPA include Maumee RAP Investigation Reports completed in 1988 and 1989; Maumee RAP Stage I Investigation Report (MRAC 1990); Maumee RAP Recommendations for Implementation (TMACOG 1991); and Activities and Accomplishments in the Maumee AOC 1991-2001 (MRAC 2002).

The Maumee RAP Stage I Investigation Report (MRAC 1990) identified the environmental problems of the Maumee AOC circa 1990. This report also identified the known sources of the pollutants and the impairments resulting from these problems. Stage I is intended as the first of three stages in the development of the complete remedial action plan trilogy as required by AOCs by the International Joint Commission. This report identified 10 beneficial use impairments (BUIs) for the Maumee Area of Concern including degraded fish and wildlife populations, fish tumors, degradation of benthos, eutrophication, and loss of fish and wildlife habitat (MRAC 1990). Assessment of BUIs is the standard approach for considering the status of water quality and related environmental issues in an AOC and the mitigation of these, also known as "delisting" is the aim of restoration by the RAP process. A 74-member MRAC spent two years working on Stage I where public participation included numerous committee meetings (with members drawn from government agencies, local planning authorities, environmental experts, the private sector, and concerned citizens), public meetings, document reviews, and consultants with a wide range of stakeholders. The resulting Stage I would become a blueprint for addressing water quality concerns as determined by the BUIs and also set the framework for subsequent community engagement in the Maumee AOC.

Following the review and approval of the Stage I report by IJC and Ohio EPA, the MRAC was divided into eight subcommittees to address specific issues and the steps needed to organize efforts to undertake solutions to resolve key water quality concerns. These subcommittees involved representation from key stakeholders such as environmental regulatory, agencies, local governments, experts from area universities, and citizen organizations. The subcommittees reflected a consideration of both the identified BUIs from the Stage I report, but also the interest of stakeholders and the targeted efforts to address the most fundamental and significant issues impacting water quality in the AOC. Examples of some of the most active and successful subcommittees (in terms of eventual actions and activities) included dredge disposal, dumps and landfills, home sewage disposal, and public and industrial discharges. It should be noted that local commitment and leadership (often by key individuals, including representatives from local governments and citizen volunteers) would be key ingredients to the success of these subcommittees. The Maumee RAP Recommendations for Implementation (TMACOG 1991) was written as a comprehensive listing of the many issues and tasks that needed to be undertaken to restore the Maumee AOC. This report led to extensive sampling throughout the Maumee AOC and guided other research and community education projects. To this day, the current Maumee RAP continues to advocate and/or directly sponsor programs and activities to address the projects and issues outlined in the Maumee RAP Recommendations for Implementation.

In 1996 the members of the Maumee RAP began to consider the progress in respect to the activities outlined in the 1991 implementation report and with the 10th anniversary of the establishment of the Maumee RAP and the first public community meeting, the MRAC undertook an exercise to prepare a strategic plan. This plan would establish future goals, guide future activities, and set a structure intended to improve the ability of the MRAC to define accomplishments in the area of addressing BUIs within the AOC. The monthly meetings of the MRIC and various subcommittees worked to redefine and expand the intended goals and actions to be undertaken by the Maumee RAP. The results of this planning exercise emphasized the importance of the Maumee RAP as a partnership and that cooperation among all groups, parties and stakeholders within the Maumee AOC (including those not directly working or affiliated with the MRAC) was essential to success. The MRAC was charged with oversight and general operational direction of the Maumee RAP while the various subcommittees were to undertake programs and activities within their defined issue or focus area. Each action group stated a purpose statement, short and long term goals, activities to be undertaken, potential barriers to success, and help needed. Action groups were organized around four issues: agricultural runoff, dumps and landfills, open space and wetlands, and urban runoff. To assist in the operational activities and program of the Maumee RAP a finance action group and public outreach and education action group were also established. In acknowledging the growing interest in watershed planning in the United States, Ohio and the Great Lakes Basin, action groups were created to focus on area specific challenges within two of the most significant (and impaired) watersheds within the Maumee AOC: the Ottawa River and Swan Creek – both located largely within the City of Toledo. Each year the Maumee RAP, and the individual action groups would prepare lists of their activities and accomplishments, highlighting projects and initiatives linked to water quality concerns within their focus area and to the BUIs for the Maumee AOC. A full series of these annual reports can be examined at www.partnersforcleanstreams.org.

Many of these programs and activities, along with those projects conducted by others in the Maumee AOC community were highlighted in the activities and accomplishments report for the Maumee Area of Concern, 1991-2001 (MRAC 2002). This report represented the first attempt to list and summarize the many projects and efforts undertaken with the Maumee AOC, by the Maumee RAP and its committees, and other community partners. More than 300 items are listed within 24 different issues identified that cross the BUIs for the Maumee AOC. In addition, for each listing an action and activity are identified, the level of Maumee RAP involved determined, date of the activity or project, the BUIs affected and a summary description. The actions ranges from small scale one time projects, such as native grass seedings, to annual activities such as educational workshops, and to the assessment and mitigation of contaminated industrial landfills sites that involved millions of dollars of local, state and federal funding and multi jurisdictional involvement and regulatory actions. Although the report clearly highlighted a high level of activities and the completion of projects, one limitation was the inability to easy link the completion of the projects and activities to efforts that were addressing or resolving BUIs.

Over the last decade of projects, planning and public outreach a number of priorities were identified the required targeted initiatives by the Maumee RAP and local partners. Notable successful examples include the Give Water a Hand program to provide information and tip cards to businesses and local residents on the importance of conserving and protecting water resources, supporting regional efforts towards creating a storm water management standards manual developed by TMACOG and Lucas County, stream dam mitigation projects, inventorying and assessing rural septic systems and their impacts on stream and ditch water quality, and assisting with various opportunities to address contaminated river sediments through proposed projects for Great Lakes Legacy Act (GLLA) funding. The Maumee RAP, and later Partners for Clean Streams (PCS) also oversaw the greatly successful Clean Your Streams annual river, stream and lake clean up activity undertaken each September in the Maumee AOC as part of the Ohio Coast Weeks Program and the International Coastal Cleanup event. Clean Your Streams grew rapidly over the years to become the largest clean up event the Ohio Coastal Weeks with more than 600 volunteers removing garbage and debris from over 50 sites within the Maumee AOC and Toledo area. Additional Maumee RAP planning and implementation efforts were targeted at addressing the loss of wetlands and developing a community based inventory and assessment tool for stemming impacts to wetlands from new development within the Lucas County, Ohio portion of the Maumee AOC (Lawrence et al. 2004, Lawrence 2005).

In January 2006, the Maumee Remedial Action Plan (RAP) Committee submitted a State II Watershed Restoration Plan for the Maumee River Great Lakes Area of Concern (AOC) to the State of Ohio for review and endorsement (www.partnersforcleanstreams.org). The plan was created in order to fulfill the requirements, needs and/or use of five water quality programs, including: Ohio Department of Natural Resources Watershed Coordinator Program (ODNR 2006, OEPA 2003); Ohio EPA Great Lakes RAP Program and state RAP delisting targets (OEPA 2005); Ohio DNR Coastal Non-point Source Pollution Control Program (ODNR 2007); Ohio EPA Total Maximum Daily Load Program; and US Fish &Wildlife Service Natural Resources Damage Program (ODNR 2000, OEPA 2003).

This plan is intended to serve as a comprehensive regional water quality improvement plan in order to provide a single resource for all jurisdictions, agencies, organizations, and individuals who are working to restore the waterways within the Maumee AOC. The plan includes information and maps regarding: Great Lakes Areas of Concern and Remedial Action Plans; an environmental background on the Maumee AOC (hydrology, geology, eco-regions, landuse, etc.); information for the six 11-digit hydrologic units; and one large river unit that comprise the Maumee AOC (MRAC 2006).

Watershed Projects Tables (WPTs) were also prepared that contain detailed project lists for each major watershed and the WPTs were organized to facilitate the delisting of areas and/or issues for the Maumee AOC. The WPTs include: the causes and sources of water quality concerns; projects; potential project partners; funding sources; timeline; status; performance/Environmental Measures; Hydrological Unit Code (HUC)/Stream Segment Addressed; and the Beneficial Use Impairment (BUI) affected (MRAC 2006).

The plan received "Full Endorsement Pending" status from the State of Ohio and it is anticipated that the plan will be fully endorsed with the completion of a Coastal Nonpoint Source Pollution Management Measures section (Lawrence 2011). As the Maumee RAP Advisory Committee begins the implementation of this plan, they are evaluating which projects will best lead them to delisting, as well as how PCS can establish itself as a new organization with community support. This strategic and organizational planning is expected to be captured in Volume 3 of the plan, which is expected to be submitted to Ohio EPA along with other requested changes in 2012.

During this planning and review stage the Maumee RAP Advisory Committee continued to develop and implement watershed projects and has maintained its community outreach efforts such as:

- Ecological and Human Health Risk Assessment for Duck and Otter Creeks;
- Highland Park Dam Decommissioning and Riparian Enhancement;
- Wetland and Riparian Inventory and Restoration Plans for Swan Creek and the Ottawa River;
- Partnering for Clean Streams Scout Patch Program;
- Get the Lead Out!; and
- Clean Your Streams.

While the above projects were underway, many other activities were also occurring. Research was being conducted and additional issues were being addressed. The Maumee RAP philosophy, process, and program all rely on the power of partnerships. Without community partners the Maumee RAP would not succeed and the Maumee AOC would not be restored. These occurred from both within the RAP structure by the action groups and outside through community partners. For more details on these various and projects the reader is referred to the website of Partners for Clean Streams (www.partnersforcleanstreams.org).

After months of discussion and a process of considered organization and capacity review, the Maumee RAP Committee determined in early 2007 that their best path forward for efficiency, effectiveness, and sustainability would be to form their own non-profit 501(c)3 organization and to leave the umbrella of the Toledo Metropolitan Area Council of Governments (TMACOG), which has been the "local host" for the Maumee RAP since its formation in 1987. In March 2007, Partners for Clean Streams (PCS) Inc. was officially created as the new "umbrella" organization for various water quality programs and initiatives within the Maumee AOC. In January 2008, the Maumee RAP Committee completed its merger into the PCS organization to become the Maumee RAP Advisory Committee. PCS was formed with an interest in supporting local and regional water quality improvements in the metro-Toledo area. PCS aims to strive for abundant open space and a high quality natural environment; adequate floodwater storage capacities and flourishing wildlife; stakeholders who take local ownership in their resources; and rivers, streams, and lakes that are clean, clear and safe (www.partnersforcleanstreams.org).

Since January 2008, the Maumee RAP Advisory Committee (MRAC) has been nestled within the committee structure of PCS, along with a Development Advisory Committee (DAC). Together they serve to assist the PCS Board of Directors to deliver a wide range of programs, projects, and community outreach efforts associated with water quality improvements that have been of long standing interest to the Maumee RAP community. Community organizations and citizen volunteers comprise the memberships of PCS and its standing committees: MRAC and DAC.

PCS continues to assist other community partners with projects such as the Ottawa River Dam Decommission and Stream Restoration Project; Ottawa River Wetland Inventory; and efforts of the Toledo Rain Garden Initiative, supporting and advocating for additional water quality monitoring and aquatic ecosystem assessments, including the completion of Total Maximum Daily Load (TMDL) studies for rivers and streams in the Maumee AOC by Ohio EPA, numerous wetland and riparian habitat inventories and restoration projects, public education and outreach efforts, community watershed tours and river canoe trips, engaging local and regional environmental non-government organizations such as the Black Swamp Conservancy, Toledo Metro Parks, the Ohio chapter of the Nature Conservancy Duck and Otter Creeks Partnership, the Boy Scouts of America, and many others. PCS has also continued the successful long history of cooperation and engagement by numerous local and regional government agencies in addressing water quality issues within the AOC, including TMACOG, Lucas County, City of Toledo, Lucas County, City of Sylvania, Sylvania Township, City of Oregon, Wood County, Ottawa County, Washington Township, and the Village of Ottawa Hills. Since 2008 many of these community based efforts have focused on habitat restoration within the Maumee AOC, leading to a grant from the 2010 Great Lakes Restoration Initiative (GLRI) awarded by the US EPA Great Lakes National Program Office to PCS that will be utilized to conduct a US $1.3 million wetland, riparian and stream corridor restoration project within the Ottawa River/Ten Mile Creek Watershed of the Maumee Area of Concern.

DISCUSSION

The history of the local remedial action planning efforts in the Maumee AOC reveals the challenges and complex nature of addressing the major water quality concerns and the struggle within a community to determine the appropriate and most effective means of organizing the citizen based participatory approach. It appears that the evolution in such efforts can be characterized by the presence of five steps of focusing local attention and actions within the Maumee AOC community to first understanding the problems and then determining the best pathways towards solutions.

Is Everyone in the Room?

With the initial formation of the Maumee RAP in 1987, a strong effort was made to develop an inclusive planning approach to ensure that all stakeholders, partners and voices were heard and that community groups, agencies and local citizens were able to participate in the formation of the guiding principles and studies to better understand the range of water quality concerns that existed within the Maumee Area of Concern. The initial MRAC included 74 members with representation from all sectors and key stakeholders: government agencies, environmental organizations, industries and the private sector, concerned citizens, experts from consulting firms and universities, and many others. For the first several years of its existence a high level of interest and support existed for the Maumee RAP as the local committee engaged the participation of the public in key essential decision making in regards to identifying and assessing the key issues that needed to be addressed and fostered a strong sense of community involvement and stewardship of the Maumee AOC. The enthusiasm and energy behind the creation of the MRAC, and the strong recognition of the need for concrete actions and solutions to address water quality concerns in the rivers and streams within the Maumee AOC, would drive much of the early success and achievements obtained by the MRAC and its community partners.

The inclusiveness and sense of community embodied by the Maumee RAP progress and the dedicated involvement of many key organizations, agencies and citizens was a remarkable undertaking and achievement of the Maumee RAP.

There are Some Big Problems Here!

With the preparation of the Maumee RAP Stage I report and Strategic Plan, it become very apparent that the vast geographic size of the AOC and the complex diverse nature of the beneficial use impairments were both presenting great challenges in planning and decision-making by the local community. In addition, the lack of basic science, monitoring, and baseline data (such as water quality measures and parameters) would mean that any efforts to immediately and swiftly resolve beneficial use impairments (BUIs) would be daunting, if not impossible. Yet the community interest and support for the Maumee RAP, although still very new with a low profile by the general public and not as well understood by many citizens, would place pressure on addressing immediate concerns and making progress on some fronts – in fact on any front – to tackle the underlying water quality concerns driving the BUIs.

The continued involvement and engagement of key local partners and stakeholders continued and the completion of the Stage I report was seen as a key "watershed" event by the Maumee RAP. The level of involvement by professionals and scientists was especially significant as it represented a continued commitment by those whose expertise and experience was essential to translating scientific data and knowledge into decision making regarding actions and solutions to address the BUIs. There was also a growing and continued effort to further the engagement of citizens and other community organizations through outreach and education activities to address the need for better understanding of the issues and challenges facing the Maumee AOC and RAP process and to reach a wider diverse representative of stakeholders drawn from various local communities and existing organizations beyond those who were initially involved with the establishment of the Maumee RAP and the completion of the Stage I report.

As is common in many planning exercises the initial excitement and enthusiasm stemming from the completion of the Stage I report would soon be met with the realization of the challenges ahead facing the Maumee RAP and local community on how to achieve the lofty goals set to "delist" the individual BUIs and by collective actions move the Maumee from an Area of Concern. Although many successful projects and activities, many noted accomplishments would be achieved, and priority actions undertaken that would sustain interest and involvement in the Maumee RAP for the coming years. Such activities were often focused in several of the Maumee RAP action groups that managed to involve and engage a strong and diverse representation of local stakeholders committed to making clear progress towards addressing key challenges and concerns, especially in the areas of rural runoff and dumps and landfills.

We Must All Work Together

During the decade following the completion of the Stage I reports, attempts were made to address the most serious and persistent problems associated with BUIs. One example was the long effort by many different partners, federal and state agencies, and local governments to address the sources of contaminants from numerous industrial landfills along the Ottawa River in the City of Toledo. At the cost of millions of dollars in federal and state funding these efforts during the 1990s led to the remediation of the most significant of these sites. During this time the Maumee RAP Committee had organize an action group on Dumps and Landfills whose active members continued to advocate and lobby for additional efforts and focus on these sites due to their ongoing contributions of PCBs and other toxic contaminants into the Ottawa River. It also became very apparent that with the organizational structure and involvement of those with the Maumee RAP, there existed a certain capacity and ability to undertake the necessary steps to achieve successes with targeted BUIs or watersheds within the Maumee AOC, although these efforts needed to extend beyond the direct oversight or involvement of the Maumee RAP to include those agencies, organizations and individuals with responsibilities, authorities and the resources (especially financial and regulatory) to tackle the most important and often most challenging and complex issues that faced the Maumee AOC.

In addition to the aforementioned industrial landfills issue, the Maumee AOC was characterized by a number of long term, regional, and complex water quality issues linked to BUIs that were the result of a multitude of causes and sources for water pollutants stemming from a variety of point and non-point sources. The urbanization portion of the Maumee AOC, especially within the City of Toledo, was plagued by aged and degraded sewer and water treatment services resulting in all too frequent combined sewer overflow (CSO) discharges of significant volumes of untreated sanitary and storm sewer waters into the Ottawa River, Swan Creek and Maumee River. This issue would require considerable political willpower, financial resources, regulatory and legal means, and commitment of local, regional, and state governments – way beyond the capacity and role of the Maumee RAP. In this case such involvement and activities in many ways superseded the role and involvement of the Maumee RAP where the fundamental actions needed to address the problem were more a function of partnerships and collaboration rather then direct activities by the Maumee RAP. The problem of CSOs within the City of Toledo would ultimately involve the negotiation of a legally binding consent agreement between the City and US EPA requiring considerable effort (and at the cost of more than US$400 million by the local taxpayers and utility ratepayers) to resolve the infrastructure improvements by the creation of another local partnership – the Toledo Waterways Initiative.

Solutions are the Solution

By 2001 it had become very apparent that the preceding 14 years had resulted in an expansive and impressive number of successful projects and efforts within the Maumee AOC by a large number of key groups, agencies, organizations and volunteers. The activities and accomplishments report by the Maumee RAP Committee highlighted hundreds of these efforts and the many dollars and volunteer hours committed to undertaking the wide range and scope of activities and initiatives within the Maumee AOC. Although a large number of activities and projects continued to be proposed, developed, implemented and completed within the Maumee AOC by the MRAC and other community partners, there was a growing concern or frustration rising over the need to move beyond planning and organization to actual solutions linked to addressing specific solutions that would resolve BUIs. The MRAC and actions groups began to expand its commitment and involvement from public outreach and education, facilitating and coordinating AOC priorities and activities, and supporting the direct work of other community organizations and agencies, to undertake more solution based efforts.

Examples would include efforts by the rural runoff action group to work with local farmers on establishing conservation buffer strips on actively cultivated farmland to reduce the input of soil from the erosion off farmland, bringing together local governments in the urban communities of the Maumee AOC to consider improvements to storm water management practices, and the work of open space and wetlands action group to secure a state project grant to complete a classification and inventory of wetlands with Lucas County through the use of remote sensing and geographic information systems (GIS) technologies (Lawrence et al. 2004). The level of involvement and activity by these action groups reached a renewed interest and effort as members became directly involved with projects from concept to design and completion. The results of these efforts would result in improvements directly linked to water quality issues within the Maumee AOC and drove interest and enthusiasm towards a move forward in successfully addressing BUIs and hope for continued progress. Alas this progress was not able to be sustained beyond the completion of the individual projects which had driven the continued public participation in the efforts of the MRAC and action groups and as these successful projects reached their natural conclusions, continued involvement and commitment by many MRAC and action groups would begin to wane. The general consensus was that after over 20 years, marked by much success and the completion of a large number of projects and activities, which collectively had achieved great progress within the Maumee AOC, the community was disappointed that measured achievements in the solutions and improvements of BUIs had yet to be gained and more fully achieved.

Let's be More Focused and Result Orientated

As the Maumee RAP approached its 20th anniversary in 2007, it had become apparent that for all the successes and achievements over the previous two decades, the fundamental goal of removing beneficial use impairments was falling short. It was no fault of the efforts, funding, partnerships, commitment, and involvement of stakeholders and volunteers, but instead a reflection of the acknowledgement of the enormity of the endeavor first expressed early during the existence of the Maumee AOC – the vast size of the area and complexity of the problems underlying the BUIs. The work on the Stage II/Watershed Restoration Plan and the formation of Partners for Clean Streams (PCS) were both in part driven by – and the result of – an effort to reformulate a more focused and strategic approach to planning efforts to address the BUIs within the Maumee AOC. These efforts were also greatly influenced by renewed and revised orientations by US EPA GLNPO and Ohio EPA to make the Great Lakes AOC and RAP progress more goal driven towards delisting AOCs by better understanding, evaluating, addressing and where possible eliminating the BUIs.

Then in 2005, Ohio EPA released comprehensive directions and guidance towards delisting BUIs by establishing set scientific targets for each BUI that if attained by the evidence of quantifiable measures would allow for delisting of BUIs. This event would in many respects revitalize the local community towards considering the means by which the Maumee RAP could identify specific goals to reach in order to begin to more directly address specific BUIs and facilitate the necessary projects and activities to mitigate BUIs to the point at which their status would led towards delisting. In addition, US EPA and Ohio EPA in assessing progress and barriers towards success with the Great Lakes Areas of Concern, would begin to encourage innovations in addressing BUIs, the functional organizational structures for AOCS, with renewed funding opportunities for operations and projects tied to BUIs, and generally support and direct RAPs, including the Maumee, to focus on those activities and projects that could be directly measured and tied to resolving BUIs. In the Maumee AOC this change and challenge to rethink approaches would result in moving to a mindset and organizational structure that would better facilitate actions directed at specific projects and in watersheds.

From 2005 to 2007, the MRAC downsized to a more lean and functional entity and discarded the long standing approach and efforts around action groups to have the MRAC serve more as an advisory function to other existing groups and organizations already in place working to address water quality issues within the Maumee AOC. It would be the role of the MRAC to link these projects and activities to BUIs and support and direct the focused efforts within the Maumee AOC towards these issues. In 2007, in order to create the necessary organizational structure, and flexibility to undertake this new role, the MRAC became a committee of Partners for Clean Streams (PCS). MRAC would transform into the role of public participation and input into the directions and activities of PCS that assumed the day to day operations and annual programs and activities formerly the responsibility of the MRAC. PCS would become the legal and financial agent for the MRAC and as a 501(c)3 non-profit organization PCS had the capacity to secure additional financial resources to support programs and initiatives with the Maumee AOC, including public outreach and education, environmental improvements, habitat restoration, and water quality monitoring. PCS would work in association with TMACOG and the Duck and Otter Creeks Partnership to complete a watershed restoration plan for the Maumee AOC, which would also take the form of a Stage II report. From 2007-2009 PCS secured a grant from the Joyce Foundation to undertake three urban restoration projects within the Maumee AOC. In recent years, PCS has continued to foster community partnerships, engage local groups in undertaking projects directed at BUIs within the Maumee AOC, supported the ongoing efforts by other organizations and agencies to address water quality concerns, and continued to raise the public awareness and involvement of the varied stakeholders who have interest in – or support for – the collective efforts towards delisting of the BUIs for the Maumee AOC.

CONCLUSION

As discussed by Downs (1972), the challenge of maintaining public interest (and the resulting community involvement and commitment) needed to address environment concerns is a real and serious barrier to achieving long term sustained success and achievement aimed to achieve significant results. The questions raised by Downs (1972), and subsequently observed in many community and volunteer based environmental organizations including with the Maumee RAP, is how to maintain and in fact foster, consistent progress towards measurable success in the light of the time and effort required for such achievement resulting in a loss or decline in the participation so necessary for such locally based community efforts. The history of community involvement and participation in the RAP process within the Maumee AOC can be characterized by the evolution of aims and approaches to engage the community and the range of stakeholders in attempts to better understand and address the fundamental water quality concerns and issues that result in the beneficial use impairments identified and addressed for the rivers and streams within the Maumee AOC.

As the 25th anniversary of the 1987 amended Great Lakes Water Quality Agreement - which designated the Great Lakes Area of Concern - is soon approaching an opportunity presents itself to reflect on the experiences, challenges and opportunities represented by the efforts and accomplishments of individual Great Lakes RAPs within the AOC and the collective history and lessons learned. The experience of the Maumee RAP as a forum of community based participatory planning and governance provides some reflection and experience, highlighted by its location as the community where the 1987 GLWQA was first signed (in Toledo in November 1987) and where one of the initial local community Great Lakes RAP organizations was formed. The steps or approaches undertaken by the Maumee RAP community over the last few decades reflect how difficult and overwhelming the challenges have been to consider - and make efforts towards - resolving the complex numerous water quality challenges that exist with an Area of Concern. The efforts have also shown a natural evolution and progression in the dedication and commitment of a community represented by agencies, organizations and citizens that can make progress towards a collective and focused efforts to again return our rivers and streams to a "drinkable, fishable and swimmable" state.

The flow of water has remained a constant natural process within the Maumee Area of Concern as the landscape leading to the drainage of these waters continues to be impacted and modified by a wide range of human activities and land uses and the need to engage and involve communities in a process of local effort and outreach to preserve and protect the valuable water resources so critical to sustaining life, the environment, and future of the watersheds, rivers and streams of the Great Lakes. The ongoing flow of water has also continued to drive the necessity for dedicated planning approaches in these local communities that watch the seasonal, annual and long term drainage of water as an essential life blood deemed so important to the sustained subsidence and survival as a collective society and as individuals with the Maumee AOC.

ACKNOWLEDGEMENTS

The projects, programs and activities discussed in this chapter have been undertaken by the Maumee RAP Advisory Committee (MRAC) and its predecessors (1987-2007) and the Partners for Clean Streams Inc (PCS), Board of Directors, Executive Director, volunteers and community partners (2007-present). The preparation of the Maumee AOC Stage II/Watershed Restoration Plan was undertaken with direction from the Maumee RAP Advisory Committee (MRAC) and volunteers serving on a Plan Development Team lead by Cherie Blair, Ohio EPA. Additional support for the plan preparation was provided by TMACOG, US EPA Great Lakes Program Office, Ohio EPA Division of Surface Water, and the Ohio Great Lakes RAP Program. The author has served as a member of the Maumee RAP Advisory Committee (2000-present), Maumee RAP Stage II Report Plan Development Team(2005-2006), and since 2006 has served as President, Board of Directors of Partners For Clean Streams Inc. The ideas expressed in this chapter are those of the author and do not necessary reflect those of the PCS, MRAC, Ohio EPA, or other agencies, individuals, and organizations cited in this acknowledgement.

REFERENCES

Beeker, J. and G. Studen, and L. Stumpe. The Cuyahoga Remedial Action Plan Coordinating Committee: A Model for Building Community Ownership of a WatershedRestoration Plan. *In:* A.A. Jennings and N.E. Spangenberg [eds.]. 1991. Surface and Ground Water Quality: Pollution Prevention, Remediation and the Great Lakes. American Water Resources Association, Bethesda, MD. USA.

Beierle, T.C. and D.M. Konisky. 1999. Public Participation in Environmental Planning in the Great Lakes Region. Resources for the Future. Washington, D.C., USA.

Downs, A. 1972. Up and Down with Ecology – The Issue Attention Cycle. Public Interest 28: 38-50.

Grima, A.P. Analyzing Public Inputs to Environmental Planning: A Summary and Discussion of Public Involvement in Great Lakes Management. *In:* G.A. Daneke, M.W. Garcia and J.D. Priscoli [eds.]. 1983. Public Involvement and Social Impact Assessment. Westview Press. Boulder, Colorado, USA.

Gurtner-Zimmermann, A. 1995. A Mid-term Review of Remedial Action Plans: Difficulties with Translating Comprehensive Planning into Comprehensive Action. Journal of Great Lakes Research 21(2): 234-247.

Gurtner-Zimmermann, A. 1996. Analysis of Lower Green Bay and Fox River,Collingwood Harbour, Spanish Harbour, and the Metro Toronto and Region Remedial Action Plan (RAP) Process. Environmental Management. 20(4): 449-459.

Hall, J.D. and K. O'Connor, and J. Ranieri. 2006. Progress Towards Delisting a Great Lakes Area of Concern: The Role of Integrated Research and Monitoring in the Hamilton Harbour Remedial Action Plan. Environmental Monitoring and Assessment 113: 227-243.

Hartig, J. and R.L. Thomas, and E. Iwachewski. 1996. Lessons from Practical Application of an Ecosystem Approach in Management of the Laurentian Great Lakes. Lakes and Reservoirs: Research and Management. 2: 137-145.

Kellogg, W.A. 1998. Adopting an Ecosystem Approach: Local Variability in Remedial Action Planning. Society and Natural Resources. 11(5): 465-484.

Krantzberg, G. 2003. Keeping Remedial Action Plans on Target: Lessons Learned from Collingwood Harbour. Journal of Great Lakes Research 29(4): 641-651.

Krantzberg, G. and E. Houghton. 1996. The Remedial Action Plan That Led to the Cleanup and Delisting of Collingwood Harbour as an Area of Concern. Journal of Great Lakes Research 22(2): 469-483.

Lawrence, P.L. 2011. Achieving Teamwork: Linking Watershed Planning and Coastal Zone Management in the Great Lakes. Coastal Management 39: 57-71.

Lawrence, P.L. A Watershed Restoration Action Strategy (WRAS) for Open Space and Wetlands in the Maumee Remedial Action Plan (RAP), Ohio. pp. 36-45. *In:* G.A. Tobin and B.W. Montz [eds.]. 2005. Papers of the Applied Geography Conferences, vol. 28. Applied Geography Conferences Inc., Kent, Ohio, USA.

Lawrence, P.L. and K. Czajowski, and N.Torbick. Policy Implication of Remote Sensing in Understanding Urban Environments: Developing a Wetlands Inventory for Community Decision-Making in the Maumee Area of Concern, Ohio. pp. 23-36. *In:* R. Jensen, J. Gatrell, and D. McLean [eds.]. 2004. Place, Meaning and the Politics of Cities: Alternative Methodologies for Using Geo-Spatial Technologies to Understanding Urban Environments. Springer Verlag Press, London, United Kingdom.

Maumee Remedial Action Plan Advisory Committee [MRAC]. 1990. Maumee River Remedial Action Plan: Stage I Investigation Report. Ohio EPA Northwest District Office, Bowling Green, Ohio, USA. (www.partnersforcleanstreams.org)

Maumee Remedial Action Plan Implementation Committee [MRIC]. 1997. Maumee River Remedial Action Plan: Strategic Plan. Ohio EPA Northwest District Office, Bowling Green, Ohio, USA. (www.partnersforcleanstreams.org)

Maumee Remedial Action Plan Committee [MRAC]. 2002. Activities and Accomplishments in the Maumee Area of Concern 1991-2001, Maumee RAP, Toledo, Ohio, USA. (www.partnersforcleanstreams.org)

Maumee Remedial Action Plan Committee [MRAC]. 2006. Maumee Area of Concern Stage II/Watershed Restoration Plan, Maumee RAP, Toledo, Ohio, USA. (www.partnersforcleanstreams.org)

Nelson, L.S. and L.F. Weschler. 1998. Institutional readiness for integrated watershed management: The case of the Maumee River. The Social Science Journal. 35: 565-576.

Ohio Department of Natural Resources [ODNR]. 2000. Ohio Coastal Nonpoint Pollution Control Program Plan. Division of Soil and Water Conservation, Columbus, Ohio, USA.

Ohio Department of Natural Resources [ODNR]. 2006. Guidance for Watershed Projects to Address Ohio's Coastal Nonpoint Pollution Control Program. Division of Soil and Water Conservation, Columbus, Ohio, USA.

Ohio Department of Natural Resources [ODNR]. 2007. Combined Coastal Management Program and Final Environmental Impact Statement for the State of Ohio. Office of Coastal Management. Sandusky, Ohio, USA.

Ohio Environmental Protection Agency [OEPA]. 2003. A Guide to Developing Local Watershed Action Plans in Ohio. Division of Surface Water, Columbus, Ohio, USA.

Ohio Environmental Protection Agency [OEPA]. 2005. Delisting Targets for Ohio Areas of Concern. Division of Surface Water, Columbus, Ohio, USA.

Toledo Metropolitan Area Council of Governments [TMACOG]. 1991. Maumee River Basin Area of Concern Remedial Action Plan: Recommendations for Implementation. TMACOG, Toledo, Ohio.

CHAPTER 14

Remedial Action Plan: A Case Study of the Presque Isle Bay Area of Concern

Lori Boughton

Pennsylvania Department of Environmental Protection,
Northwest Regional Office, 230 Chestnut Street, Meadville, PA 16335
E-mail: lboughton@state.pa.us

INTRODUCTION

Located on the southern shore of Lake Erie in Pennsylvania, Presque Isle Bay is formed by a seven-mile long, re-curved, "flying" sand spit named Presque Isle (Fig. 14.1). This peninsula is home to the most visited park in Pennsylvania, Presque Isle State Park. To the south, the City of Erie and a portion of Millcreek Township border the 1,505 ha bay. The Presque Isle Bay drainage basin is about 6,475 ha in area, and is primarily urbanized with heavy manufacturing industries coexisting within residential and commercial neighborhoods. The geography and geology of the bay make it a natural "settling" basin for solids and pollutants entering the bay and generally become entrapped in the sediments.

In 1984, local fishermen reported the presence of external sores and lesions on brown bullhead catfish caught in Presque Isle Bay. These reports served as the catalyst for local citizens to band together with the common goal of restoring and protecting the "crown jewel" of Pennsylvania's piece of the Great Lakes. It was largely due to their efforts that Presque Isle Bay became the 43rd Area of Concern (AOC) under the Great Lakes Water Quality Agreement and was the first American AOC to receive the Recovery Stage designation in 2002.

The Great Lakes Water Quality Agreement does not specify how to involve the public nor does it define who constitutes the "public". Annex 2 merely

Fig. 14.1 Aerial view of Presque Isle Bay
Color image of this figure appears in the color plate section at the end of the book.

instructs that the public be "consulted in all actions undertaken pursuant to this Annex" (United States and Canada 1978). This flexibility allows each AOC to tailor involvement in a way that best meets the needs of the local community and engage those most impacted by the environmental conditions. A review of the AOC program in 1994 concluded among other things that "the process of involving stakeholders and securing broad-based support is at least as important as the technical and scientific aspects of RAPs" (Hartig 1997). This was the case for the Presque Isle Bay AOC, where an engaged and active citizenry, in concert with local and state government, accelerated action and led to consensus decisions and positive environmental outcomes. Community participants and government officials came to recognize that the AOC designation offered the opportunity for long-term restoration planning that would leave a sustainable legacy. Presque Isle Bay's restoration story illustrates how regional collaborations and participatory decision-making necessitated by local environmental and economic issues can work successfully.

FROM DEGRADATION TO RESTORATION: RECOUNTING LOCAL KNOWLEDGE

The City of Erie, founded in 1792, grew around Presque Isle Bay and its Lake Erie harbor. As was the practice in the 18th and 19th centuries, much of the wastewater from the City of Erie's industrial and domestic sources was discharged directly to the bay or its tributary streams. Urban streams in Erie were looked at more as sewers than natural resources, and much of the bay's largest tributary, Mill Creek, was converted to an enclosed combined sewer (i.e., Mill Creek Tube). At the same time, the bay was the primary drinking water source for the city. Following a typhoid outbreak in the early 1900s, the city extended its water intake pipe out into Lake Erie and constructed its first wastewater treatment plant. Upgrades to the original plant to install primary and secondary treatment began in the 1930s (Fig. 14.2) and continued through the early 1980s. In spite of these improvements, problems remained as a result of discharges to the bay and lake from combined sewer overflows, unregulated industrial releases, and the treatment plant itself. During heavy storm events, raw sewage flowed directly into the bay, drawing the attention of bay front businesses and raising concerns among local citizens.

Fig. 14.2 Erie's Wastewater Treatment Plant, 1940. Image courtesy of John L. Scott, Scott Historical Aerial Photos

The Great Debate: To List or Not to List

Erie has a long history of citizen involvement in local decision-making. Advisory committees in one form or another have played an integral role in water quality issues, particularly related to Presque Isle Bay. One of the first of these groups was the Erie County Environmental Coalition (Coalition), which began meeting in 1983. The Coalition consisted of representatives from a diverse group of stakeholders, including the League of Women Voters, Benedictine Sisters, Erie County Sportsmen Association, and Eastside Federation of Polish American Societies. Eventually, the Lake Erie Sierra Club, Presque Isle Audubon Society, American Lung Association, Zero Population Growth, and Earth Day Coalition also became members. In the mid-1980s, Coalition members began traveling to other Great Lakes states to share the concerns they had and to collect stories and ideas from other parts of the basin. The Coalition was responsible for initiating the actions that led to the petition to the United States Department of State to designate Presque Isle Bay as an AOC.

The story, however, is not that simple. Along with the urging of local citizens to designate the bay as an AOC, other events were taking shape that would align the priorities of citizens, regulators, and the business community and eventually bring them together with a common goal of restoring the bay. Like many other Great Lakes cities, the 1980s brought challenges to Erie in terms of revitalizing its economy and addressing environmental issues. Industry had moved away from the bay front and the number of ships visiting Erie had greatly decreased, leaving the area neglected and in disrepair. The City's wastewater infrastructure was in need of major upgrades and the bay itself was considered too contaminated for recreational uses. These three factors and the events that followed illustrate the depths to which a motivated group of citizens can affect positive and sustainable change.

In 1986, Great Lakes United (GLU), an international environmental citizens group, working with the Coalition, held one of its 19 citizens' hearings on pollution problems in the Great Lakes in Erie. After hearing from Coalition members and others, GLU adopted a resolution for the International Joint Commission (IJC) to designate Presque Isle Bay as an AOC. Pennsylvania's regulatory community opposed this designation citing insufficient water quality evidence to support it.

Local government officials and businessmen also opposed the AOC designation, fearing it would hamper development efforts and private investments in the bay front. Seeing a resurgence of waterfront growth in cities like Baltimore and Boston, local companies and developers saw the potential for Erie's bay front. From their viewpoint, future economic stability was contingent on the tourism and recreation industry. Planned projects in Erie included marinas, townhouses, condominiums, hotels, commercial shops, and a maritime museum requiring an estimated US $225 million private investment. Plans were underway to replace railroad tracks with a parkway connecting the city and bay front with south, west, and east bound interstates. Local leaders worried that labeling the bay an AOC was a very public and negative statement about the bay's environmental condition that would harm existing businesses and dissuade potential bay front investors. Already concerned with the need to raise funds for extremely expensive wastewater infrastructure improvements, local government viewed the AOC designation as an additional burden for citizens and businesses. With its limited budget, the city saw funding for studies and remediation reducing the amount available for basic services unless local taxes increased to provide for both. Local government officials also worried that an increase in taxes and services charges would deter new businesses from selecting Erie.

Members of the Coalition, in turn, raised concerns about the environmental impact of bay front development on already impaired water quality. The Coalition saw the AOC designation and RAP process as a way to secure state and federal funding, raise awareness of the bay's degraded water quality, and pressure public and private groups to allocate more resources toward restoring the bay.

Citizens List an Area of Concern

In January 1988, members of the Coalition consulted with the IJC's Science Advisory Board on the appropriateness of designating the bay as an AOC. Recognizing local business concerns, the Coalition jointly sponsored a workshop with the Chamber of Commerce and the IJC's Science Advisory Board. Coalition members used this opportunity to bring all the potentially impacted parties together, emphasize the immediate need for action to correct the flow of raw sewage into the bay, and garner support from Erie's City and County Councils. A key theme of the discussion was the need to link environmental quality and the economy with an emphasis on how a clean bay will bring more business to Erie. With support from the Board and GLU, the Coalition was encouraged to move forward with a petition to list the bay as an AOC. Based upon the recommendation of the Science Advisory Board and at the urging of city government officials and the County's Water Resources Commission as a result of the workshop, both the City of Erie and Erie County Councils passed resolutions in March 1988 supporting the designation.

Pennsylvania Governor Robert Casey came to Erie in June 1988 for a town meeting on the Great Lakes issues. The conversation covered topics like the potential AOC designation, regulations banning phosphates in detergents, and funding for environmental work in the bay and Lake Erie on par with the state's commitment to the Chesapeake Bay. Members of a local ad hoc water quality study committee appointed by the County's Water Resources Commission created the "Swimmable Bay in 20 Years" objective and introduced this slogan at the Governor's meeting (Fig. 14.3).

Fig. 14.3 Swimmable Bay slogan from a 1988 Town Meeting on Great Lakes issues

In November 1988, Erie County and City government officials created a new entity, the Erie Harbor Improvement Council (Council), to serve as the vehicle for engaging representatives of business, academia, development, government, civic, and environmental groups. Local government appointed Council members to ensure representation by all sectors. The Council in essence began the Remedial Action Plan (RAP) process developing a plan to clean up the bay and convening public meetings to engage both regulators and the community in discussion of how to implement it. Additionally, the Council invited experts and representatives from other AOCs to Erie to provide information on the effect of the AOC designation, funding resources, environmental issues, and their experiences working through the RAP process.

Concurrent with the Council and Coalition's efforts to list the bay, the City of Erie was discussing the regulatory requirement to update its sewage management plan under the Pennsylvania Sewage Facilities Act (Act 537 of 1965, P.L. 1535; 35 P.S. §§ 750.1 et seq.). Local citizens, particularly Council and Coalition members, expressed their concerns to the county health department, state environmental agencies, the Governor's office, and legislators. They pressed for upgrades to the city's sewer system, which was a key component of the action plan for the bay. Members of city government and local businesses reiterated their concerns that the AOC designation and any sort of agreement to upgrade the wastewater collection, conveyance, and treatment system within the city would blemish the community's image and hinder the region's possibilities for future growth and tourism. Council meetings provided a forum for this debate. While concerned about funding such a major infrastructure project, members of city and county government as well as the local sewer authority saw an opportunity to act and implement large-scale improvements with public support.

In 1989, the city and the Pennsylvania Department of Environmental Resources (PADER), the predecessor to the Department of Environmental Protection (PADEP), entered into a "Consent Decree" to address multiple sewage issues. The City of Erie and the Erie Sewer Authority agreed to attain and maintain compliance with Pennsylvania's Clean Streams Law (Act 394 of 1937, P.L. 1987; 35 P.S. §§ 691.1 et seq.) and any current or future National Pollutant Discharge Elimination System (NPDES) permits for the facilities under their control. At the time the Consent Decree was signed, there were 52 permitted overflow locations in the Erie sewer system; many which experienced dry and wet weather overflows that discharged into local receiving waters. In addition, there were numerous illegal sanitary sewer connections directing untreated wastewater into the storm sewer system. The Consent Decree required Erie to determine the extent of pollutants discharged from the sewage collection system, storm sewer system, and treatment plant into the Mill Creek Tube, Presque Isle Bay, and Lake Erie and to evaluate both structural and nonstructural alternatives to reduce these pollutants.

Although state government agencies believed there was insufficient evidence to support it and local business owners remained concerned about the economic impact, in 1991 the United States Department of State listed Presque Isle Bay as an AOC. The Council expanded its objectives to ensure Pennsylvania met its responsibilities under the Great Lakes Water Quality Agreement, including development and implementation of the RAP. The Council also worked to address concerns regarding the perceived negative impact of implementing environmental quality goals on the community's efforts to revitalize the local economy. To that end, the Council conducted a survey of the existing 42 AOCs to determine how that designation had impacted the local economy and restoration actions. The survey provided compelling evidence that the designation had positive or no effect on the local tourism economy or ability to attract new business. Of the 26 respondents, 73% reported that the AOC designation had no effect and more than 50% said cleanup actions were accelerated (EHIC 1991). The survey results supported both the Coalition and Council's assertion that designating Presque Isle Bay an AOC was a positive step for both the environment and the economy.

Life after Listing

As part of the RAP evolutionary process, the Council dissolved and its members became the Public Advisory Committee (PAC) formed to provide advice to state and local government entities. The PAC membership included representation from both economic and environmental interests. Members created a process for engagement by establishing PAC meetings as a forum for active participation in government decision-making that would affect the Erie community. PAC members viewed their role as much larger than providing advice. While decision-making authority remained clearly the responsibility of government entities, every effort was made to achieve consensus amongst all parties. In many cases conflicts arose between individual PAC members and the government role became one of facilitator.

The PAC and PADEP, the lead state agency for implementing the Presque Isle Bay RAP, agreed that good science would be the foundation for decision-making. To achieve that goal, the PAC was integrated into the evolving RAP process. Strategic planning sessions gave definition to the problems, identified long-term goals for the ecosystem and the PAC itself, and prioritized research needs. The government and the community together drew the roadmap for restoration. Each gained an understanding of, or at least heard, the other's priorities, concerns, and constraints in terms of regulatory authorities, responsibilities, and resources.

By 1993, the PADEP and the PAC had reviewed and analyzed existing data to identify contaminants of concern, assessed impaired uses in the bay, and evaluated the potential sources of the problems (PADEP 1993). Contaminants of concern included 10 heavy metals, nutrients, cyanide, oil and grease, and polycyclic aromatic hydrocarbons (PAHs). Two of the 14 beneficial uses identified in the Great Lakes Water Quality Agreement were found impaired: (1) fish tumors or other deformities and (2) restrictions on dredging activities.

The three pieces of the RAP process, science, technical, and stakeholder support (Hartig 1997) came together in tackling the beneficial use impairments. Individual PAC members and their organizations supplemented government-sponsored investigation with research of their own. PAC members participated in fieldwork and critical review of the results. A key role for the PAC was in outreach and education within the community and the formation of watershed groups to undertake specific restoration, monitoring, and data gathering projects. Slowly, over the years, a process evolved that merged the three prongs of the RAP process and became the template for work on the impairments and within the watershed.

The centerpiece of this process was identifying and convening both nationally and locally recognized experts to function as an advisory group on science and technical issues. The experts represented federal and state agencies as well as academia from within and outside the Great Lakes Basin. Through a series of workshops, the experts provided advice to PADEP and the PAC on study design, sample analysis, interpreting results, and setting delisting targets for the impairments in Presque Isle Bay. PAC members interacted directly with the experts, participated in sampling, and worked through the development of delisting and ecosystem health targets. Each series of workshops included a public meeting to share the information with the broader community. This collaborative approach between the PAC and the government allowed for the assimilation and consideration of the community's needs at every decision point in the RAP process. The opportunity to fully engage directly with experts mitigated the perception that regulators exercised selective disclosure, releasing only the information that supported their position.

After more than 20 years of study, investigations showed that concentrations of contaminants in the sediment were well below standards and the incidence rate of liver tumors on fish was stabilizing (PADEP 2002). In conjunction with the PAC, PADEP determined that monitoring the natural attenuation of contaminants would provide the most cost effective and practical method for addressing the sediment-related impairment. While no sediment management actions were implemented in the bay, a number of active interventions in the watershed, most importantly the upgrades to the sewer system were responsible for the improvements in environmental quality. The decision was premised on the commitment of both regulators and the PAC to focus attention and resources on continuing to reduce the amount of sediment entering the bay from its watershed. In recognition of the progress made, Presque Isle Bay became the first American AOC to be redesignated as an "Area of Recovery" in 2002.

As part of the long-term monitoring plan for the bay, in 2005, the PADEP and the United States Environmental Protection Agency (USEPA) undertook a comprehensive sediment sampling effort. Data showed that the contaminant concentrations in the sediment met both the ecosystem health and delisting targets. With the PAC's concurrence, PADEP requested the delisting of the restrictions on dredging activities beneficial use impairment (PADEP 2006). The USEPA granted that request in 2007.

Results of Participatory Governance

Without a doubt, the citizen-supported AOC designation and urging for the Consent Decree between Erie and PADEP served as a turning point for cleaning up the bay. Both of these actions were driven in large measure by the members of the Erie County Environmental Coalition and Erie Harbor Improvement Council. Without the leadership of these groups, it is unlikely that the bay would be in the healthy condition it is today. As a result of the AOC designation and the RAP program, more than US $1 million in federal and state resources have been focused on investigating, understanding, and addressing beneficial use impairments. Additionally, the Consent Decree resulted in close to US $100 million being spent on upgrades to the city's wastewater treatment, collection, and conveyance system. While bonds funded a good portion of the work, the state provided grant resources to offset the cost of upgrades. The number of combined sewer overflows (CSOs) was reduced from 70 to five. All five CSOs are fitted with screening devices to remove debris and are flow monitored. These improvements have been made to the sewer system while customers pay some of the lowest service rates in Pennsylvania.

Contrary to the initial concerns of some local businesses and city government officials, the AOC designation and the massive infrastructure improvements have both had a positive effect on the local economy. It is likely that neither would have happened without the urging and support of the PAC. The city completed construction of the highway (i.e., Bayfront Parkway) connecting the bay front to east, west, and south interstates in the mid-1990s and new commercial ventures including a convention center were built. Erie's waterfront underwent a transformation from an industrial center to a tourist attraction (Fig. 14.4). The wastewater system has excess capacity and the city can and does solicit new businesses by marketing this commodity. In 2008, the PAC celebrated achievement of this goal of a "swimmable bay" set 20 years ago by sponsoring a mile swim across Presque Isle Bay. Close to 200 swimmers from Erie and across the Great Lakes participated in the event.

Fig. 14.4 The City of Erie's Bay Front Today

Color image of this figure appears in the color plate section at the end of the book.

LESSONS LEARNED

Public involvement is a cornerstone of the Great Lakes environmental movement. RAPs and other restoration experiments, as offered by Sproule-Jones (2002), promise a way in which resource users, regulators, and those interested in restoring the local ecosystem can collaborate towards a common purpose. They promise to enable local stakeholders to determine their own solutions to ecological degradation and open new venues for collaboration.

There is ample evidence that communities have been instrumental in helping governments be more responsive to and responsible for restoring beneficial uses in AOCs. Broad-based partnerships among diverse community constituencies as described by Hartig and Zarull (1992) are a movement towards grassroots ecological democracy in the Great Lakes Basin. Their observations are equally valid today. This ongoing process of human interaction is seen by Hartig and Zarull as essential to ensure all parties are committed to a common cause. From its inception, the RAP program was intended to change the environmental restoration paradigm in the Great Lakes to a more inclusive, community driven process. The empowerment of the community resulted in diverse and direct involvement in decision-making, coordinating societal response to environmental problems, as evidenced by Krantzberg (2006). The process of democratic, deliberative and shared decision-making results in ownership of those decisions, and the determination, then, to ensure implementation proceeds. Such was the case in the Presque Isle Bay AOC where the public involvement and local ownership expectations of the RAP program were fully realized.

The Presque Isle Bay story provides evidence of the impact citizens can have on restoring AOCs. Those who sought the AOC designation became members of the PAC and worked cooperatively with PADEP and local government to communicate with and address the concerns of local citizens. The PAC served its function of representing the diverse Erie community at the decision-making table.

One difficulty with participatory decision-making is creating a system that allows both forward progress and meaningful involvement. There have to be realistic expectations on the part of the PAC members and the organization responsible for developing and implementing the RAP. There also has to be a decision point and an entity responsible for making it. One of the biggest challenges and points of confusion was defining roles in decision-making. To address this, the PAC worked with PADEP on a set of by-laws, which provided structure, definition, and clear expectations for both organizations. The PAC provides recommendations; however, the PADEP makes the final decisions on technical, scientific, and policy issues. The key to success in many cases was allowing room for negotiating and incorporating the needs and concerns of all parties to the maximum extent practicable.

A good example of how divergent opinions came together is the exercise to define the AOC boundaries. Arguments for inclusiveness came from the environmental community while the state and local government as well as industry believed that the boundary should reflect the original listing description modified by results of the beneficial use impairment studies. The discussion went on for months with both sides attempting to convince the other. The two groups reached a compromise by including study areas outside the AOC boundary. The PADEP committed to incorporating those areas in all future studies and remedial efforts.

Participatory decision-making works in the case of Presque Isle Bay for a number of reasons. First, is the relationships built over the years between the PAC and PADEP and the shared commitment to restoring the bay. That commitment is evidenced by the fact that many of the original PAC members continue to participate today nearly two decades after the bay became an AOC (Fig. 14.5). Second, the citizens took ownership and exercised leadership in the decision to list the bay as an AOC. Third, both sides recognize the value of the other and appreciate the tools available to each; the PADEP understands the

Fig. 14.5 Members of the Presque Isle Bay Public Advisory Committee

importance of public support and the PAC knows that government agencies are required to take or prevented from taking action due to laws, regulations, or agreements they implement. Fourth is the diversity of the PAC membership. Balancing of economic and environmental needs occurred around the PAC table as members of both communities worked hand-in-hand with regulators as part of the decision-making process. The composition of the PAC itself provided the opportunities to integrate the different yet dependent goals of economic prosperity and environmental restoration in the RAP process. While not always in agreement nor able to be equally balanced, PAC members considered both perspectives. A willingness of members to compromise after being heard and the assurance of their role in decision-making prevented either perspective from bringing the RAP process to a standstill.

CONCLUSION

To achieve sustainability ecology, economics and society must be integrated. Costanza (2001) defines sustainability as a balanced relationship between the dynamic human economic systems and the dynamic, but generally slower-changing ecological systems. In deliberative democracy, political leaders, observe Ohja and Timsina (2007), do not make decisions on their own; rather they seek opinions from concerned citizens by engaging in an open public process. Such an approach to collaborative decision-making was reflected in the Presque Isle Bay process. Progress in restoring the AOC was a direct result of civic ownership and leadership leading to cooperative partnerships between government and citizens. This point is crucial to understanding how management of public resources might be effectively carried out. Presque Isle Bay is a healthier ecosystem today as a direct result of the leadership and support of the community through the Coalition, Council, and PAC, as well as the vision and willingness to act exhibited by the local and state government.

ACKNOWLEDGEMENTS

The author wishes to acknowledge the contributions to this chapter from Mark Kwitowski of the City of Erie, David Sterrett of HRG, Inc., Sr. Pat Lupo of Earth Action, Mr. Martin Visnosky, Freda Tarbell and Kelly Burch of the Pennsylvania Department of Environmental Protection. I also thank the Presque Isle Bay Public Advisory Committee and the citizens of Erie, Pennsylvania for their dedication to excellence.

REFERENCES

Costanza, R. 2001. Visions, values, valuation and the need for an ecological economics, BioScience 51: 459-468.

Erie Harbor Improvement Council [EHIC]. 1991. In Erie Harbor Improvement Council Public Information Workshop Report: Presque Isle Bay, our place in the Great Lakes Basin.

Hartig, J.H. 1997. Great Lakes remedial action plans: fostering adaptive ecosystem-based management processes. American Review of Canadian Studies. Fall. Association for Canadian Studies in the United States.

Hartig, J.H. and M.A. Zarull [eds.]. 1992. Under RAPs. Toward Grassroots Ecological Democracy in the Great Lakes Basin. University of Michigan Press. Ann Arbor, Michigan, USA.

Krantzberg, G. 2006. Sustaining the Gains Made in Ecological Restoration. Environment, Development and Sustainability 8: 413-424.

Ojha, H.R. and N.P. Timsina. From Grassroots to Policy Deliberation: The Case of the Community Forest Users' Federation in Nepal. *In:* H.R. Ojha, N.P. Timsina, R.B. Chhetri and K.P. Paudel [eds.]. 2007. Knowledge Systems and Natural Resources: Management, Policy, and Institutions in Nepal. Foundation Books/IDRC 2007.

Pennsylvania Department of Environmental Protection [PADEP]. 1993. Presque Isle Bay Stage I Remedial Action Plan. Northwest Regional Office. Meadville, PA.,USA.

Pennsylvania Department of Environmental Protection [PADEP]. 2002. Presque Isle Bay Remedial Action Plan Update. Office of the Great Lakes. Meadville, PA., USA.

Pennsylvania Department of Environmental Protection [PADEP]. 2006. Delisting the Restrictions on Dredging Activities Beneficial Use Impairment in the Presque Isle Bay Area of Concern. Office of the Great Lakes. Meadville, PA., USA.

Sproule-Jones, M. 2002. The Restoration of the Great Lakes. University of British Columbia Press. Vancouver BC., Canada.

United States and Canada. 1972. Great Lakes Water Quality Agreement. Windsor, Ontario, Canada.

CHAPTER 15

Sustaining Restoration in Light of Climate Change

Sommer Abdel-Fattah

92 Melbourne St., Hamilton, Ontario, L8P2A6

E-mail: abdelfs@mcmaster.ca

INTRODUCTION

The ecosystems and human economies of the Great Lakes region, profoundly shaped by human activities, are now further threatened by climate change (Millerd 2007). Climate change can be seen in the region through the temperature increases in the northern hemisphere by more than 0.5°C, lengthened growing seasons, and precipitation increases by 5 to 10% since 1900 (Kling et al. 2003). Other indicators that the climate is warming include documented losses of alpine glaciers, sea ice, and seasonal snow cover (Chao et al. 1999). Changes in precipitation patterns and temperature changes will cause increased evaporation and evapotranspiration, and more coastal storms. These changes will affect many ecosystem components including water temperatures, water levels, soil moisture, seasonal flow and timing, vegetation type and animal habitats as well as other systems.

As we look to the future, consensus is building that the biggest threat to the Great Lakes Basin ecosystem is climate change (Krantzberg 2008). Adapting management tools, such as the Canada-US Great Lakes Water Quality Agreement (GLWQA) of 1972, to address the threat of climate change is vital. The GLWQA expresses the commitment to restore and maintain the chemical, physical and biological integrity of the waters of the Great Lakes Basin Ecosystem (United States and Canada 1972, 1978, 1987).There have been no changes to the programs, policies, or priorities within the GLWQA since 1987, although it is currently under renegotiation. By no means has science, technology, and public policy remained static in that time. Although the GLWQA has had substantial influence on the remediation and restoration of features of the regional ecosystem, threats to the Great Lakes in the face of climate change have been overlooked in the legislation and thus demand a renewal of program and policy approaches to the restoration of beneficial uses.

Restoration of beneficial uses under Annex 2 of the GLWQA should be better adapted to anticipate climate change impacts on ecosystem structure and function. Already delisted beneficial uses will need to be included in policy recommendations to ensure that restoration is restored or sustained as the climate changes. Management responses related to beneficial uses need to accommodate uncertainties anticipated regarding climate change impacts on ecosystem structure and function.

CLIMATE CHANGE IN THE GREAT LAKES BASIN

The Intergovernmental Panel on Climate Change (IPCC 2001) defines climate change as "statistically significant variations of the mean state of the climate or of its variability, typically persisting for decades or longer." Climate change occurs by the greenhouse effect that has for millions of years warmed the Earth, by naturally occurring gases in the atmosphere that trap heat within the atmosphere. Many studies report that human-induced climate change, particularly the burning of fossil fuels, has played a more significant role in climate fluctuation.

Climatologists using the General Circulation Model (GCM) have been able to determine the manner in which the increase of carbon dioxide emissions will affect the climate in the Great Lakes Basin. Climate scenarios developed from GCMs produce changes in the mean values of climatic elements by applying $1XCO_2$ and $2XCO_2$ ratios to historic base-case climate.

There are many basin-wide changes that are expected to affect the entire Great Lakes region. An increase in annual and seasonal temperatures for the Great Lakes watershed is predicted. The greatest range in projected temperature increase is in the spring (0.75 to 5.0°C) while summer had the least (1.5 to 4.0°C) (Great Lakes Water Quality Board 2003). Summer daily temperatures are projected to gradually increase towards 2030 and then a more rapid increase could have daily average summer temperatures 10°C higher than the 1960-1990 average by 2100 (Kling et al. 2003). Numerous GCM experiments have determined that the daily temperature range, the difference between the daily high and daily low temperature, tends to decrease with increasing greenhouse gas (Stone and Weaver 2002).

Overall, annual average precipitation is expected to increase. Winter and spring modelling scenarios show consistent increases in precipitation while autumn scenarios show precipitation decreases. Analyses of precipitation extremes in GCMs and regional climate models (RCMs) indicate more heavy precipitation events, fewer moderate events and more dry days or days with light precipitation (Cubasch 1995, Hennessy et al. 1997, Jones et al. 1997, Trenberth 1999). The distribution of precipitation throughout the year will be altered. Sharif and Burn (2006) estimate that only the months of January, March and October will have increased monthly precipitation while the other months may see a decrease in precipitation, including the months between April and September when water demand is the highest. Changes in precipitation, combined with temperature increases, will influence soil moisture, ground water recharge, and runoff in the Great Lakes watershed. Warmer air temperatures in winter and early spring affect the form of precipitation. It is expected that increased frequency of winter temperatures rising above the 0°C threshold will cause precipitation that previously fell as snow to fall as rain. As a result, more runoff may occur in winter and become intermittent and less snow may accumulate (Great Lakes Water Quality Board 2003). There will be slightly different changes for each of the Great Lakes dependent on the differences in the physical characteristics of the lakes and local climatic differences.

It is important to investigate these changes, and identify trends in order to accurately assess the ecological, social and economical impacts.

Ecological Impacts due to Climate Change

Many attributes of the climate system in the Great Lakes watershed are projected to change, specifically air temperature, precipitation events, evaporation rates, water levels, ice cover, water quality, natural ecosystems and biodiversity. A changing climate may put added stress on beneficial uses, especially those that are already impaired (see Fig. 15.1), creating heightened sensitivities as well as new risks within ecologically healthy areas.

Environmental restoration of sensitive areas could be hindered by a changing climate and an altered hydrological regime (Rhodes and Wiley 1993). Twelve of the 14 beneficial uses were identified as having a potential vulnerability to climate change. The two beneficial uses impairments' that were considered not directly affected by climate change are fish tumours or other deformities and bird or animal deformities or reproduction problems (IJC 1991). These impairments however, may be related to climate change through harmful runoff, and excessive algae blooms which are both increased in predicted scenarios. In an attempt to identify the potential risks posed by climate change to the beneficial uses of the Great Lakes, an evaluation of ecosystem impairments particularly at risk in light of projected impacts of climate change are examined below. The most significant ecosystem changes discussed are temperature, precipitation, and evaporation.

	Lake Basin/ Region	AOC (Canadian and Binational)	Beneficial Use Impairments	Common Basin Impairments
Upper	Superior	Thunder Bay	A,B,C,D,F,G,J,K,M,N	A,C,F,G,K,N
		Nipigon Bay	A,B,C,F,G,H,K,N	
		Jackfish Bay	C,D,F,G,K,N	
		Peninsula Harbour	A, C, F, G, I, N	
Middle	Huron	Severn Sound	A,C,F,G,H,K,N	A,C,F,G,H,J,K,N
	Middle Connecting Channels	St. Mary's River	A,C,D,F,G,H,J,K,N	
		St. Lawrence River (Cornwall)	A,C,F,G,I,J,N	
		Niagara River	A,C,E,F,G,H,I,J,K,N	
Lower	Erie	Currently no Canadian AOCs - Wheatley Harbour recently delisted.		A,C,F,G,H,J,K,N
	Ontario	Hamilton Harbour	A,C,D,E,F,G,H,J,K,M,N	
		Bay of Quinte	A,C,F,G,H,I,J,K,M,N	
		Port Hope	G	
		Metro Toronto	A,C,F,G,H,J,K,N	
	Lower Connecting Channels	St. Clair River	A,D,F,G,H,J,K,L,N	
		Detroit River	A,D,F,G,J,K,N	

Beneficial Uses

A	Restrictions on Fish and Wildlife Consumption
B	Tainting of Fish and Wildlife Flavour
C	Degraded Fish and Wildlife Populations
D	Fish Tumours or Other Deformities
E	Bird or Animal Deformities or Reproductive Problems
F	Degradation of Benthos
G	Restrictions on Dredging Activities
H	Eutrophication or Undesirable Algae
I	Restrictions on Drinking Water Consumption or Taste and Odour
J	Beach Closings
K	Degradation of Aesthetics
L	Added Costs to Agriculture or Industry
M	Degradation of Phytoplankton and Zooplankton Populations
N	Loss of Fish and Wildlife Habitat

Fig. 15.1 Beneficial use impairments in Canadian AOCs

Increased Temperature

An increase in annual and seasonal temperatures for the Great Lakes watershed is expected. Air temperature plays an important role in defining the length of the growing season, the frost-free season and thus also species-breeding seasons. Increases in air temperature drive water surface temperatures and affects freeze and thaw cycles, the rate of chemical reactions, and biological productivity (Great Lakes Water Quality Board 2003). Since the moisture-holding capacity of the atmosphere is enhanced with increased temperature, the potential for loss of water to the atmosphere through evaporation and evapotranspiration demand is expected to also increase. In a humid region such as the Great Lakes, atmospheric moisture content is a very important factor. Mean annual lake surface evaporation is expected to increase by +6 to +39% due to an increase in lake surface temperatures (Lofgren et al., 2002) which will impact water levels, soil moisture content and wetlands.

Rising surface temperatures may amplify the negative consequences of eutrophication, as well as hypoxia (Kennedy 1990) and their subsequent pressures on benthic communities. Benthic communities are vital to lake health. Many species of *chironomids* and *tubificidoligochaetes* are tolerant to low dissolved oxygen, such that these become the dominant in lakes with hypoxic hypolimnia. As hypoxia becomes more severe tubificids can become dominant over chironomids. In cases of prolonged anoxia, the common variety of benthic organisms might disappear entirely (Mackie 2004).

Warming will also result in water temperature thresholds being reached for certain species. Earlier warming of spring temperatures and later cooling of temperatures in autumn contribute to an earlier start for plant growth and a longer growing season. Natural adaptation of flora and fauna by migration assumes that species are able to move rapidly enough to remain within their preferred climate zones even with natural and human-caused barriers. Temperature changes can influence the timing of breeding and bird migration (Kerr and Packer, 1998); rare and endangered species will be particularly vulnerable to this change. These populations have more specific habitat requirements or habitat niches and thus are the most susceptible to changes and have the least adaptation options (Great Lakes Water Quality Board 2003).

Cold-water fish may lose important habitat as temperatures rise above their thermal thresholds (Meisner 1990). Climate change impact assessments have projected that fish ranges could move more than 500-600 kilometres northward leading to invasions of warmer water fish and extirpations of colder water fish (Magnuson et al. 1997). Breeding windows may be compressed or shifted. Optimum temperatures for lake trout spawning in the autumn is 8°C to 11°C and spring spawning for northern pike is 4°C to 12°C; warming could affect biological cycle timing. Egg development is largely temperature dependent, as is the period following hatching when juveniles rely on the local nutrients. For some fish, such as salmon, juveniles remain in the gravel for several weeks surviving on the nutrients in their yolk sacs before beginning exogenous feeding and emerging. The timing of emergence from the gravel has likely evolved in response to selection for a presumably optimal time for survival (e.g. temperature and flow), food availability and competitive advantages. If winter temperatures increase following climate change, egg incubation will be shortened and result in earlier juvenile emergence from the gravel, in the future. Such a shift away from optimal timing for juvenile emergence is likely to result in increased mortality. Moreover, if climate change results in increased winter flows, as many models predict, gravel shifts caused by flooding may become more frequent and larger, increasing the susceptibility of incubating eggs to destruction (Fleming and Jensen 2002).

Species currently limited to southern states may be able to extend northward into the Great Lakes region. Thus, the spread of invasive species could be exacerbated. Species such as carp, zebra mussel, purple loosestrife, curly-leaf pondweed, and Eurasian milfoil are examples of species whose introductions have affected the Great Lakes ecosystem. At present, zebra mussel is primarily found in the lower Great Lakes while the coldwater of Lake Superior limits its expansion (Easterling and Karl 2001). Warmer water temperatures would allow zebra mussel to become more widespread on Lake Superior. Changes to the natural ecosystems and the inhabitants of the region will ultimately affect the entire dynamics of the system.

Increased temperatures may better allow for bacteria and viruses to survive and disease local plants, trees and animals. Dutch elm disease contributed to an almost extinction of this species throughout the region in the 1970s. Oak wilt is another disease that exists in many parts of the region, although it mostly affects the northern portion. Other forest-related diseases and pests that are found in the region in non-epidemic numbers include spruce budworm, cankerworms, forest tent caterpillar, white pine blister rust, white pine weevil, basswood thrips, butternut canker, and Asian long-horned beetle (Great Lakes Water Quality Board 2003). New diseases may become prevalent in the region and retard plant growth or cause animal illness.

Climate change is expected to directly and indirectly influence water quality (Murdoch et al. 2000). Warmer water temperatures reduce dissolved oxygen concentrations. Higher temperatures change the rate of chemical reactions in the water column, sediment-water interface, and the water-atmosphere interface. Water temperatures also influence phytoplankton growth; blue-green algae dominate at the highest temperatures, followed by green algae, then flagellates and finally diatoms at lowest temperatures (Magnuson et al.

1997). Blue-green algae have been associated with taste and odour problems in drinking water (Anderson and Quartermaine 1998).

As temperatures rise, the ability of the atmosphere to hold moisture increases allowing for increased and more intense precipitation events.

Precipitation

Overall, annual average precipitation is expected to increase. Winter and spring modelling scenarios show consistent increases in precipitation while autumn scenarios show precipitation decreases. Changes in precipitation, combined with temperature increases, will influence soil moisture, ground water recharge, and runoff in the Great Lakes watershed. Whether projected increases in precipitation will be offset by more water loss due to higher evaporation is the concern especially during summer and autumn. In the basin, these seasons are typically characterized by low stream flow. Low flow conditions could become more extreme and more vulnerable to water level decreases (Great Lakes Water Quality Board 2003). Winter in the Great Lakes region is a period with reduced runoff due to ice coverage; most precipitation falls as snow and is stored in snow pack. Warmer winter temperatures could lead to more rainfall events, which create runoff. Often infiltration is reduced at this time due to frozen ground conditions and more overland runoff may occur. In response, flow in rivers and streams will increase, causing a higher risk for flooding. Peak flows can be high if the snow cover melts quickly. Earlier spring warming may bring on an earlier melt. Since there may be less snow stored in the snow pack, the amount of water available to runoff and contribute to the peak flow will be less. These changes could cause an increase in mean flow which will ultimately influence Great Lakes water levels, instream assimilative capacity changes affecting water quality, habitat deterioration, and access to water for irrigation, drinking water, and hydropower.

Extreme precipitation and high runoff is expected to increase causing the risk of soil erosion, land and water quality degradation, flooding, and infrastructure failure. With more intense precipitation events, more pollutants reach watercourses directly and rapidly through surface transport by runoff. In urban areas, longer periods of dry weather between rainfall events allow more pollutants to accumulate on road and land surfaces. Spring can be a high pollution-loading period because of pesticide application combined with little vegetative cover (Soil and Water Conservation Society 2003). Although benthic macroinvertebrates are moderately long-lived, contamination and toxicity of sediment caused by runoff will affect those sensitive benthic organisms (Mackie 2004).

Wildlife is susceptible to extreme climate events. Droughts and periods of excessive wetness can affect habitat as well as food supply. Extreme weather also plays a role in regional forest destruction. Lightning, high winds, hail, or tornadoes can destroy trees. Exceptionally cold or hot weather can also retard growth or kill trees depending on the duration and location of such weather (Great Lakes Water Quality Board 2003). Wildlife will also be highly susceptible to increased evaporation that may cause wetlands, streams and rivers to dry-out and reduce animal access to fresh water.

Evaporation and Evapotranspiration Increases

The frequency of both droughts and flooding are expected to increase due to fluctuations of stream flow, lake levels, and ground water. Increased evaporation and evapotranspiration may cause summer and fall low flow to become even lower due to reductions in ground water base flow. Dry periods reduce recharge and water levels in aquifers. Shallow aquifers consisting of unconsolidated sediment, weathered or fractured bedrock are more vulnerable to these changes. Although a changing climate may result in higher total annual precipitation, ground water levels will likely decrease (Nastev et al. 2002). Development of dry, crusty tops on soils due to increased temperature will impede infiltration reducing ground water levels (Piggott et al. 2001). As a result, the low water levels experienced in the Great Lakes during 1999 to 2001 could occur more frequently. Minimum levels occur in late autumn and early winter; and rise in spring with snowmelt and reach a maximum from July to September. Then, they decline in autumn due to evaporation and reduction in runoff. Distinct shifts are being detected in the seasonal cycle of water levels for the Great Lakes. In Lakes Erie and Ontario, during the period 1860 to 1990, the annual rise and fall of levels occurred one month earlier; spring levels are becoming higher while fall levels are becoming lower sooner (Lenters 2001). Lake Huron water levels have declined due to a combination of lower precipitation, higher air temperatures, and increased evaporation over the past eight years. As cold, dry air masses pass over warmer Lake Huron waters in the fall and winter, the potential exists for lake levels to plummet to their lowest point in 36 years (Schwartz 2001). Michigan-Huron exhibited higher levels in winter and reflected a shift, since 1965, to more runoff in winter and less spring runoff. Studies have indicated that Lake Huron water levels are projected to decline by as much as 1 m by 2050 (Mortsch et al. 2000). In Lake Superior, summer and fall levels are decreasing (typically the high period) while spring and winter levels remain roughly the same. Declining water levels will produce wider beaches and may influence shoreline erosion (Schwartz 2001).

Wetlands would experience changes including reduction in water levels; changes in timing and amount of water flowing through a wetland affecting flushing, sedimentation, nutrient input; and length of ice cover. Indirect effects of lower lake levels include oxidation of wetland bottoms (Mortsch 1998).

Increasing air temperatures combined with evaporation and alterations to precipitation amount, timing, and duration could lead to more variability of water supply in the Great Lakes region. Thus it can be seen that the terrestrial and aquatic ecosystems that make up the biosphere provide vital goods and services to humanity, including food, water, recreation, and natural beauty.

SOCIAL AND ECONOMIC IMPACTS CAUSED BY CLIMATE CHANGE

The economy of the region is largely diversified and includes manufacturing, tourism and recreation services, agriculture, and forestry (Sousounis and Glick 2007). The region is also a major shipping corridor, with traffic generating US$3 billion business revenue annually and 60,000 jobs (Kling et al. 2003). The Great Lakes themselves are a recreation and tourist attraction with more than 15 million people in the Great Lakes states participate in fishing, hunting, or wildlife watching; a revenue of US $18.5 billion in sales annually (Kling et al. 2003).With a changing climate there would be a corresponding change in social dynamics and economy of the Great Lakes. Particularly, commercial fisheries, recreation and hunting, local industry including water purification and environmental conservation will be economically and socially affected.

Commercial Fish Market and Recreational Fishing and Hunting

Due to warmer water temperatures there would be an increased warmer water species and aquatic life that is tolerant to warmer temperatures. There would also be a shift away from cold-water fisheries (Atkinson et al. 1999). A predicted warming of 2.5°C over the next 70 years would decrease the habitat space of brook, rainbow, cutthroat and brown trout by one-fourth, possibly even down to one-third (Dempsey et al. 2008). Climate change will also alter growth rates, mortality rates, reproductive capacity and distribution of fish which may cause a shift in contaminant uptake leading to greater restriction on fish consumption (Brander 2006). Some smaller, localized species may undergo severe population reductions or even extinction. These changes may affect which types of fish can be commercially viable. This may mean that some fish need to be farmed to reduce contamination. Recreational fishing will need to be reduced or monitored if fish populations become at risk, or contaminated. If wildlife habitats are put at risk or diminished, hunting may need to be reduced or monitored to protect native species.

Warmer water temperatures may also lead to the invasion of non-native species. This invasion of both exotic species and or warm water fish is difficult to prevent. Fish populations have been impacted heavily by the invasive parasitic sea lamprey which is expected to flourish in warmer temperatures (Holmes et al. 1994). These problems will impact the fisher's ability in catching healthy fish to sustain revenue and local fish prices could rise as a consequence. In Lake Ontario, scientists have identified invasive mussels as a main culprit responsible for the collapse of the lower food web and mussels are thought to be part of the causal mechanism of the botulism outbreaks that are responsible for the death of thousands of fish and fish-eating seabirds. This could increase the cost habitat cleanup and conservation efforts. Many of these impairments can also be very cost intensive for industry.

Industry and Water Sanitation and Purification

Climate change is expected to directly and indirectly influence water quality (Murdoch et al., 2000). As air temperature increases, water temperatures mirror that rise.

Warmer water temperatures predicted with climate change will enhance productivity of algae blooms (Poff et al. 2002). These compounds occur in lakes, reservoirs and rivers. Blue-green algae have been associated with taste and odour problems in drinking water (Anderson and Quartermaine 1998). They possess musty, earthy odours at very low thresholds. Moreover, these compounds tend to accumulate in fish flesh where they significantly alter fish taste. They can also markedly reduce the quality of drinking water. An increase in algal blooms may result in added costs for drinking water treatment. Healthcare costs could rise if contamination enters drinking water causing human sickness.

Within the Great Lakes region, the zebra mussel infestation has created a severe economic problem. Zebra mussels have an impact on facilities that use raw surface water, such as utility plants, factories, and water treatment plants. These mussels colonize inside water intake pipes, wells, and can attach to many types of surfaces. In addition to clogging pipes and reducing the flow, zebra mussels infiltrate interior plant structures, causing obstruction of pump valves and leading to failures of some components. The cost of cleaning intake pipes and repairing failed components can be massive. Mussels have caused millions of dollars of damage in the Great Lakes region alone (The National Atlas of the United States of America 2010).

With water level declines and possible water scarcity there could be less potential for hydropower, and less water for industry operations. Farming industries may also be affected and possibly need to rely on less water intensive techniques or crops that are not as water dependent.

Recreational use of Great Lakes Beaches, Cottages and Marinas

The intensity of precipitation that is expected with climate change will have an impact on beach closings due to increased runoff as well as pollution or sewage waste that would result from overflows and affect the quality of beaches. Human sewage is a major source of bacteria in surface waters, and can come from combined sewer overflows caused by an abundance of stormwater. Runoff can also carry dangerous substances into the lakes. Heavy metals such as copper, lead, aluminum and zinc can be carried into the lakes and at higher concentrations; these substances can be toxic to aquatic life.

Changes in the timing and duration, height and elevation of annual and seasonal water levels may also lead to beach closings. Lower lake levels will impact beaches, with the amount of new exposure a function of water depth, lakebed composition and slope, and water level decline (Wall 1998), such that larger beach surfaces could affect recreation space. Furthermore, exposed mud flats could reduce shoreline aesthetics, and there is the potential that exposed lakebeds could include toxic sediment (Mortsch et al. 2000). Pollution problems may also be exacerbated on some beaches due to low water levels, resulting in more beach closures. With reduced water levels, there could be decreased depth of navigation channels, stranded docks and harbours impacting marinas and shipping ports.

Recent and ongoing global environmental changes—including climatic change, habitat fragmentation, pollution, and the spread of invasive species—are affecting the functioning of many ecosystems, and therefore the goods and services that they provide.

Governance

Clearly, the changes expected with a changing climate cannot be ignored. Polices must be developed to highlight these changes, and build adaptation strategies. Restored beneficial uses may be better adapted to cope with these changes in climate, but over time could become re-impaired. This problem will result in increased and advanced impairments in beneficial uses; whether currently impaired or restored. Many of the beneficial use impairments are affected by climate change determinants that are common. For example, algae blooms will respond to increased nutrient loadings associated with severe storm effects, which also effects drinking water, beach closings, and added costs to treat water for agriculture or industry. Further, water quantity issues associated with lake evaporation and increased temperatures will exacerbate the deterioration of fish and wildlife populations and loss of fish and wildlife habitat which are among the most widespread impairments. These are highly vulnerable beneficial uses. Figure 15.2 identifies how ecosystem changes are could result in multiple beneficial use impairments.

Beneficial Use Affected	*Climate Variability and Climate Change Effects*				
	No. of Canadian AOCs Impaired (% of 14)	*Air Temperature Increases*	*Intensity of Precipitation and Storm Events, Increased Runoff*	*Evaporation/ Evapo-transpiration (Water Levels)*	*Shortened Snow Seasons and Reduced Ice Cover*
1. Restrictions on fish and wildlife consumption	11 (79%)	X	X		
2. Tainting of fish and wildlife flavour	0 (0%)	X	X		
3. Degradation of fish wildlife populations	9 (64%)	X	X	X	X
4. Fish tumours or other deformities	3 (21%)	-	-	-	-
5. Restrictions on Dredging Activities	13 (93%)	X	X	X	
6. Bird or animal deformities or reproduction problems	2 (14%)	-	-	-	-
7. Degradation of benthos	13 (93%)	X	X	X	
8. Eutrophication or undesirable algae	6 (43%)	X	X		
9. Restrictions on drinking water consumption, or taste and odour problems	3 (21%)	X	X	X	
10. Beach closings	8 (57%)	X	X	X	
11. Degradation of aesthetics	9 (64%)	X	X	X	
12. Added costs to agriculture or industry	1 (7%)	X	X		

Fig. 15.2 contd. ...

Fig. 15.2 contd. ...

Beneficial Use Affected	*Climate Variability and Climate Change Effects*				
	No. of Canadian AOCs Impaired (% of 14)	*Air Temperature Increases*	*Intensity of Precipitation and Storm Events, Increased Runoff*	*Evaporation/ Evapo-transpiration (Water Levels)*	*Shortened Snow Seasons and Reduced Ice Cover*
13. Degradation of phytoplankton and zooplankton populations	3 (21%)	X	X	X	
14. Loss of fish and wildlife habitat	12 (86%)	X	X	X	X

Fig. 15.2 Beneficial uses affected by climatic change

In order to highlight the most sensitive beneficial uses, an examination of common beneficial use impairments, forecasted ecosystem changes and the specific properties of the individual Great Lakes allows for insight into the future of the Great Lake Basin to aid in building policies that combat the main issues affecting the individual lakes (see Fig. 15.3).

Due to the fact that climate is constantly changing, restored beneficial uses also need to be highly monitored and adaptation responses directed at these beneficial uses that are highly sensitive to projected effects of climate change and generally widespread across Ontario's AOCs should aid in achieving more resilient end points to sustain delisting activities.

A policy framework must be developed, but will need to be very adaptable as the climate changes, and include monitoring and assessments tools. Although there is much uncertainty in predicting the future, by looking at the past one can see that the climate has already changed. It is important to present possible future scenarios and policy goals around social and environmental impacts to push for a policy that aims to better our future. Thus, a renewed Great Lakes Quality Agreement, or efforts towards its implementation need to include two important considerations:

- Options for consideration to ensure restored beneficial uses remain restored
- Potential adaptation responses to enhance the resilience of beneficial uses

Maintaining the Great Lakes ecosystems in light of climate change is vital in preserving the surrounding habitats as well as our access to clean water and the beneficial uses the Great Lakes provide. A development of adaptive strategies to reduce the vulnerabilities of the conservation targets to actual or potential climate change effects is needed. It implies adjustments in ecological, social, or economic systems in response to actual or expected climate, for example, adaptation techniques to limit runoff or nutrient loading in the system.

One way of approaching the identification of policy options is to question what ecological, social or economic adjustments can be made to reduce the magnitude of the threats to a particular conservation target and thus reduce the vulnerability of beneficial uses and make them more resilient to climate change? For example, how might the threats of urban and agricultural runoff, dams and invasive species be reduced and ameliorate Great Lakes vulnerability to climate changes? What are some ecological, social or economic alternatives to be considered? This will allow for sequential consideration of the individual conservation targets for impaired beneficial uses and create a policy solution for including climate change in decision making in the Great Lakes. Adapting to the changes now, could reduce the potential impacts in the future. Progress will come from realizing that changes in global climate will have significant impact on local and regional hydrological regimes, which will in turn affect ecological,

	Lake Basin/ Region	Ecosystem Characteristics Vulnerable Impairment	Resultant of Impairments
Upper	Superior	High Retention Time	A,B,C,D,E,F,G
		Shoreline Length	A,B,C,D,E,F,G,I,J
		Largest Temp Increase	A,B,C,F,G,H,I,J,K,N
Middle	Huron	Exposed Shorelines	A,B,C,D,E,F,G,I,J,K
		Low Water Levels	A,B,C,D,E,F,G,I,J,K
		High Biodiversity	C,N
Lower	Erie	Low Water Levels	C,D,G,H,I,J,K,N
		High Biodiversity	C,N
	Ontario	Variable Spring/Fall Changes	A,C,N
		Urbanization	A,B,C,D,E,F,G,H,I,J,N

Beneficial Uses

A Restrictions on Fish and Wildlife Consumption
B Tainting of Fish and Wildlife Flavour
C Degraded Fish and Wildlife Populations
D Fish Tumours or Other Deformities
E Bird or Animal Deformities or Reproductive Problems
F Degradation of Benthos
G Restrictions on Dredging Activities
H Eutrophication or Undesirable Algae
I Restrictions on Drinking Water Consumption or Taste and Odour
J Beach Closings
K Degradation of Aesthetics
N Loss of Fish and Wildlife Habitat

Fig. 15.3 Sensitivities

social and economical systems. The overall aim is to create suggestions for adaptations, and policy goals to include climate change into planning, remedial programs and improve the resilience and sustainability of the Great Lakes and its beneficial uses.

CONCLUSION

Climatic changes will result in changing vulnerabilities and sensitivities in the Great Lakes region. Synergies between development pressures and environmental change have resulted in weather events that have the potential to have negatively impact the Great Lakes. Changes in regional climate may accelerate or exasperate the degradation of beneficial uses. Climate-induced changes must be assessed in the context of impacts on water quantity and quality and the corresponding ecosystem changes. Relating climatic change predictions to local impacts will result in a holistic to Great Lakes conservation and preservation. Slight changes in regional climate such as more intense precipitation can result in water-quality problems due to increased harmful runoff and sewage overflows during storms which can be detrimental to fish and wildlife health. Warmer temperatures result in altered lake mixing regimes that reduce the availability of fish habitat, change the magnitude and seasonality of runoff regimes that alter nutrient loading and limit habitat availability through wetland reduction through evaporation. These changes have potential ecological risks and costs and thus improvements are required to strengthen mitigation efforts and to evaluate management options. The challenge will be to find effective adaptation strategies in managing resources within a changing climatic regime. This will also require assessment of the effects of water management activities on organisms and ecosystem processes, including development and testing of management strategies designed to counteract changes in climate. It is hoped that this will be accomplished through revisions of the Great Lakes Water Quality Agreement. A warming climate presents an important opportunity to revitalize the Great Lakes and add strength to current regimes and remedial action plans through innovative adaptation mechanisms and policy implementation.

REFERENCES

Anderson, B.C. and L-K Quartermaine. 1998. Tastes and odors in Kingston's municipal drinking water: a case study of the problem and appropriate solutions. J. Great Lakes Res. 24: 859–867.

Anderson, J. and I. Craine, A.W. Diamond, and R. Hansell. Impacts of climate change and variability on unmanaged ecosystems, biodiversity and wildlife. pp. 121-188. *In:*

G. Koshida and W. Avis [eds.]. 1998. The Canada Country Study: Climate Impacts and Adaptation, National Sectoral Volume. Environment Canada, Canada.

Arnell, N. 2002. Climate Change due to Global Warming. pp. 212-249. *In:* Prentice Hall [eds.] Hydrology and Global Environmental Change. Pearson Education Limited, Essex, UK.

Atkinson, M.J. and H. Barnett, H. Aceves, C. Langdon, S.J. Carpenters, T. McConnaughey, E. Hochberg, M. Smith and B.D.V. Marino. 1999. The Biosphere 2 coral reef biome. Ecol. Eng. 13: 147-171.

Blumeberg, A.F. and D.M. DiToro. 1990. Effects of climate warming on dissolved oxygen concentrations in Lake Erie. Transactions of the American Fisheries Society 119: 210-213.

Brander, K. 2006. Assessment of Possible Impacts of Climate Change on Fisheries. Berlin: WBGU. ISBN 3-936191-13-1. Available at: http://www.wbgu.de/wbgu_sn2006_ex02.pdf

Brandt, S.B. and D.M. Mason, E.V. Patrick, R.L. Argyle, L. Wells, P.A. Unger, and . J. Stewart. 1991. Acoustic measures of the abundance and size of pelagic planktivores in Lake Michigan. Canadian Journal of Fisheries and Aquatic Sciences 48: 894-908.

Canada and the United States.1972, 1978, 1987. Great Lakes Water Quality Agreement. Retrieved from www.ijc.org

Chao, P.T. and B. Hobbs, and V. Boddu. 1999. How Climate Uncertainty should be Included in Great Lakes Management: Modeling Workshop Results. Journal of the American Water Resources Association 35(6). http://www3.interscience.wiley.com/journal/119085779/abstract-fn2

Cubasch, U. and J. Waszkewitz, G. Hegerl, and J. Perlwitz. 1995. Regional climate changes as simulated in timeslice experiments.Climate Change 31: 273-304.

Dempsey, D. and J. Elder, and D. Scavia. (May 2008). Great Lakes Restoration and the Threat of Global Warming. A Report by the Healing Our Waters - Great Lakes Coalition.

Easterling, D.R. and T.R. Karl. 2001. Potential consequences of climate variability and change for the Midwestern United States. pp. 167-189. *In:* National Assessment Synthesis Team Climate Change Impacts on the United States: The Potential Consequences of Climate Variability And Change. Report for the U.S. Global Change Research Program. Cambridge University Press, United Kingdom.

Fleming, I.A. and A.J. Jensen, Fisheries: effects of climate change on the life cycles of salmon. pp. 309-312. *In:* I. Douglas [ed.]. 2002. Encyclopedia of Global Environmental Change, vol. 3, Causes and Consequences of Global Environmental Change. John Wiley and Sons Ltd., Chichester.

Great Lakes Water Quality Board. 2003. Climate Change and Water Quality in the Great Lakes Basin. International Joint Commission, Ottawa, Ontario. Canada.

Hennessy, K.J. and J.M. Gregory, and J.F.B. Mitchell. 1997. Changes in daily precipitation under enhanced greenhouse conditions. Climate Dynamics 13: 667-680.

Hofmann, N. and L. Mortsch, S. Donner, K. Duncan, R. Kreutzwiser, S. Kulshreshtha, A. Piggott, S. Schellenberg, B. Schertzer, and M. Slivitzky. 1998. The Canada Country Study: Climate Impacts and Adaptation. vol. 7. National Sectoral Volume. Climate Change and Variability: Impacts on Canadian Water. Environment Canada.

Holmes, J.A. and F.W.H. Beamish, J.G. Seelye, S.A. Sower, and J.H. Youson. 1994. Long-term influence of water temperature, photoperiod, and food deprivation on metamorphosis of sea lamprey, *Petromyzon marinus*. Can. J. Fish. Aquat. Sci. 51: 2045–2051.

International Joint Commission. 1991. List/delist criteria for Great Lakes areas of concern, International Joint Commission, March–April 1991, Windsor, Ont. - Detroit, Michigan. Available online at http://www.ijc.org/focus/listdelist

IPCC [Intergovernmental Panel on Climate Change]. 1996. Climate Change 1995: Impacts, Adaptation and Mitigation of Climate Change: Scientific-Technical Analysis. Contribution of Working Group II to the Second Assessment Report. Cambridge University Press. Cambridge, UK.

Intergovernmental Panel on Climate Change [IPCC]. 2001. Climate Change 2001: Impacts, Adaptation, and Vulnerability. Contributions of the Working Group II to the Third Assessment Report of the IPCC. Cambridge University Press, Cambridge, UK.

Jones, P.D. and T.J. Osborn, and K.R. Briffa. 1997. Estimating sampling errors in large-scale temperature averages. Journal of Climate 10: 2548-2568.

Kennedy, V. 1990. Anticipated Effects of Climate Change on Estuarine and Coastal Fisheries. American Fisheries Society. Fisheries, vol. 15, Number 6, November, Horn Point Environmental Laboratories, University of Maryland, Cambridge, USA.

Kerr, J.T. and L. Packer. 1998. The impact of climate change on mammal diversity in Canada. Environmental Modeling and Assessment 49: 263-270.

Kharin, V.V. and F.W. Zwiers. 2000. Changes in the extremes in an ensemble of transient climate simulations with a coupled atmosphere-ocean GCM. Journal of Climate 13: 3760-3787.

Kilgour, B.W. and R.C. Bailey, and E.T. Howell. 2000. Assessing changes in the nearshore benthos of Lake Ontario. J. Great Lakes Res 26: 272–286.

Kling, G. and K. Hayhoe, L. Johnson, R. Lindroth and J.J. Magnuson. 2003. Confronting Climate Change in the Great Lakes Region.Union of Concerned Scientists; Ecological Society of America. Retrieved from http://www.ucsusa.org/greatlakes/glchallengetoc. Html

Krantzberg, G. 2008. The Great Lakes, A 35th Year Anniversary: Time to Look Forward. Electronic Green Journal, Issue 26, Spring 2008.

Kunkel, K.E. and S.A. Changnon, T.E. Croley II, and F.H. Quinn. 1998. Transposed climates for study of water supply variability on the Laurentian Great Lakes. Climatic Change 38: 387-404.

Lehman, J.T. 2002. Mixing patterns and plankton biomass of the St. Lawrence Great Lakes under climate change scenarios. Journal of Great Lakes Research 28(4): 583-596.

Lenters, J.D. 2001. Long-term trends in seasonal cycle of Great Lakes water levels. Journal of Great Lakes Research, 27(3): 342-353.

Lofgren, B.M. and F.H. Quinn, A.H. Clites, R.A. Assel, A.J. Eberhardt, and C.L. Luukkonen. 2002. Evaluation of potential impacts on Great Lakes water resources based on two GCM climate scenarios. Journal of Great Lakes Research 28: 537-554.

Mackie, G.L. 2004. Applied Aquatic Ecosystem Concepts. 2nd edn. Kendall/Hunt Publishing Company, Dubuque, Iowa. ISBN 0-7575-0883-9

Magnuson, J. and K.E. Webster, R.A. Assel, C.J. Bowser, P.J. Dillon, J.G. Eaton, H.E. Evans, E.J. Fee, L.D. Mortsch, D.W. Schindler and F.H. Quinn. 1997. Potential effects of climate changes on aquatic systems: Laurentian Great Lakes and Precambrian Shield Region. Hydrological Processes 11: 825-871.

McCormick, M.J. 1990. Potential changes in thermal structure and cycle of Lake Michigan due to global warming. Transactions of the American Fisheries Society 119: 183-194.

McGeehin, M.A. and M. Mirabelli. 2001. The potential impacts of climate variability and change on temperature related morbidity and mortality in the United States. Environmental Health Perspectives, vol. 109, Supplement 2, May, 185-189.

Meisner, J.D. and J.S. Rosenfeld, and H. Regier. 1988. The role of groundwater in the impact of climate warming on stream salmonines. Fisheries 13: 2-8.

Meisner J.D. 1990. Effect of climatic warming on the southern margins of the native range of brook trout, Salvelinus fontinalis. Can. J. Fish. Aquat. Sci. 47: 1065-1070.

Millard, F. 2007. Climate Change and Great Lakes International Shipping. Wilfrid Laurier University Waterloo, Ontario, Canada.

Mortsch, L.D. and F.H. Quinn 1996. Climate change scenarios for Great Lakes Basin ecosystem studies. Limnology and Oceanography 41: 903-911.

Mortsch, L.D. 1998. Assessing the impact of climate change on the Great Lakes shoreline wetlands. Climatic Change, 40, 391-406. ON, Canada, 193 pp.

Mortsch, L.D. and H. Hengeveld, M. Lister, B. Lofgren, F. Quinn, M. Slivitzky, and L. Wenger. 2000. Climate change impacts on the hydrology of the Great Lakes-St. Lawrence system. Canadian Water Resources Journal 25(2): 153-179.

Mortsch, L.D. and M. Alden, and J. Klaassen. 2005. Climate Change Scenarios for Impact and Adaptation Studies in the Great Lakes-St. Lawrence system. Report prepared for the International Joint Commission, International Lake Ontario-St. Lawrence River Study Board, Hydrologic and Hydraulic Modelling Technical Working Group. Adaptation and Impacts Research Group, Waterloo, Ontario, 28pp.

Murdoch, P.S. and J.S. Baron, and T.L. Miller. 2000. Potential effects of climate change on surface-water quality in North America. Journal of the American Water Resources Association 36(2): 347-366.

Nastev, M. and M.M. Savard, D. Paradis, R. Lefebvre, M. Ross, and A. Rivera. 2002. Regional Hydrogeologic Characterization of the Fractured Aquifer System in South-Western Quebec- Part 2: Quantitative Study of the Groundwater Resource, Technical Report 37, Geological Survey of Canada.

Piggott, A. and D. Brown, B. Mills and S. Moin. 2001. Exploring the dynamics of groundwater and climate interaction. In Proceedings of the 54th Canadian Geotechnical and 2nd Joint IAH-CNC and CGS Groundwater Specialty Conferences. Canadian Geotechnical Society and the Canadian National Chapter of the International Association of Hydrogeologists, pp. 401-408.

Poff, N.L. and M.M. Brinson, and J.W. Day. 2002. Aquatic ecosystems and global climate change. Pew Center on Global Climate Change, Arlington, VA, USA.

Proceedings of a Binational Symposium. 1998. Mortsch, L. Assessing the impact of climate change on the Great Lakes shoreline wetlands. 13-15 May, Toronto Climatic Change 40: 391-416.

Rhodes, S. and K. Wiley. 1993. Great Lakes toxic sediments and climate change: implications for environmental remediation. Global Environmental Change 3(3): 292-305.

Schwartz, R.C. 2001. The potential impact of climate change on the Lake Huron shoreline at Goderich, Department of Geography, Faculty of Environmental Studies, University of Waterloo, Waterloo, Ontario, Canada.

Sharif, M. and D.H. Burn. 2006. Simulating climate change scenarios using an improved K-nearest neighbor model. Journal of Hydrology, 325: 179-196.

Sousounis, P. and P. Glick. 2007. The Potential Impacts of Global Warming on the Great Lakes Region. Retreived from www.climatehotmap.org/impacts/greatlakes.html

Stone, D.A. and A.J. Weaver. 2002. Daily maximum and minimum temperature trends in a climate model. Geophysical Research Letters, 29(9), 70: 1-70:4, 1356, doi:10.1029/2001GL014556

The National Atlas of the United States. "Zebra Mussles". United States Department of the Interior. http://www.nationalatlas.gov/articles/biology/a_zm.html retrieved online December 2010.

Trenberth, K.E. 1999. Conceptual framework for changes of extremes of the hydrological cycle with climate change. Climatic Change 42: 327-339.

Wall, G. and M. Sanderson [eds.]. 1990. Proceedings of an International Symposium/ Workshop on Climate Change: Implications for Water and Ecological Resources. Department of Geography Publication Series, Occasional Paper No. 11, University of Waterloo. Brunswick.

Wall, G., 1998. Implications of global climate change for tourism and recreation in wetland areas. Climatic Change, 40: 371-389.

WQB [Water Quality Board]. 2003. Climate change and water quality in the Great Lakes Region. Report to the International Joint Commission.

Section 4

Comparative Analysis of the Great Lakes

CHAPTER **16**

A Holistic Approach to Natural Resource Management: A Case of Lake Victoria Basin

Tom O. Okurut and Doreen M. Othero

Lake Victoria Basin Commission Secretariat, P.O Box 1510- 40100,
Kisumu, Kenya
E-mail: lvbc@lvbcsec.org

INTRODUCTION AND BACKGROUND

Lake Victoria (LV), with a surface area of about 68,870 km^2, is Africa's largest and the world's second largest freshwater lake. It lies at an altitude of 1134 m above sea level and relatively shallow with a maximum depth of about 80 m, and an average depth of about 40 m. Figure 16.1 shows the extent of coverage of the lake and its entire catchment. The catchment area is 180,950 square kilometres with Tanzania occupying 44%, Kenya 22%, Uganda 16%, Rwanda 11% and Burundi 7% (Table 16.1). Direct rainfall into the lake is its main source of water accounting for 82% while 18%, comes in through rivers. Evaporation accounts for 76% of water loss and 24% loss are through the River Nile outflow at Jinja, Uganda. The lake's water residence time is 23 years, while its flushing time is 123 years.

The lake is one of the most important shared natural resources for three of the East African Community Partner States; it is a major source of water and fisheries in the region (Shared Vision for Lake Victoria 2004). The vegetation cover around the lake basin comprised savannah and wetlands. Its biodiversity and ecosystem provide a wide range of species of aquatic life, plant and forest cover.

Fig. 16.1 Map of Lake Victoria and its catchment

Color image of this figure appears in the color plate section at the end of the book.

The population within the Lake Victoria Basin was approximately 35 million as of 2005 (RTDA 2007). Most of this population depends on the natural resources in the basin for their livelihoods. Activities such as agriculture and livestock production, hydropower generation, forestry, fisheries, wildlife and tourism, and mining constitute major sources of income and employment to the population within the basin.

The socio-economic importance of Lake Victoria to the Eastern Africa region is associated with the fact that, it is

(i) the largest inland water fishing sanctuary
(ii) a major inland water transport linkage for the three East African countries
(iii) a source of water for domestic, industrial and commercial purposes
(iv) a major reservoir for hydroelectric power generation
(v) a major climate modulator in the region and
(vi) rich in biodiversity

Considering the whole basin, the socio-economic importance widens to include the vast natural resources such as wildlife, forests and minerals together with fertile soils that sustain the main economic activity of the region – agriculture. However, despite these stated potentials, investments by both the local and international entrepreneurs are still low. The approximately 35 million people living within this basin are still bedevilled with poverty with average per capita incomes of between US$ 90- 270 (EAC 2003a).

Table 16.1 Description of Lake Victoria and its Basin

Country	*Lake Surface Area*		*Catchment Area*		*Lake Shoreline*	
	Km²	%	Km²	%	Km	%
Tanzania	33,756	49	79,570	44	1150	33
Uganda	31,001	45	28,857	15.9	1750	50
Kenya	4,113	6	38,913	21.5	550	17
Rwanda			20,550	11.4		
Burundi			13,060	7.2		
Total	68,870		180,950		3,450	

(Adapted from the Lake Victoria Environmental Management (LVEMP I) Project document, 1994)

MAIN DRIVERS FOR ADOPTION OF A HOLISTIC APPROACH FOR MANAGEMENT OF LAKE VICTORIA BASIN

The structure of management of the lake and its catchment prior to the arrival of colonialists was the responsibility of the local rulers of the time and their communities. Because of the low population and minimal interference with the catchment land based resources, the lake and the entire catchment remained nearly pristine.

However, national independence in the early 1960s by the three riparian countries of Lake Victoria and the subsequent adoption of development agenda for the new nations focussing on economic empowerment of the populace through improved agricultural activities coupled with increasing population density (RTDA 2007) triggered off a continuous pressure build up on the resources of the basin from then to date causing a severe environmental decline both in the catchment and in the lake itself. Typical examples include, the periodic spells of lack of oxygen (Hecky 1993); increase in algal concentrations and changes in the abundant species (Mugidde 1993) (Fig. 16.2); the reduction of the fish species diversity, the emergence of water hyacinth in the Lake (Twongo 1996) (Fig. 16.3) and the pollution loading from the municipal and urban centres spread in the entire basin (Okurut et al. 2003). In the catchment, major environmental and conservation shocks have arisen out of the high

Fig. 16.2 Algal blooms in Lake Victoria

Fig. 16.3 Water Hyacinth in Lake Victoria (March 2006)

Color image of this figure appears in the color plate section at the end of the book.

deforestation rates, poor agricultural practices, poor soil and water conservation practices and drainage of critical wetlands and silt loading to the lake has increased by two-fold (LVEMP 2001). It may be worth noting that each of the EAC countries has over time enacted several legislations that are intended to mitigate some of these pressures (NEMA Statute, Uganda, 1995, l EMCA, Kenya 1999; Environment Act, Tanzania 2006). However, the deterioration of the Lake Victoria Basin environment continues to be a challenge and this state was one of the key points for changing approach of managing the basin resources.

The consequence of the resultant environmental decline constitutes a formidable threat to the economic benefits hitherto derived from the national exploitation of natural resources of the Basin and the potential benefits thereof. Key among these resources affected include: (a) the fishery industry totalling over 1 million MT with an annual beach value of US$400 million annually as of 2006 (LVFO, Acoustic Survey Reports 2007); (b) Lake Victoria Inland water transport industry with annual turnover of just over US$10 million but with a potential of over US$100 million once the industry gets fully developed (LVBC LV Trade statistics 2008); (c) reliable water sources sustaining human and livestock populations as well as sustaining ecosystem services associated with it. The total revenues by the water service providers based in the region are over US $60 million but this could increase severally given the increasing demand and population (RTDA 2007). The water resources of the lake constitute and in some of the rivers provide sources of hydropower, in Uganda the potential amount to over 2000 MW (EAC Power Mater Plan 2005); and (e) the biodiversity in the lake, the insurance policy of resource sustainability. Two world renowned biodiversity and subsequently rich tourists' destinations are located in the basin – the Mara and Serengeti – but all driven by the Mara River ecosystem and have a combined annual revenue of over US $200 million (EAC Statistics 2007)

In addition to the environmental decline and its resultant effects, the other important driver for adoption of the holistic approach arose from the management challenges. First, many stakeholders have been attracted to the lake region due to its rich natural resource base but most of them have diverse interests especially in the lake itself. The motive of those involved is driven by the need to maximize exploitation of the natural resources in whatever manner but with minimum or no regard for the impact of their activities. A survey undertaken in the lake region showed that there are several programmes and projects that are being implemented by over 2000, associations and Non-Governmental Organizations (NGOs) with or without assistance from development partners and other interest groups and most of them did not effectively contribute towards the attainment of the common goals (EAC 2003, unpublished data).

Another characteristic feature that existed in the Basin was the minimal interaction between and among the various groups/associations/NGOs implementing various projects in the Lake Basin and often acted as adversaries to each other. This state of affairs was not healthy and as a result, most of the projects/programmes implemented are often duplicated even within the same areas and most often in the different Partner States. The concept of "a starting syndrome" was in play where each project was initiated as if no other one has existed before it and to learn from. Often donor funding is sought from same sources. It was therefore not surprising that poverty and degradation of the environment are still endemic in this region in spite of the interventions and the funding quoted.

These economic threats enhanced by unregulated actions on the population coupled with increased levels of both general and scientific awareness and capacity in the Partner States were and are still motivators of the political leaderships in the countries to appreciate ecosystem approaches for tackling the challenges. Further, governments have recognized the futility and risks of, unilateral actions by governments on basins' resources that have immediate consequences for all. The most recent controversial example was the drop in water levels in the lake due to the alleged over-releases at the outlet in Jinja; this has motivated all riparian to strengthen regional management arrangements.

Overall, the sum total of the threats opened up the thinking and led to the eventual decision to establish LVBC to coordinate holistically, the initiatives by the government, CSO and private sector including the development partners.

ROADMAP FOR INTRODUCTION OF HOLISTIC APPROACH AND PRACTICE IN LAKE VICTORIA RESOURCES

The genesis of the current management of Lake Victoria Basin (LVB) was a bold political decision by the Heads of State of the EAC Partner States (EAC Development Strategy 1997) as a long term to designate LVB as a "Regional Economic Growth Zone" which is to be exploited jointly in a coordinated manner so as to maximize its economic and social benefits as well as the environmental concerns. This political declaration which was informed by the output of the World Earth Summit of 1992 sparked the process of establishing structures that led to the entrenchment of holistic management of the LVB.

Regional Management Initiatives

Initial steps included commencement of mechanisms for addressing the concerns of poverty and environmental degradation in the Lake Victoria Basin. One first such mechanism was the establishment of the Lake Victoria Fisheries Organization (LVFO) in 1994. The key objective of this organization was to harmonize measures for sustainable utilization of the living resources of the lake and to develop and adopt conservation and management measures thereof (LVFO 1996). The second major intervention was the development and joint implementation of a regional project: the Lake Victoria Environmental Management Project (LVEMP), (LVEMP 1994). The implementation of this project started effectively in 1996 with a holistic approach in environmental management for the Lake Victoria Basin. Both LVFO and LVEMP have contributed enormously towards the knowledge base of the Lake Victoria environment as well as contributing towards poverty reduction in various ways.

The achievements and challenges gained from LVEMP and LVFO propelled conceptualization of a framework for a coordinated governance and management of this shared ecosystem. In addition, the experience learnt from management scenarios from other shared water bodies, such as the Baltic Sea and the American Great Lakes strengthened the argument and justification for the establishment of a central coordination body for the activities in the Basin at the East African Community level.

The strategy adopted for the management of Lake Victoria and its Basin was based on the identification of critical requirements that must be put in place to support all the initiatives. Key among these are:

(i) Sustained political commitment at the highest levels for the Lake Basin affairs.
(ii) Establishment of a regional institutional mechanism to coordinate the process of development.
(iii) Strengthening existing organizations that have overlapping mandates in the Lake Basin affairs.
(iv) Involvement and participation of the multi-stakeholders in Lake programmes in a coordinated manner.
(v) Partnerships with interest parties – local and international.
(vi) Funding by the Partner States governments.

RESULTS AND ACTIONS

To date the EAC Strategy for development and management of Lake Victoria and its Basin has revolved around responding to these critical factors as highlighted above. The sections below show the outcome or results of response to each of these factors.

Political Actions

The Political commitment by the EAC Partner States in the development and rationale management of the Lake and its Basin has been demonstrated and sustained from 1996 to date. This is clearly evident from the several decisions taken by the Summit (Heads of State) and the Council of Ministers at their various meetings that have been the basis for undertaking various interventions in the Lake. Typical examples of these include:

(a) The designation of the Lake Victoria and its Basin in the 1977-2000 EAC Development Strategy as an "Economic Growth Zone" and "An Area of Common Economic Interest" to be exploited jointly in a coordinated implementation process in order to maximize its economic and social benefits, while taking into consideration the need to provide an effective environmental management regime for posterity. These provisions were further crystallized in The Treaty for the Establishment of the East African Community (EAC 1999a) under various chapters and more specifically in Chapter 19. Arising from this decision several follow-up actions have been undertaken in the development of the management framework for the Lake ecosystem.

(b) The Establishment of Committee for Lake Victoria Development Programme (CLVDP) in 1999 (EAC 1999b). This Committee is composed of Permanent Secretaries from the relevant sector Ministries and is charged with the responsibility of providing the necessary policy guidance for the activities of the projects and programmes related to the Lake Victoria Basin. This Committee has actively steered the development initiatives of the Basin through their effective consultative fora.

(c) The Establishment of the Sectoral Council for Lake Victoria Basin in 2002 (EAC 2002) to provide overall policy directions for the development and management of the Lake Victoria Basin in accordance with the provisions of the Protocol. The Sectoral Council is now the top policy organ for the newly created Lake Victoria Basin Commission.

(d) The signing and ratification of the Protocol for Sustainable Development of Lake Victoria Basin in 2003 and 2004 respectively by all the three Partner States (EAC 2003b, 2004b). The Protocol provides the legal status of the Lake Victoria Basin Commission and policy guidance to the national initiatives related to the Lake Victoria Basin management. The application of this Protocol has already begun at both national and regional levels.

(e) The preparation, adoption of the Shared Vision and Strategy Framework (SVSF) for the Management and Development of Lake Victoria Basin and its use as a Planning Tool by all stakeholder in 2004 (EAC 2004a,b). The SVSF clusters priority areas for intervention under five policy areas. The use of this document by the various stakeholders is now gaining momentum among government, stakeholders and development partners.

(f) The preparation and development of a single legislation for regulation of transportation in Lake Victoria from 2002-2004 and the eventual enactment of the Lake Victoria Transport Act (2007) and the enabling Regulations concluded thereafter. This Act took effect in November 2009.

Regional Institutions and Mechanisms

As noted earlier on, the management of Lake Victoria and its Basin is very complex because of the different stakeholders undertaking different activities. The EAC Strategy was the sequential establishment of regional management mechanisms notably, the Lake Victoria Basin Commission (LVBC) is established as an institution of the East African Community (EAC). The institutional set-up of LVBC is unlike that in most of the River/Lake Basin organizations in Africa that have been formed outside the jurisdiction of existing regional governance mechanisms. The LVBC was uniquely set up to facilitate the fulfilment of the EAC strategic objectives for spurring economic development in the Lake Victoria region using the massive shared natural resources of Lake Victoria and its Catchment using a regionally coordinated approach (1st EAC Development Strategy 1997). The declarations in the Agenda 21 of the Rio conference (1992) informed the processes and justification for the establishment of the LVBC. Specifically, it was noted that shared natural resources such as Lake Victoria could no longer be sustainably managed using only national approaches but by considering their Transboundary nature and hence involvement of all the riparian states. Further, the rivers or lakes are continuously influenced by the activities in the catchment and therefore the sustenance of the water resource can only be secured by integrating whole catchment land-based activities in any management intervention identified. The current management of Lake Victoria and its Basin under the auspices of LVBC is now wholly based on an ecosystem and holistic approach.

Organizational Structure of LVBC

The current organizational structure for LVBC is based on Articles 34-42 of the Protocol for Sustainable Development of Lake Victoria Basin. The structure comprises the Sectoral Council, Coordination Committee, Sectoral Committees and the Secretariat. The LVBC Secretariat in carrying out its mandate works with the Development partners, Civil Society and Private Sector institutions as detailed in the sections that follow. Figure 16.4 shows the current structure of inter-linkages of the Commission.

As stated earlier, the political declaration by the EAC Heads of State to designate Lake Victoria Basin an Economic Growth Zone spurred the process of establishing strategies and structures for a holistic management for the basin as described below.

(a) *Legal and institutional arrangements:* These are defined in the Treaty for the establishment of the East African Community (1999a) and its annex, the Protocol for the Sustainable Development of Lake Victoria Basin (2004). The latter defines the broad mandate of the LVBC namely, to promote equitable economic growth; measures aimed at eradicating poverty; promote sustainable utilization and management of natural resources; the protection of environment

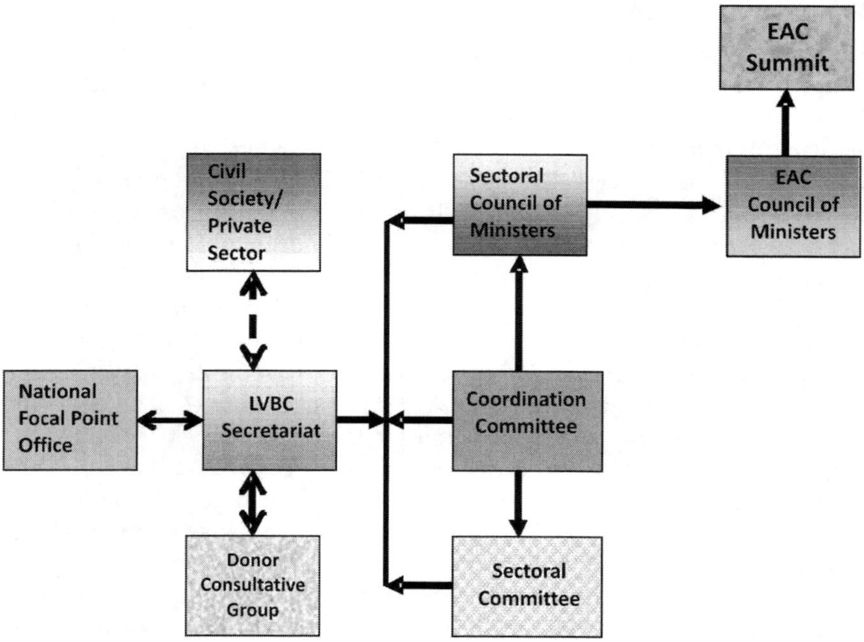

Fig. 16.4 Linkages of LVBC to other Organs/Institutions

with the Lake Victoria Basin; and compliance on safety of navigation. This broad mandate is to be implemented in the 14 areas of cooperation specified in the Protocol.

Setting of a long-term development agenda for 25 years for the basin by defining through an extensive consultative process, a Shared Vision and a development framework that is to guide all actors in the basin; this is the Vision and Strategy Framework (VSF) for the Management and Development of Lake Victoria Basin (2004). The VSF is anchored on three basic tenets: economic development, poverty reduction and environmental sustainability.

(b) *Mainstreaming the LVB development agenda into the EAC five-year development strategies*: This is in consideration of the fact that the LVBC drives the development agenda as an institution of the EAC and hence its programmes continue to be aligned to the overall development agenda of the Community.

Strengthening active involvement and participation of the non-governmental stakeholders in the management and development of the Lake Victoria Basin as provided for in the Protocol for sustainable development of Lake Victoria Protocol (2004). The regional Civil Society Organization and Private Sector Organization form part of the LVBC. Further, strong partnerships were continuously forged with international partners and funding agencies and this has enhanced focus on interventions that deal directly with issues defined in the VSF.

(c) *Public monitoring and evaluation of the Commission activities as well as the activities of all other LV stakeholders*: This is undertaken annually through the Annual Stakeholders Forums (ASF). The forum provides an opportunity for all participants to evaluate if all the activities undertaken have made us realize the intent of the political leaders that the lake Victoria Basin is an economic growth zone. To date the political commitment in the Lake Basin Affairs at the highest levels has been sustained and at each of the annual meetings, the LVBC has to report on the progress.

Current Coordination Arrangements of LVBC in the EAC Partner States

The National Focal Point Ministries for the Lake Victoria Basin Commission as provided for under Article 37(4) of the Protocol for Sustainable Development of Lake Victoria Basin provide a link between the LVBC Secretariat and Partner States. Currently, the Ministry of Water, Tanzania; Ministry of Environment and Mineral Resources, Kenya; Ministry of Water and Environment, Uganda; Ministry of Water, Environment, Land and Urbanization, Burundi and Ministry of Environment and Lands in Rwanda are the designated Focal Ministries.

Each of the Partner States has designated a Senior Officer as the National Focal Point Officer who on a regular basis liaises with different Government Ministries on common issues of the Basin development. They also facilitate stakeholder involvement in the development process, coordinate the activities of the Commission at country level, and seek to harmonize the implementation of programmes and projects within the Lake Victoria Basin.

In each Partner State, the coordination mechanism is such that the National Focal Point set up a National Steering Committee, which is a forum that brings on board various sectors of Government including private sectors and civil society organizations. The National Steering Committees have performed at different levels of success. However, it is realized that once fully developed it is a good forum for bringing on board key players that are relevant to the achievements of the mandates of the Commission.

The critical role of the NFP as seen in the extensive list of functions requires full-time staff. Under the current structure, the National Focal Points Officers in addition to the Lake Victoria Basin Commission have also other national responsibilities and hence are seriously overstretched and accordingly need beefing up. Currently, National Focal Points are financed by the Partner States.

Strengthening Existing Institutions

There are several national and regional institutions with mandates extending to the Lake. Since 1996, the EAC has strongly collaborated with these institutions in various ways but more so in promoting networking and data sharing. The most significant aspect for the national institutions was the strengthening/ building up of the human resource and infrastructure capacity through the

Lake Victoria Environment Management Project from 1994-2005 (LVEMP 1996–2004, unpublished data). The Lake Victoria Fisheries Organization (LVFO), which is an institution of the EAC responsible for fisheries management on Lake Victoria, has been strengthened in its working ethics especially in financial management and legal aspects. The East African Development Bank (EADB), which is also an institution of EAC has provided critical funding to the LVDP through the Partnership Funding and through this interrelation, EADB is now poised to fund a number of investment ventures in the Lake Basin.

The Inter University Council of East Africa is an institution of the Community responsible for higher education and research. Its capacity has been enhanced through fund mobilization from the government of Sweden for specific research activities focussing on the Lake Victoria Basin (VicRes). This programme started in 2003 and is operational to date. This programme is of great significance to various stakeholders who need research outputs for informed decision-making process.

EAC has strategically also supported the Nile Equatorial Lakes Subsidiary Action Program (NELSAP) which is responsible for implementation of three River Basin Integrated Water Resources Management Projects covering Kagera, Mara and Sio-Malaba-Malakisi. The implementation of these projects now has the necessary political clout to successfully drive the entire process.

Partnership and Cooperation with other Stakeholders

The role played by other stakeholders in the development of the Lake Victoria Basin is an undisputable factor that is recognized by the EAC. To tap on the resource base from these Partners, as well to refocus their initiatives to the same priority areas of LVBC/LVBC, the EAC, entered into partnerships with the following governments and institutions:

The Governments of France, Norway and Sweden, the World Bank and the East African Development Bank through a Partnership Agreement on Sustainable Development of Lake Victoria Basin (Partnership Agreement, 2001, unpublished data). The Agreement provides for a partnership to promote sustainable development in the Lake Victoria Basin on the basis of agreed objectives and principles as spelt out in the Agreement.

The World Conservation Union, Eastern Africa Regional Office (IUCN-EARO); the World Wide Fund for Nature (WWF), Eastern Africa Regional Programme Office (WWF-EARPO) and the International Centre for Research in Agroforestry (ICRAF) through Memoranda of Understandings (MoUs) concluded in 2002 and 2003 (MoUs: EAC-IUCN, 2002; EAC-WWF, 2003 and EAC-ICRAF 2003, unpublished data). Under the partnerships, these international NGOs are now also able to plan and develop their interventions in the different communities in a coordinated manner and along the same priority areas of the EAC with minimal duplication.

The LVBC works very closely with the regional Private Sector and Civil Society Organizations whose interests have a bearing in the development of the Lake Victoria and its Basin. Notable among these are: the East African Business Council (EABC) and the East African Communities' Organization for Management of Lake Victoria Resources (ECOVIC), the East African Law Society and the Lake Victoria Regional Local Authorities Cooperation (LVRLAC). The cooperation with LVRLAC for instance has culminated into the development of a joint project proposal with the EAC on water and sanitation. This proposal has now been concretized in cooperation with UNHABITAT under its realm of Lake Victoria water and sanitation initiative.

Funding by EAC Partner States

The political commitment highlighted earlier needed to be backed up by funding directly coming from the Partner States and not only from the donors. With regard to the development of the Lake Victoria and its Basin, the EAC Partner States have since creation of the LVBC fully funded the Recurrent Expenditure of the Unit. The countries contributions have steadily been increasing on an annual basis starting from US$ 66,950 in 2001 to US$ 1,911,380 in 2009. Besides, the Partner States have allocated budget lines towards meeting the costs for the National Focal Points offices in each country. This funding commitment is the critical indicator of the support the Partner States are giving towards the management and coordination modalities of Lake Victoria and its Basin.

Funding from the Development Partners

The LVBC development budget is mainly derived from the generous contributions it has received from various Development Partners. The main contributors are: (i) The Partnership Fund: this fund was established in 2002 following a Partnership Agreement signed between EAC and five partners, viz., the Governments of France, Norway and Sweden; the World Bank and East African Development Bank. The fund supports several capacity building projects in the Commission and Partner States; (ii) The Norwegian Government; an Agreement signed with EAC to support the implementation of the Mt. Elgon Regional Ecosystem Conservation Program (MERCEP); (iii) The French Government supports the Safety of Navigation project on Lake Victoria; (iv) The World Bank has supported the preparation of Lake Victoria Environmental Management Project (LVEMP II); (v) Sida supported the implementation of the LVEMP I project, LVEMP II bridging phase activities, the LVEMP II preparation activities; and the EAC regional HIV&AIDS Project in targeting mobile populations; (vi) The European Union supported the LVEMP II bridging phase activities.

CONCLUSION

The experiences from EAC activities in Lake Victoria and its Basin indicate that the management and coordination of a trans-boundary lake system requires a holistic approach anchored on a committed and focussed management body to coordinate all initiatives therein with a view of building synergies among different stakeholders. On the other hand, for this to be achieved, a high-level of political commitment is a prerequisite since it creates and stimulates a sense of ownership among the different actors as well as gaining the stakeholders willingness to cooperate in managing the resource. Strengthened institutions play a crucial role in developing and implementing shared programmes in the Lake. For these institutions to operate effectively, a funding mechanism from the governments is a necessity so as to justify and attract additional support from Development Partners. Finally, it is worth noting that the management and coordination of a shared lake ecosystem is not restricted to country administrative boundaries. A partnership with countries that benefit from or contribute to the destruction/disturbances of an ecosystem is an important commitment to ensure all countries are fully involved.

ACKNOWLEDGEMENTS

The authors acknowledge the contribution of the East African Community Partner States and her Secretariat, who contributed a great deal in the process and implementation of the Lake Victoria Ecosystem management and coordination mechanism. The authors, further express their appreciation to the Development Partners mainly the Governments of Sweden, Norway and France, and the World Bank and the East African Development Bank among others for their financial contribution that made this achievement a reality. The contributions from the stakeholders of Lake Victoria who have been supportive partners in building up the new management and coordination framework are acknowledged.

REFERENCES

EAC. 1997. First East African Cooperation Development Strategy; East African Community Publication.

EAC. 1999a. 13th Meeting of the Permanent Tripartite Commission Report. East African Community Secretariat Publications EAC/C13/02/99

EAC. 1999b. The Treaty for the Establishment of the East African Community. East African Community Secretariat Publications No. I, Arusha

EAC. 2002. 4th Meeting of Council of Ministers Report. East African Community Secretariat Publications EAC/CM/04/2002

EAC 2003a. EAC Statistics, facts and figures, 2002. East African Community Secretariat Publications No. III, Arusha.

EAC 2003b. The 6th Meeting of the Council of Ministers Report. East African Community Secretariat Publications EAC/CM6/CC/2003

EAC 2004a. The 7th Meeting of the Council of Ministers Report. East African Community Secretariat Publications EAC/CM7/CC/2004

EAC 2004b. The Protocol for Sustainable Development of Lake Victoria Basin. East African Community Secretariat Publications No. V.

EAC Statistics. 2007. East African Community Statistics- Facts and Figures; a Publication of the East African Community.

Hecky, R.E. 1993. The Eutrophication of Lake Victoria. Verh. Internal. Verein. Limnol. 25: 39-48.

LVBC. 2008. Lake Victoria Trade Statistics; a Publication of the Lake Victoria Basin Commission.

LVEMP 1994. The Tripartite Agreement on the implementation of the Lake Victoria Environmental Management Project and the project document. Available from URL: http://www.lvemp.org/. Accessed on September 9th 2005.

LVEMP 2001. Proceedings of the LVEMP Regional Scientific Conference, Kisumu, Kenya, 3-7 December 2001. Summary Available from URL: http://www.lvemp.org/. Accessed on 2nd September 2005.

LVFO 1996. Convention for the Establishment of Lake Victoria Fisheries Organization. Lake Victoria Fisheries Organization Secretariat Publications.

LVFO Acoustics Study Reports. 2006. Regional Catch Assessment Survey Reports April 2005- November 2006; December 2006.

Mugidde, R. 1993. The increase of phytoplankton primary productivity and biomass of Lake Victoria (Uganda). Verh. Internal. Verein. Limnol. 25: 846-849.

Okurut, T.O. and F. Marron, R.C. Kaggwa, and L.Okwerede. 2003. Assessment of industrial and municipal effluent and; urban runoff loads discharged into Lake Victoria from Uganda catchment. African Journal of Tropical Hydrobiology and Fisheries, vol. II.

RTDA. 2007. Regional Transboundary Diagnostics Analysis Report; a Publication of the Lake Victoria Basin Commission.

Twongo, T. The growing impact of water hyacinths on near-shore environments of Lake Victoria and Kyoga (East Africa). pp. 633-642. *In:* T.C. Johnson and E. Odada [eds]. 1996. The Limnology, Climatology and Paleoclimatology of the East Africa Lakes. Gordon and Breach Publishers.

CHAPTER 17

Top-Down and Bottom-Up Approaches in the Management of the Laurentian Great Lakes and Lake Victoria Fisheries: A Comparison of Two Shared Water Bodies

Marc Gaden,[1] Oliva C. Mkumbo,[2] Ted Lawrence[3] and Chris Goddard[4]

[1,4]Great Lakes Fishery Commission, 2100 Commonwealth Blvd. Ste. 100, Ann Arbor, MI 48105
E-mails: [1]marc@glfc.org; [4]cgoddard@glfc.org

[2]Lake Victoria Fisheries Organization, Plot No. 7B/7E Busoga Square, Belle Avenue, P.O. Box 1625, Jinja 256, Uganda
E-mail: ocmkumbo@lvfo.org

[3]School of Natural Resources and Environment, University of Michigan, 440 Church Street, Ann Arbor, Michigan, 48109
E-mail: ted@glfc.org

INTRODUCTION

The Laurentian Great Lakes of North America and Lake Victoria of East Africa support thriving fisheries that provide millions of people with jobs, food, income, subsistence, and recreational opportunities. The Great Lakes[1] comprise

[1]Throughout this chapter, the term "Great Lakes" refers to the Laurentian Great Lakes: Lakes Superior, Michigan, Huron, Erie, Ontario, and the connecting channels. The Great Lakes discussed in this chapter are not to be confused with the "Great Lakes of Africa," a term used to refer to Lakes Victoria, Malawi, Tanganyika, Albert, Kivu, Turkana, Edward, and others as a single group.

the largest system of freshwater lakes on the planet and Lake Superior and Lake Victoria rank first and second, respectively, in surface area as the world's two largest lakes. Fishery management authority in both regions is diffuse, with two nations, eight states, the province of Ontario, and U.S. tribes involved in Great Lakes fishery management and three partner states – Kenya, Tanzania, and Uganda – and scores of local entities called "Beach Management Units" involved in Lake Victoria fishery management. Officials in both regions coordinate their fishery management activities amongst themselves, while also understanding and incorporating stakeholder interests and needs into management actions.

This chapter discusses fishery management in both the Great Lakes and Lake Victoria Basins, with a focus on the primary structures and institutions established to facilitate interjurisdictional cooperation, collaborative processes, or co-management. Fishery management in both regions involves both "top-down" and "bottom-up" elements; top-down is defined here as the execution of fishery management by central government agencies, and bottom-up is defined here as resource-user or stakeholder involvement in the development of policy and engagement in management and enforcement. In the Great Lakes region, top-down authority is exercised by each of the states, the province of Ontario, and U.S. tribes in that they develop and implement their fishery regulations on behalf of stakeholders, while bottom-up coordination is done at the discretion of each jurisdiction and, thus, varies across the region. Cooperative inter-jurisdictional management occurs as well, coordinated by the Great Lakes Fishery Commission through a governance arrangement called *A Joint Strategic Plan for Management of Great Lakes Fisheries* (GLFC 1997), which serves to moderate unilateral, parochial action on a shared resource. On Lake Victoria, the Lake Victoria Fisheries Organization (LVFO) – the fisheries management arm of the East African Community – harmonizes fisheries regulations between the three partner states. The LVFO also coordinates fishery management responsibilities between each partner state and their respective community-level resource users in a co-management program (LVFO 2001). In both the Great Lakes and Lake Victoria regions, the established coordination structures were developed deliberately and over time to meet the specific needs of the respective regions.

DIMENSIONS OF TOP-DOWN AND BOTTOM-UP

Management of the Great Lakes and Lake Victoria fisheries ultimately comes down to issues of sovereignty and how best to respect that sovereignty when resources are shared among jurisdictions. A "sovereign" government has a defined territory, a defined population, autonomy, governmental powers, a legal identity, and fiscal independence (Holloway 1972). Sovereignty has two basic elements:

the government's ability to control its own domestic activities and its ability to interact autonomously and equally with other sovereigns (Haas and Sundgren 1990). Simply, sovereignty allows jurisdictions to pursue their own interests (Weiss 1999). A jurisdiction's sovereignty and its ability to exercise its will is limited by its ability to defend itself from the encroachments of other jurisdictions, its ability to expend resources to encroach upon the sovereignty of others, or its willingness to give up some sovereignty to achieve a collective goal. Conversely, preventing the exercising of another jurisdiction's sovereign will is difficult in the absence of binding mechanisms that include enforcement or punishment. Nevertheless, as this chapter will discuss, sovereign, top-down execution of a jurisdiction's will can be tempered by strategic governance approaches.

In fishery management, sovereign actions affect and in turn are affected by fishers' behavior, attitudes, and compliance. A management decision involving a quota, for instance, will influence how much a commercial fisher will harvest, while that commercial fisher's level of compliance with the quota will likely influence future management decisions. Because the relationship between those who govern the fishery and those who are governed is somewhat symbiotic, the process about how decisions are made and ultimately who has the authority for the decisions are central to management.

Decisions and subsequent management actions are regularly considered to be "top-down" or "bottom-up." Top-down (also called "command-and-control" or "managerial") decisions descend from a central authority—usually a government agency with prevailing regulatory or management responsibilities—and contain often broad policies that are to be applied to specific cases (Rachlinski 2006). Top-down management is driven by the sovereign will, goals, and responsibilities of the governments entrusted to allocate the fishery in the name of the public good; compliance on the part of the governed for rules and regulations is sought, regardless of whether the governed agree with the jurisdiction's goals (Bryan 2004). Throughout the world, sovereign authorities generally have the power requisite to carry out fishery management and, in fact, top-down approaches to management practices like quota-setting, law enforcement, and bag limits are the models generally followed (Acheson et al. 1998, Kooiman and Chuenpagdee 2005, Townsend and Pooley 1995). Townsend and Pooley (1995) say

> not only does the government determine the appropriate level of rights to allocate at each point in time, but it also has the sole responsibility for all other decisions that determine stock condition (such as closed seasons or minimum mesh sizes). The only responsibility that is distributed to individuals is the right to organize fishing activity within the constraint of the input or quota allocation.

Important here is the jurisdiction's ability to exercise its sovereign authority and the limited conveyance of discretion to the individual fishers. A jurisdiction's laws, regulations, and other fishery management mechanisms do not always include provisions mandating fisher involvement or consultation in decision-making (Bryan 2004), thus creating the inherent tendency toward top-down management.

In practice, however, government decisions are not made in a vacuum, absent of an understanding or consideration for socio-economic dimensions. Indeed, while government authorities may make fishery policies, laws, and regulations, they usually do so in consultation with the governed (Kooiman and Chuenpagdee 2005). Given the myriad interest in fisheries – whether stakeholders are interested in maintaining their livelihoods, preserving or enhancing recreational opportunities, or protecting subsistence fishing rights – governments rarely are afforded a blank cheque to impose unfettered power.

The incorporation of stakeholders into the process of fishery management is a "bottom-up" approach, and such approaches have been on the rise in fishery management for decades (Rettig et al. 1989, Townsend and Pooley 1995). Rettig et al. (1989) say: "What is . . . surprising is the increasing number of forums in which fishermen are seriously discussing new approaches [to management]." Bottom-up decisions emerge from the resource users themselves and reflect case-specific policies that might be applied broadly (Rachlinski 2006). This public participation could spawn from citizens who, through political processes, lobbying, or coalition-building demand to be heard, or it could come from government actions to facilitate input (Head 2007). This chapter focuses on the latter – the establishment of formal processes to ensure that interests beyond the individual jurisdiction's are incorporated into management. These processes take many forms, from formal to informal, from enforceable to non-binding, from narrow in focus to lakewide or basinwide. The point of bottom-up approaches is not to relinquish the government's public trust responsibilities to non-governmental actors, rather, it is to create some mechanism to ensure that the public's interests are heard and, to the greatest extent practicable, incorporated into management decisions.

Management processes used in Great Lakes and Lake Victoria fishery management fall into four broad categories: (1) interjurisdictional cooperation, (2) collaborative management, (3) decentralization, and (4) co-management. These four general approaches illustrate progressively stronger movement toward true stakeholder empowerment. These approaches define how people work together and stem from the notion that stakeholder involvement legitimizes and enhances management decisions (Jentoft et al. 1998, Krueger and Decker 1999).

Interjurisdictional Cooperation

Sovereign entities, whether nations or sub-national units, retain authority over their natural resources. When a common pool resource is involved, independent jurisdictions choose either to compete or cooperate when conflict arises (Kohn 1992). Sovereigns are "jurisdictionally equal" (Young 1994) and, as such, can erect barriers to intervention in their affairs. "Interjurisdictional cooperation" results when jurisdictions must and do work together to protect and sustain the shared resource. Undoubtedly, sovereign jurisdictions will face conflict over their fishery management activities, as jurisdictions often have differing management philosophies, needs, constituent pressures, and political dynamics. Managers may feel conflicted over whether to satisfy their parochial interests or to consider the interests of other jurisdictions. In the absence of communications, conflicting management philosophies could cause jurisdictions to work at cross purposes; to permit selfish, unsustainable harvest; or to distrust others' motivations or intentions.

To transcend competition and facilitate cooperation, a typical tool is the use of an institution or an agreement. On the international level, treaties are the strongest, most legalistic way nations cooperate. Treaties are enforceable through international law and often are supported by a commission or a secretariat to facilitate compliance. Agreements among non-federal governments (e.g. states, provinces, tribes) are also common. In the United States, for example, "interstate compacts" are agreements among states that bind them to a shared set of laws addressing a shared policy. In a non-binding or less-formal setting, governments discuss routinely interjurisdictional matters with each other and seek ways to harmonize regulations, share information, and establish reciprocal practices. In many cases, a commission or some other body is formed to ensure compliance with the agreement. The nature and level of formality of an agreement is, of course, highly dependent on the unique circumstances that exist and what the jurisdictions hope to accomplish. The ultimate goal is to facilitate interjurisdictional cooperation among sovereign entities.

Collaborative Management

The term "collaborative management" is often used synonymously with terms like "collaborative decision making," "citizen participation," "cooperative management," or "public consultation," but regardless of the term, such a process allows people to come together and address issues of mutual interest. Collaborative management occurs when citizens, either by government prompting or their own initiative, have an avenue to participate in the development and implementation of natural resource policies (Ansell and Gash 2008, Head 2007). Although the ability of citizens to be involved in bottom-up processes ranges broadly, collaborative processes go beyond simple information exchange and instead give participants a role in the decision-making process (Ansell and Gash 2008). Collaborative management as a concept, however, generally stops short of the government delegating responsibilities or authorities to stakeholders. Instead, collaborative management is a way for government to incorporate some bottom-up mechanisms into their decision-making processes.

Decentralization

"Decentralization" (also called "devolution") of natural resource management occurs when a government formally cedes authority and power to lower orders of government, institutions, or the resource-extractors themselves (Ribot 2002). The logic behind decentralized management is that local-level institutions and resource-users can often better determine and respond to local needs and aspirations because of their intimate knowledge of the resource and their proximity to it (Jentoft 2004, Ribot 2002). Indeed, some say the most effective governance is the arrangement closest to the resource, where local knowledge will be greatest and users will have an enhanced ability to interact frequently with local officials (see Chapter 10 for a discussion on the role of municipalities in governance of the Great Lakes). Decentralization goes beyond delegation of responsibilities to local governments or users; it is designed to increase the responsibility of the resource-user in hopes of better resource management. It creates a sense of ownership, investment, and deep responsibility for the resource (Jentoft 2004); it is an "institutionalized form of community participation" (Ribot 2002).

Co-management

"Co-management" is a broad, widely applied term that describes a cooperative arrangement whereby a traditional government agency shares power with other agencies, with local institutions, or even with the stakeholders themselves (Armitage et al. 2005, Berkes 2007, Castro and Nielsen 2001, Kooiman and Chuenpagdee 2005, Sen and Nielsen 1996, Townsend and Pooley 1995). Embedded in co-management is the sharing of authority and responsibilities (either voluntarily or forced, as in by a court), as multiple entities manage the same resource or the same waters, and share accountability. Co-management is a "horizontal" relationship, less focused on hierarchical control and more focused on the joint development and implementation of fisheries policies between the governed and government (Kooiman and Chuenpagdee 2005, Pinkerton 1989, Townsend and Pooley 1995). A jurisdiction could choose to create a co-management arrangement or could have that arrangement forced upon it (i.e. because another jurisdiction has responsibilities over the same waters or a court imposes it). Commonly, co-management arises after significant conflict and serves as a way to overcome competition (Castro and Nielsen 2001).

A key aspect of co-management is that no one jurisdiction has complete autonomy over the resource (Pomeroy and Berkes 1997). The literature discusses a wide range of co-management arrangements covering a variety of natural resource policies. Most commonly in fisheries, co-management refers to arrangements made between governments and local users as a way to involve the stakeholders at all levels of policy development, and often including implementation (Gough 2006, Kearney 1989). For instance, co-management could involve the government establishing broad policies and allowing local users to work out details equitably amongst themselves (Pinkerton 1989). Co-management is also discussed in terms of the relationship between governments and native peoples, where management authorities and expected responsibilities are carefully negotiated and relatively clear (Castro and Nielsen 2001). In many instances, treaties, agreements, court orders, or other legal mechanisms affirm or afford native peoples autonomous rights, which might include management responsibilities (e.g. Busiahn 1989; Doubleday 1989). The important point in cases of co-management between government and indigenous communities is the fact that, in some instances, both authorities manage the same waters and that co-management arrangements recognize such things as who has the right to harvest, who can restrict access to the fishery, and who can participate in management (Doubleday 1989).

One major challenge of co-management is the potential lack of balance in the power relationship between the government agency and the other entities involved in co-management (Townsend and Pooley 1995). In cases where the government devolves management to a local institution or a set of users, the government still retains management authority and, as such, can still dominate management. Often the local participants will need central governments to enforce laws and regulations or even to preserve the rights of the fisher (Folke and Berkes 1995). In other cases, such as arrangements between native peoples and the government, the co-management regime is the result of court cases or binding negotiations (e.g. Busiahn 1989, Doubleday 1989). Additionally, central governments can devolve management power to lower-levels of authority and not provide enough resources for effective management to persist. Unless co-management exists under law or court order, governments can exercise their ability to ignore or supersede many management decisions made through a co-management process, given governments often have more power than local institutions.

Scholars have identified benefits to a bottom-up approach. Berkes and Folke (1998), who discuss bottom-up actions in terms of co-management, hold the view that humans and the natural system are so closely linked that management decisions made without local knowledge or citizen input lean toward being arbitrary. Others stress that policies made through collaborative processes have the potential to be more balanced than top-down policies and convey a greater sense of ownership among the participants in the policies; citizens feel they have an enhanced stake in the matter (Bryan 2004, Hanna 1995, Krueger and Decker 1999, Pinkerton 1989, 2009, Pinkerton and John 2008). Because of a sense of shared ownership, policies originating at a local level are likely more accepted than top-down policies that come from a distant source, thus reducing conflict, enhancing compliance, and lessening the costs of implementation and enforcement (Acheson 2003, Berkes and Folke 1998, Bryan 2004, Pinkerton 1989, Pinkerton and John 2008, Townsend and Pooley 1995).

While many espouse the benefits of a bottom-up approach, others are not sure it serves to mobilize citizens effectively or even produce better policies. For instance, while a problem might generate strong, local interest, stakeholders often participate with a special, parochial agenda and a lack of broad knowledge on the topic. Moreover, those who participate in a bottom-up process are often those who are the most passionate about the issue; the bottom-up process, therefore, does not necessarily incorporate the range of views (Abel and Stephan 2000, Manthesian 1998, van Sittert 2002). A policy that fails to take the range of views into consideration is, some argue, an undemocratic reflection of parochial interests; special interests are allowed to have an inordinate affect on policy to the detriment of those interested in larger-scale ecosystem protection (Manring 2005, McCloskey 2000). If the government relies on collaborative processes to legitimize policies, those at the table will dictate the policy, regardless if the policy is sound (Lange 2001). In an attempt to establish bottom-up policies that are more participatory, strong local interests could work to prevent change, divert culpability, and even allow managers to avoid making tough decisions (Coggins 2001).

FISHERY MANAGEMENT ON THE LAURENTIAN GREAT LAKES

The Great Lakes are the major natural feature on the North American continent. From east to west, the lakes span more than 1,200 kilometers, a distance that is doubled if the Saint Lawrence River is included. Today, as historically, the Great Lakes fishery is important to the region economically and culturally. The region has a rich history of native fishing from the time of human inhabitation, commercial fishing since European settlement, and recreational fishing starting in the late 1800s and burgeoning the middle of the 20th century. Currently, the Great Lakes fishery provides at least $7 billion (ASA 2008, 2373) in economic return to the people of Canada and the United States annually, and hundreds of thousands of people earn income from fish-related business including commercial fishing, fish processing, charter fishing, and related industries such as tourism, supplying bait, and boat manufacturing.

The lakes' large geographic range spreads the fishery resources across many political jurisdictions, thus also diffusing fishery management responsibilities. In the Great Lakes region, two nations, eight states, the province of Ontario, and Native American Tribes border the lakes, and the non-federal governments exercise sovereign control over their fisheries, including migratory fishes. Because the state and provincial boundaries extend to the international border (Bogue 2000, Piper 1967), and because tribal fishing areas are defined by treaties (Busiahn 1985, 1989, Flanagan 2000, Zorn 1989), jurisdictional authorities are usually clear. Each jurisdiction formulates and executes its own policies in its own waters, illustrating that a jurisdiction will be motivated by its legal needs and political desires (Francis and Regier 1995). Through enumerated powers, ownership rights, court cases, precedent, and legislation, each of the non-federal jurisdictions retained or attained the authority to manage its section of the resource, though with some federal involvement as well. These authorities are generally understood and accepted, though they are not always exclusive (Gaden et al. in press). In the Great Lakes region, the states, in managing their fisheries, behave like nations.

The non-federal jurisdictions operate through their own agencies (e.g. Departments of Natural Resources) to carry out fishery management. These jurisdictions maintain the authority to manage in their own waters and, in fact, the Great Lakes fisheries are highly regulated. The individual jurisdictions establish harvest regulations, issue fishing licenses, stock fish, enforce the law, and conduct fisheries assessment. The individual state, provincial, and tribal legislative and regulatory bodies have full control over their own waters; they govern the behavior of their own citizens and have their own processes to develop, promulgate, and enforce their fishery regulations (Bence and Smith 1999, Brown et al. 1999). As such, Great Lakes fishery management has inherent top-down tendencies, since sovereignty over fishery resources allows each authority to impose its own regulations. Individual fishers in the Great Lakes region do not share the resource with each other *per se*, as fisher behavior is governed by the will of government; fishers have little ability themselves to make allocation decisions beyond those determined by their personal ethics (e.g. willingness to break the law, belief in harvest versus catch-and-release, etc.). In other words, fishers' behavior is governed by the top-down will of the individual jurisdictions, as determined and implemented through the management agencies.

How that authority is exercised varies from jurisdiction to jurisdiction. For example, in the State of Michigan, a Natural Resources Commission (made up of individuals appointed by the governor) and the Department of Natural Resources (DNR) are each involved in fishery matters, but the commission is consultative only, with the DNR retaining by law the ability to issue fishery rules and regulations. In the State of Ohio, the Department of Natural Resources has the authority to issue administrative rules that establish such things as bag limits, open and closed seasons, and fish size limits, though any rule change requires formal hearings and economic analyses. In Ontario, the Minister of Natural Resources maintains complete authority to issue commercial fishing regulations, while sport regulations are developed by the minister and formally adopted by the federal government. Individual tribes each retain the authority over on-reservation fishing and shares responsibilities with the states in ceded waters. To develop management rules, each tribe has its own natural-resource committee, but with varying levels of authority. The two intertribal organizations work to coordinate the activities of the member tribes, but the intertribal organizations do not have authority beyond that which is delegated to them. The other states each carry out their fishery management based on variations of these examples.

In practice, however, top-down management does not occur without considerable consultation with stakeholders. Indeed, the diffuse nature of jurisdictional authority and a heavily engaged constituency in the Great Lakes demand bottom-up management. Great Lakes fishery managers from the non-federal jurisdictions face myriad interests and organized stakeholder groups. These stakeholders, who tend to be well-organized, range from those promoting a Pacific salmon fishery to promoting commercial fishing to protecting native biodiversity. The fishery managers must balance those interests if they wish to make their policies defensible with both the public and with politicians (Ferreri et al. 1999, Gaden 2007, Krueger and Decker 1999).

To balance interests and incorporate bottom-up actions into individual management decisions, each of the non-federal jurisdictions, in one way or another, has established a robust regime for consulting stakeholders and incorporating their desires into management. These consultations vary widely in formality, from advisory-committee-type structures (whereby the jurisdiction assembles citizens to discuss proposed rules and regulations) to direct outreach with fishing clubs or business leaders to hear concerns, float new ideas, and receive other types of input. The cooperative arrangements between government and stakeholders that exist throughout the Great Lakes Basin influence, to varying degrees, the top-down decisions that are made in each jurisdiction. While some of the jurisdictions are compelled by law to consult stakeholders before, say, the promulgation of an administrative rule, none are required to accept stakeholders' input.

Bottom-up mechanisms are designed to create some degree of transparency in the decision-making process and allow stakeholders direct access to fishery officials. In Michigan and Ontario, for instance, lake-by-lake advisory committees and zone councils, respectively, are convened regularly as a way for DNR and ministry officials to meet directly with stakeholders; both are relatively formal processes. In Wisconsin, a "Conservation Congress," comprising elected citizens, meets annually to propose fishery rules. Approved rules go before the Wisconsin Natural Resources Board, though neither the board nor the Wisconsin Department of Natural Resources must accept the decisions of the Conservation Congress. In both New York and Ohio, public comment periods are required for the establishment or change of administrative rules.

Some individual U.S. tribes retain positions on their natural resource boards for commercial fishers. Fishery management staff in all jurisdictions attend annual meetings of sportfishing clubs, commercial fishing associations, and tribal natural resource committees for one-on-one conversations in informal settings. Overall, lines of communication between managers and stakeholders are open, regular, robust, and near-constant; each jurisdiction has its own process and culture dictating how those communications occur.

Stakeholder input is an integral component of fishery management throughout the Great Lakes Basin. Fishery managers interviewed for this chapter indicated that their consultations with users, whether required by law or not, frequently influence policy decisions, because managers have the ability to better gauge stakeholder acceptance and, thus, potential compliance. While none of the managers suggested they compromise the fishery by accepting suggestions that would be unsustainable, they do use the consultations to understand socio-economic needs and desires and then to formulate policies that are consistent with those needs. Said one fishery manager "Stakeholder groups need to know their role in helping to attain goals, but managers need to know the stakeholders' desires so that alternative solutions can be proposed and evaluated." Indeed, consultation with stakeholders is one way to facilitate the development of alternative solutions that might be both desirable to the stakeholders and contribute to achieving the management goals.

Fishery managers also report considerable communication benefits from a bottom-up, consultative process. Such a process is a "two-way street," said one manager, as it not only allows the managers to hear from stakeholders, but it also allows the stakeholders to hear from the managers about the agency's strategies and goals. Stakeholders, by participating in the process, feel more vested in the outcome than with a straight top-down approach. Managers, by consulting the stakeholders, have the opportunity to explain their policies and promote them directly with stakeholders. In fact, fishery managers generally hold that a major benefit of consultation is it heightens the chances that stakeholders will accept and embrace the agencies' policies and then communicate that acceptance to other members of their community (Gaden 2007). Stakeholders who have been in regular contact with the management agency are stakeholders who are more informed and better educated on the rationale behind the agency's decisions. Such participants are more likely to talk-up the policies with others and eschew political routes that might circumvent the jurisdiction's decisions. Broad acceptance also heightens the chances that fishers' behavior will be consistent with the agency's regulations.

Unlike the ability to enforce the laws and regulations upon their own citizens in a top-down manner, the individual jurisdictions cannot coerce their neighboring jurisdiction's behavior. Thus, a process of "interjurisdictional cooperation" needs to exist to help the jurisdictions coordinate fishery management amongst themselves. For much of the early history of Great Lakes fishery management, the many jurisdictions actually showed very little interest in harmonizing their fishery policies and instead exercised their sovereign will, regardless of the actions of other authorities (Bogue 2000, Gaden et al. in press, Gallagher et al. 1942, Piper 1967). This parochial use of top-down authority resulted in conflicting and unsustainable practices, which in turn led to serious declines of fishery resources (Bogue 2000, Dempsey 2001). This untenable fishery governance regime began to change in the 1960s when the binational Great Lakes Fishery Commission – urged by its enabling treaty to establish working relationships – formed "lake committees" as a place for state, provincial, and federal agencies to discuss matters and share information. By 1981, noting the need to be more strategic in policy and more defensive in fending off federal intrusion, the jurisdictions created a regional arrangement – *A Joint Strategic Plan for Management of Great Lakes Fisheries* – to help them identify and work toward their shared objectives (Gaden et al. in press). The plan is non-binding and, as such, is only as effective as the signatories wish it to be. The process is based on consensus and all members must accept (or at least be able to live with) a decision before the decision can move forward.

"Lake committees" are the mechanisms used to achieve interjurisdictional cooperation and to implement the plan. Under the plan, high-ranking officials from the non-federal agencies on each lake meet as a group to coordinate their fishery management responsibilities. To facilitate the generation of science and integrate the work of the field biologists into management, each lake committee has at least one technical committee, which consists of field-level professionals who undertake such tasks as deciding on data needs and then gathering and interpreting the data. Unlike the lake committees, whose members represent agencies with management authority, technical committees and task groups include as members federal officials and, by invitation, outside experts, such as academicians. Entities that have relevant data to share and that have suitable biological expertise are usually invited to participate in the technical committees. Through the Joint Strategic Plan, fishery managers meet regularly to engage in three primary collective action activities: to collect and share information; to decide on tasks and who carries them out; and to develop shared plans, policies, and objectives. The Joint Strategic Plan helps the fishery managers achieve shared goals through a cooperative process.

The deliberations through the Joint Strategic Plan process reflect elements of bottom-up management, whereby the jurisdictions take more than their individual will into account as they formulate their fishery policies. Although deliberations under the Joint Strategic Plan influence each jurisdiction's management decisions, the Joint Strategic Plan has served to coerce the jurisdictions, gently, into doing things they might not otherwise have done absent a common forum (Gaden 2007). The result of this interjurisdictional cooperative management is a process that facilitates collective action to manage a common property resource. The lake technical committees further facilitate the Joint Strategic Plan process. The technical committees, by ensuring joint collection and interpretation of data, ensure that no one entity has a monopoly of data and, therefore, can have an inordinate influence over decisions. Moreover, as the data have been collected and analyzed jointly, disagreements about the underlying science and the status of fish stocks do not occur. The work of the technical committees reflects the shared understanding of the field-level scientists because the information needs were identified and then collected and analyzed in a highly coordinated fashion.

Finally, in some areas of the Great Lakes, an arrangement that combines "interjurisdictional cooperation" and "co-management" exists between the States of Michigan, Wisconsin, and Minnesota; the U.S. federal government; and Native American tribes. The role of U.S. tribes in Great Lakes fishery management is relatively strong[2]: tribes have exclusive fishing rights on-reservation, where other fishing laws (e.g. state laws) are generally not applicable (Busiahn 1985, Cohen 1988). Off-reservation (areas known as "ceded" or "sold" lands, which include adjacent waters), tribal fishing rights in the Great Lakes were recognized in two treaties - one in 1836 and one in 1842 - between the tribes and the U.S. federal government (Anon 1873, GLIFWC 2006, Kappler [1904] 2006, Mattes et al. 2005).

While in the past, Native fishers were subjected to state fishery laws and regulations, court cases in the 1970s and 1980s re-affirmed tribal fishing rights; rights that could be exercised without a state license (Busiahn 1989, Chiarappa and Szylvian 2003, Zorn 2003). The major case affirming Tribal rights is *U.S. v. Michigan* of 1979 (United States District Court 1979). In this case, the court ruled that the state does not have exclusive management authority in the ceded waters. The consequence of this and other rulings has been to allow tribes to govern their own fishing activities, such as issuing licenses to their members, conducting law enforcement, and otherwise managing their fisheries, though the tribes must still manage consistently with state conservation goals (Busiahn 1985, 1989, Schlender (undated), Zorn 1989).

Tribal responsibilities in some cases, thus, overlap with state authorities, and both states and tribes regulate fisheries in the same waters. To manage a shared resource in this setting, the states, the federal government, and the U.S. tribes have developed an arrangement that maintains a significant level of autonomy for tribes to manage their fisheries, delineates respective state and tribal activities in the same waters that they each manage, allows for deliberation to occur between the states and the tribes, and brings tribes to

[2]In Canada, although treaties between the First Nations and the federal government also include tribal rights to fish and hunt, the courts have ruled that federal and provincial regulations do not themselves deny tribes access to fish, a conclusion that has often resulted in federal and provincial management on behalf of the tribes. In other words, co-management between the Canadian First Nations and the federal and provincial governments does not occur nearly to the same degree as it does in the upper Great Lakes region between the tribes and the state and federal governments.

the same interjurisdictional table that is used to facilitate interstate/provincial management on shared lakes. In the Great Lakes, the states and the tribes work together to formulate and adhere to agreements that allow licensing, open and closed seasons, species-specific management, and law enforcement, among other things. They collaborate on the development of shared policies. The "Consent Decree," an agreement between tribes, Michigan, and the federal government, governs the co-management relationship in northern Lakes Michigan and Huron and in eastern Lake Superior, and court-ordered tribal consensus on state management decisions and formal state-tribal committees govern co-management in western Lake Superior (United States District Court 2000, Zorn 2003).

Overall, each jurisdiction is responsible for its own waters, deciding on its own regulations, enforcing its own laws, and consulting its own citizens. The process to consult citizens varies by jurisdiction, and some processes are more formal than others. Nevertheless, all jurisdictions have mechanisms in place to meet with and consult stakeholders. Even with this relatively strong ability to manage parochially, the jurisdictions have also recognized the shared nature of the fishery and established a cooperative process that at once respects sovereignty while at the same time facilitates cooperative fishery management. That cooperative process is driven by the Joint Strategic Plan, the lake committees, and the technical committees. This structure is intended to promote the emergence of science through the technical process and to allow the independent jurisdictions to work together, avoid parochialism, and develop shared interjurisdictional management policies. It is a way for the independent jurisdictions to engage in interjurisdictional, cooperative management on a common property resource. Native American tribes are closely involved in management, participating in the process through interjurisdictional settings like the Joint Strategic Plan and exercising their rights in a co-management-like way in treaty waters.

FISHERY MANAGEMENT ON LAKE VICTORIA

Lake Victoria is the second largest freshwater lake by surface area in the world, second only to Lake Superior. A multi-use resource, it is known for its highly productive fishery, with about one million tons of annual fish production. The fishery employs more than three million people and brings in US$600 million annually to the East African Community (Njiru et al. 2008). Though only a single tropical lake, it produces more fish catch than all of the five commercial fisheries on the Laurentian Great Lakes combined. Like many other freshwater systems around the world, Lake Victoria faces numerous challenges including environmental degradation, introduction of exotic species, eutrophication, and over fishing. Recent stock assessments, for instance, have indicated a serious decline in the Nile perch (*Lates niloticus*) due to overfishing (Kayanda et al. 2009, Mkumbo and Mlaponi 2007, Mkumbo et al. 2007, Njiru et al. 2008, Ojuok et al. 2007, Witte et al. 2007).

Like the Laurentian Great Lakes, the management of Lake Victoria is multi-jurisdictional; three countries – Kenya, Tanzania, and Uganda – share the lake, 6, 51, and 43% respectively (Njiru et al. 2008). Whereas fishery management authority on the Great Lakes is vested in the non-federal governments, authority on Lake Victoria is decided at the national political level. Compared to the Great Lakes, where bottom-up processes are the prerogative of the management agencies and are largely *consultative*, Lake Victoria, while vesting management authority in national governments, nevertheless has developed a much more robust bottom-up *process* to decentralize authority and facilitate considerable co-management between the government and the local-level users. In fact, the social, political, and economic differences between the Laurentian Great Lakes and Lake Victoria create dynamics which change what "bottom-up" means and how it is implemented in each region. Because the governments see the Lake Victoria resources as a key factor in alleviating poverty, the success and failure of efforts to achieve sustainable fisheries might be considered a much higher priority in that region than in the Great Lakes. Moreover, where interest groups and local collective action can influence fisheries policies in the Great Lakes, in Lake Victoria, inherent weakness in community organization disenfranchises a large segment of the Lake Victoria users; users who depend on the resource as a relief from poverty. Thus, it is necessary to empower Lake Victoria communities in a "bottom-up" arrangement, such as formalizing co-management institutions between the national-level government and the communities. The fisheries governance regime that has emerged on Lake Victoria was designed to address many of the challenges of fisheries management on a common property resource by applying tenets of decentralization and co-management.

Decentralization of Lake Victoria fishery management largely evolved from a lack of compliance on the part of the resource users and the inability of the national governments to institute policies at a basin-wide level. Historically, a centralized, top-down approach to natural resource management in East Africa, exercised by sovereign nations, was based on both resource and economic extractive policies that neither benefited the resource nor the local populations (*see* Stein 2008, Wiebe and Dodge 1987). This led the local resource users to distrust the government and it created a culture of fishing activity that resembled Hardin's (1968) "Tragedy of the Commons." Further, a lack of agreement on measures to enforce shared management recommendations increased the level of government distrust and heavy skepticism of authority by the people who depended on the resource for their income (Abila et al. 2006, Mkumbo 2006, van der Knaap and Ligtvoet 2006, Wilson 2002). Unsustainable, parochial fishing practices contributed to the fisheries' exploitation at more than two times its sustainable level (Hecky 2003), devastating the resource, diminishing economic benefits, and harming livelihoods.

The three countries that border the lake have acknowledged that actions taken by users in one partner state will have impacts on other users in the lake's partner states. To coordinate their sovereign actions, the three nations formed the Lake Victoria Fisheries Organization by convention, which entered into force on May 24, 1996. The LVFO plays two important roles: first, the LVFO coordinates and harmonizes national fisheries management and research measures and second, it develops and adopts conservation and management measures for the sustainable utilization of the fishery resources of the Lake.

The LVFO is the central, harmonizing organization within the Lake Victoria fishery management institution. Uganda, Kenya, and Tanzania are the partner states, or contracting parties within the institution, and the other organizations included in the fishery management institution on Lake Victoria include the three partner states' fisheries departments, research institutions, and the local fisher communities that exist around the lake. Like *A Joint Strategic Plan for Management of Great Lakes Fisheries*, the LVFO is a way to achieve "interjurisdictional cooperation" between the partner states.

The LVFO achieves its regional coordination through the organization's statutory organs (LVFO 2001). The Fisheries Management and Fisheries Research committees (constituted by directors of fisheries departments and research institutions from the three contracting parties) are the technical arms of the LVFO, which receive fisheries management and research information from the organization's working groups. The working groups conduct research and socio-economic studies related to the status of the stocks and the welfare of the fishing communities. The two committees recommend measures to the LVFO's Executive Committee (made up of the six directors of fisheries management and research institutions) which in turn considers and recommends appropriate management measures to the Policy Steering Committee, comprising chief executives of the ministries dealing with fisheries matters in each of the contracting parties. The Policy Steering Committee submits the recommendations to the supreme body, the Council of Ministers (comprising ministers responsible for fisheries matters in the contracting parties), which adopts management and conservation measures. The LVFO committees hold one session annually to review progress and compliance in implementing the agreed measures. The measures are also revised as needed. The Council of Ministers meets biennially, though special sessions are held as the need arises.

At the regional level, the LVFO plays the role of harmonizing measures, advocating for the resources and the users, and disseminating the agreed measures to the contracting parties after adoption by the Council of Ministers. The convention establishing the LVFO demands the contracting parties provide the organization with such laws, regulations and any other related documents and information for the purpose of assessing compliance or any disparities which will warrant adjustments and harmonization.

Top-down exertion of authority in the Lake Victoria region is essentially used only to develop broad policies through the LVFO's Council of Ministers. The contracting parties adopt the agreed regional measures and formulate their national laws and regulations to fit their situations and enforce them in their territorial waters. Each contracting party exercises direct (top-down) authority on matters regarding enforcement of laws and regulations. Like the Joint Strategic Plan in the Great Lakes region, the LVFO committee structure provides the mechanism to seek and implement shared objectives among the parties. However, also like the Great Lakes region, each of the national governments separately controls policy-making, implementation, and enforcement within their area of jurisdiction. While the LVFO as an institution is non-binding, the ministers in each of the three contracting parties sharing Lake Victoria depend on the LVFO to provide independent advice, develop sound science, and facilitate the development of policies that maximize the benefits of Lake Victoria stakeholders. Additionally, *Article XIII, Sect 4* of the convention establishing the LVFO states that "[E]ach Contracting Party shall remain free to adopt national laws and regulations more stringent or extensive than those required to fulfill its obligations" (LVFO 2001). The three partner states report their progress in implementing agreed measures to the executive secretary of the LVFO.

Two overarching premises influence community involvement in natural resource management: first, the need for communities to have the opportunity to make decisions about management of the resource on which they rely, and second, effective resource management through the use of local knowledge (Coffey 2005). The LVFO is designed to transcend resource issues across national borders and to overcome the differences in fisheries management that comes with varying degrees of decentralization between the three governments themselves. At the same time, local-user or community participation in natural resource management programs in underdeveloped countries is seen as a necessary component for effectively managing natural resources, in part because of the past failure of centralized resource management efforts and in part because of the realization that communities are essential for effective management (*see* Kaimowitz and Ribot 2002).

While the three partner states sharing Lake Victoria exercise their sovereign will and use established mechanisms to work together, they also have ways to engage directly with the stakeholders, primarily the subsistence fishers of Lake Victoria, in a bottom-up way. Lake Victoria fishing is highly localized and community-based; fishing communities are spread throughout the lake's shoreline. To ensure participation of the fishing communities and other stakeholders in resource management and development, the LVFO adopted in

2007 a co-management approach called "Beach Management Units," or BMUs. BMUs, which are legally recognized by each contracting party, are community-run governing bodies which manage the fishery and share policy development, enforcement, and research duties with each national government's fisheries department. The fishers themselves throughout the Lake Victoria Basin help standardize rules of fisheries extraction. The BMUs enforce the lake's fishing rules, serve as resource-data collection points for better fisheries management and monitoring, and increase members' capacity to manage their fishing profits (Ebong et al. 2003). BMUs have been established at all Lake Victoria landing sites, totaling 1,069 BMUs, with each BMU containing no fewer than 30 fishing boats. Consequently, landing sites with fewer than 30 boats are required to amalgamate with other smaller sites to form a single BMU. These beach-level management institutions operate using Harmonized Beach Management Guidelines, which were approved by the LVFO's Council of Ministers in 2006 (LVFO 2007). According to the BMU guidelines, each BMU has a structured executive committee consisting of a wide range of local stakeholders (e.g. fishers, boat owners, traders, gear repairers); each set of guidelines is formally legalized. Collaborative action between neighboring BMUs is facilitated by the formation of higher level BMU committees aligned to government administrative systems in each country. The BMU committees at district and national levels are represented regionally by a regional committee (called the Regional BMU Network), which is offered membership on the statutory technical fisheries management and scientific committees of the LVFO. BMUs' operational funding comes from levying equipment and fishing licensing fees, fish-catch taxes, and illegal fishing fines on individuals involved in the fishery (LVFO 2005). Fees, taxes, and fines are levied by the BMU committee, vetted through district-level governments, and returned to the BMU as a percentage of the district total. In some cases, fishers make voluntary, in-kind contributions toward BMU operations, often in the form of fish to be sold, with the proceeds used to pay workers for cleaning the surroundings at a given landing site, local security, or other social programs.

To include other key stakeholders in fishery management and to ensure compliance, the Fish Processors and Exporters Associations of the three partner states established a collaborative process, known as the East African Industrial Fishing and Fish Processor's Association (EAIFFPA). The association is given membership on the statutory technical committees of the LVFO. The (EAIFFPA) has imposed strict guidelines parallel to those of the LVFO's to ensure sustainable harvest of Nile perch in Lake Victoria. Severe penalties are imposed on processors who fail to follow the guidelines. The (EAIFFPA) has also established a self-monitoring team in each partner state to ensure compliance by fish processing factories to the minimum slot sizes for Nile perch processing and have moved ahead to establish a regional self monitoring team which is allowed in any fish factory in the three nations for inspections. Any factory not complying is reported to the government authorities, which can deny a factory to operate for a given time as per the agreed guidelines governing self-monitoring.

The (EAIFFPA) and the BMUs play active roles in self-policing, enforcement, and sensitization of their respective communities. They are allowed to develop their own by-laws so long as the by-laws are consistent with the national fisheries laws or regulations. The authority is, therefore, with the central government, but with delegated powers to the different stakeholder groups like the BMUs. BMUs have a mandate to protect and conserve fishery resources in their areas of jurisdiction, in collaboration with the fisheries officers from the local and district levels. National fisheries committees, which are formed pursuant to the convention, bring in all the relevant stakeholders at the national level to deliberate on national fisheries matters before they are taken to the regional level for harmonization through LVFO structures.

The Lake Victoria fishery management institution is a co-management approach to managing Lake Victoria's fisheries. The institution is based on decentralization, which includes a formal relationship between the national government and local communities through the BMUs. This co-management arrangement incorporates the concepts that active citizen participation at all levels of fisheries governance can protect the diverse interests of those affected by environmental problems (Lemos and Agrawal 2006, Nunan 2006); that natural resources are a source of livelihood and income, and, therefore, are best managed by the local people (self-governance) (Kaimowitz and Ribot 2002, Ostrom 2009); and, that rather than being an expense for central governments, natural resources become a major source of revenue and poverty alleviation when the appropriate property rights and management scheme is instituted (Kaimowitz and Ribot 2002).

Bottom-up processes of fishery management are indeed stronger and more formalized in the Lake Victoria region than in the Great Lakes region. Through the fisheries governance institution on Lake Victoria, the three nations include all fisheries stakeholders in decision-making, incorporate the needs of the poor and disenfranchised, and employ bottom-up, community-based development in management plans. These design elements are intended to motivate and reinforce legal fishing behavior and, therefore, a sustainable fishery. Like the non-federal jurisdictions on the Great Lakes, the three nations of Lake Victoria have also developed processes for interjurisdictional cooperative management. The three partner states have agreed to work together to achieve shared objectives, but in a way that respects their sovereignty and their individual constitutional procedures and national laws. And, like the Great Lakes jurisdictions, they agree to implement any decisions that arise through the consensus-based, cooperative process.

CONCLUSION

Neither the Lake Victoria nor Laurentian Great Lakes fisheries are strictly "open access" resources; rather, governments hold the fisheries in public trust, determine the rules and regulations governing fishers' behavior, carry out fishery management actions, and enforce the rules. Authorities in both the Great Lakes and Lake Victoria regions decide on the level of public involvement or devolution to communities in management. The Lake Victoria and Great Lakes fisheries are "common property resources" to the extent that nations and sub-national jurisdictions, with sovereign rights and responsibilities, share the fisheries. Resource-extractors are the jurisdictions themselves who, in the case of the Great Lakes, decide how much of the resource they will allocate to individual fishers[3] and in the case of Lake Victoria, control harvest by limiting the type of gear, harvest techniques, and fishing in areas of spawning habitat.

In both the Great Lakes and Lake Victoria Basins, multiple jurisdictions and stakeholders share the resource, each with varying views about how the fisheries should be managed. Governments in both basins – whether federal or non-federal – strongly retain their sovereignty over their fishery resources and exert top-down authority as they manage fishers' behaviors. Even with multi-jurisdictional agreements like *A Joint Strategic Plan for Management of Great Lakes Fisheries* and the LVFO's Fisheries Management Plan, day-to-day management decisions that affect fishers behavior – decisions related to licensing, harvest, gear – remain the prerogative of the governments, and the governments generally remain willing and able to exert their will as they see fit.

Nevertheless, both basins to varying degrees maintain cooperative management between the governments and the stakeholders. In the Great Lakes region, interjurisdictional cooperation and co-management exists between government and the U.S. tribes. Interjurisdictional cooperation also exists among the non-federal agencies through the Joint Strategic Plan, and each of the non-federal jurisdictions incorporates bottom-up collaborative processes into their management decisions through the use of advisory council-type arrangements. On Lake Victoria, the three partner states have established a formal co-management process administered by the Lake Victoria Fisheries Organization. Local-level participation is incorporated into policy through the legally recognized Beach Management Units, which reflect decentralization of authority and co-management. Indeed, in Lake Victoria, local governance arrangements serve as the foundation of influencing fishers behavior and to instill a sense of ownership in the resource.

---

[3]In the Great Lakes, jurisdictions establish individual commercial fishing quotas or individual bag limits for recreational anglers. In Lake Victoria, there is no system of allocation to individual fishers, but once a fisher has a fishing license, he fishes throughout the year.

In both the Great Lakes and Lake Victoria, the management structures reflect the history, composition of the fisheries, and particular socio-economic needs of the fishers. In the Great Lakes region, non-federal sovereignty allows each jurisdiction to promulgate its own fishery regulations, while the shared nature of the fishery mandates co-management with the U.S. tribes and cooperation among the non-federal entities. The growth of educated, capable stakeholders, representing myriad interests, mandates a high level of consultation with the stakeholders so that policies are better accepted. In Lake Victoria, a lack of financial resources at the national level to enforce fishery regulations has led to the emergence and sanction of the local Beach Management Unit as the primary community-level fishery management arrangement. Moreover, the nature of dependence on fisheries by local-level users on Lake Victoria necessitates a more-robust co-management relationship between the national government and fishers than on the Great Lakes.

REFERENCES

Abel T.D. and M. Stephan. 2000. The limits of civic environmentalism. American Behavioral Scientist 44: 613-627.

Abila R. and P. Onyango, and K. Odongkara. 2006. Socio-Economic Viability and Sustainability of BMUs: Case Study of the Cross-Border BMUs on Lake Victoria. Paper presented at Great Lakes of the World IV; February 21, 2006, Bagamoyo, Tanzania.

Acheson J.M. 2003. Capturing the commons. Devising institutions to manage the Maine lobster industry. Univesity Press of New England Hanover and London.

Acheson J.M. and J.A. Wilson, and R.S. Steneck. Managing chaotic fisheries. pp. 390-413. *In:* F. Berkes, C. Folke, and J. Colding [eds.]. 1998. Linking Social and Ecological Systems: Management Practices and Social Mechanisms for Building Resilience, Cambridge University Press, Cambridge, U.K. New York, NY, USA.

Anonymous. 1873. A compilation of all the treaties between the United States and the Indian tribes. U.S. Government Printing Office. Washington, DC, USA.

Ansell C. and A. Gash. 2008. Collaborative governance in theory and practice. Journal of Public Administration Research and Theory 18: 543-571.

Armitage D. and F. Berkes, and N. Doubleday. Introduction: Moving beyond co-management. pp. 1-15. *In:* T.S. Gray [ed.]. 2005. Participation in Fisheries Governance, vol. 4 Springer, Dordrecht, The Netherlands.

ASA. 2008. Today's angler: A statistical profile of anglers, their targeted species and expenditures. American Sportfishing Association. Alexandria, VA, USA.

Bence J.R. and K.D. Smith. An overview of recreational fisheries of the Great Lakes. *In:* W.W. Taylor and C.P. Ferreri [eds.]. 1999. Great Lakes Fisheries Policy and Management: A binational perspective, Michigan State University Press, East Lansing. Michigan, USA.

Berkes F. Adaptive co-management and complexity: Exploring the many faces of co-management. pp. 19-37. *In:* D. Armitage, F. Berkes, and N. Doubleday [eds.]. 2007.

Adaptive Co-management: Collaboration, learning, and multi-level governance, UBC Press, Vancouver. Canada.

Berkes F. and C. Folke. Linking social and ecological systems for resilience and sustainability. pp. 1-25. *In:* F. Berkes, C. Folke, J. Colding J [eds.]. 1998. Linking Social and Ecological Systems: Management Practices and Social Mechanisms for Building Resilience, Cambridge University Press, Cambridge, U.K. New York, NY, USA.

Bogue M.B. 2000. Fishing the Great Lakes: An Environmental History, 1833-1933. University of Wisconsin Press. Madison, USA.

Brown R.W. and M. Ebener, and T. Gorenflo. Great Lakes commercial fisheries: Historical overview and prognosis for the future. *In:* W.W. Taylor and C.P. Ferreri [eds.]. 1999. Great Lakes Fisheries Policy and Management: A Binational Perspective, Michigan State University Press, East Lansing. USA.

Bryan T.A. 2004. Tragedy averted: The promise of collaboration. Society & Natural Resources 17: 881-896.

Busiahn T.R. 1985. An introduction to native peoples' fisheries issues in North America. Renewable Resources Journal 3: 23-26.

Busiahn T.R. The development of state/tribal co-management of Wisconsin fisheries. pp. 170-185. *In:* E. Pinkerton [ed.]. 1989. Co-operative Management of Local Fisheries: New Directions for Improved Management and Community Development, University of British Columbia Press, Vancouver, Canada.

Castro A.P. and E. Nielsen. 2001. Indigenous people and Co-management: implications for conflict management. Environmental Science and Policy 4: 229-239.

Chiarappa M.J. and K.M. Szylvian. 2003. Fish for All: An Oral History of Multiple Claims and Divided Sentiment on Lake Michigan. Michigan State University Press. East Lansing, USA.

Coffey, C. 2005. What role for public participation in fisheries governance? Participation in Fisheries Governance (Reviews: Methods and Technologies in Fish Biology and Fisheries). T. S. Gray. Dordrecht, The Netherlands, Springer. 4: 27-44.

Coggins G.C. 2001. Of Californicators, quislings, and crazies: Some perils of devolved collaboration. pp. 163-171. *In:* P. Brick, D. Snow, and S. Van de Wetering [eds.]. 2001. Across the Great Divide: Exploration in Collaborative Conservation and the American West, Island Press, Washington, DC. USA.

Cohen F.S. 1988. Cohen's handbook of federal Indian law. United States Government Printing Office. Washington, DC, USA.

Dempsey D. 2001. Ruin and Recovery: Michigan's Rise as a Conservation Leader. University of Michigan Press. Ann Arbor, Michigan, USA.

Doubleday N. Co-management provisions of the Inuvialuit Final Agreement. pp. 209-230. *In:* E. Pinkerton [ed.]. 1989. Co-operative Management of Local Fisheries: New Directions for Improved Management and Community Development, University of British Columbia Press, Vancouver, Canada.

Ebong I. and M. Lwanga, and J. Scullion. 2003. Beach Management Units and Integrated Lake Management. International Workshop on Community Participation in Fisheries Management on Lake Victoria. Kisumu, Kenya.

Ferreri C.P. and W. Taylor, and J,M. Robertson. Great Lakes fisheries futures: Balancing the demands of a multijurisdictional resource. *In:* W.W. Taylor and C.P. Ferreri [eds.]. 1999. Great Lakes Fisheries Policy and Management: A Binational Perspective, Michigan State University Press, East Lansing, USA.

Flanagan T. 2000. First Nations? Second thoughts. McGill-Queen's University Press. Montreal and Kingston.

Folke C. and F. Berkes. Mechanisms that link property rights to ecological systems. pp. 121-138. *In:* S. Hanna and M. Munasinghe [eds.]. 1995. Property Rights and the Environment: Social and Ecological Issues, Beijer International Institute of Ecological Economics and The World Bank, Washington, DC., USA.

Francis G. and H.A. Regier. Barriers and bridges to the restoration of the Great Lakes basin ecosystem. pp. 239-291. *In:* L.H. Gunderson, C.S. Holling and S.S. Light [eds.] 1995. Barriers and Bridges to the Renewal of Ecosystems and Institutions, Columbia University Press, New York, USA.

Gaden M. 2007. Bridging jurisdictional divides: Collective action through a Joint Strategic Plan for Management of Great Lakes Fisheries. Ph.D. diss. University of Michigan. Ann Arbor, Michigan, USA.

Gaden M. and C. Goddard and J. Read (in press). A history of multi-jurisdictional management of the shared Great Lakes fishery: Transcending conflict and diffuse political authority. *In:* W. Taylor, A.J. Lynch and N. Leonard [eds.]. Great Lakes Fishery Management and Policy, Michigan State University Press, East Lansing, USA.

Gallagher H.R. and A.G. Huntsman, D.J. Taylor and J. Van Oosten. 1942. Report and Supplement, U.S. Government Printing Office. Washington, DC, USA.

GLFC. 1997. A Joint Strategic Plan for Management of Great Lakes Fisheries. Great Lakes Fishery Commission. Ann Arbor, Michigan, USA.

GLIFWC. 2006. A guide to understanding Ojibwe treaty rights. Great Lakes Indian Fish and Wildlife Commission. Odanah, WI. (online at www.glifwc.org/pub/treaty_rights.pdf)

Gough J. 2006. Managing Canada's Fisheries: From Early Days to the Year 2000. McGill-Queens University Press. Georgetown, ON, Canada.

Haas P.M. and J. Sundgren. Evolving international environmental law: changing practices of national sovereignty. pp. 401-430. *In:* N. Haigh and F. Irwin [eds.]. 1990. Integrated Pollution Control in Europe and North America, The Conservation Foundation, Washington, DC., USA.

Hanna S. 1995. User participation and fishery management performance within the Pacific Fishery Management Council. Ocean & Coastal Management 28: 23-44.

Hardin G. 1968. The Tragedy of the Commons. Science 162: 1243-1248.

Head B.W. 2007. Community engagement: Participation on whose terms? Australian Journal of Political Science 42: 441-454.

Hecky R. 2003. What does the future hold for the Great Lakes of Africa? African Journal of Aquatic Science 28: iii-vi.

Holloway W.V. 1972. Intergovernmental relations in the United States. MSS Information Corp. New York, USA.

Jentoft S. The community in fisheries management: Experience, opportunities and risks. *In:* B. Hersoug, S. Jentoft and P. Degnbol [eds.]. 2004. Fisheries Development: The Institutional Challenge, Eburon, Delft, The Netherlands.

Jentoft S. and B.J. McCay and D.C. Wilson. 1998. Social theory and fisheries co-management. Marine Policy 22: 423-426.

Kaimowitz, D. and J.C. Ribot. 2002. Services and infrastructure versus natural resource management: building a base for democratic decentralization. Conference on Decentralization and the Environment, Bellagio, Italy, Washington, DC: World Resources Institute.

Kappler C. [ed.] [1904] 2006. Treaty with the Chippewa, 1942. *In:* Oklahoma State University (online at digital.library.okstate.edu/kappler/), ed. Indian affairs: laws and treaties, vol. II, Treaties U.S. Government Printing Office, Washington, D.C., USA.

Kayanda R. and A.M. Taabu, R. Tumwebaze, L. Muhoozi, T. Jembe, E. Mlaponi and P. Nzungi. 2009. Status of the major commercial fish stocks and proposed species-specific management plans for Lake Victoria. African Journal of Tropical Hydrobiology and Fisheries 12: 15-21.

Kearney J.F. Co-management or co-optation? The ambiguities of lobster fishery management in southwest Nova Scotia. pp. 209-230. *In:* E. Pinkerton [ed.] 1989. Co-operative Management of Local Fisheries: New Directions for Improved Management and Community Development, University of British Columbia Press, Vancouver, Canada.

Kohn A. Cooperation: What it means and doesn't mean. pp. 3-11. *In:* A. Combs [ed.] 1992. Cooperation: Beyond the Age of Competition, vol. 4, Gordon and Breach, Philadelphia, USA.

Kooiman J. and R. Chuenpagdee. Governance and governability. pp. 325-349. *In:* J. Kooiman, M. Bavinck, S. Jentoft and R. Pullin [eds.]. 2005. Fish for Life: Interactive Governance for Fisheries, Amsterdam University Press, Amsterdam, The Netherlands.

Krueger C.C. and D.J. Decker. The process of fisheries management. pp. 31-59. *In:* C.C. Kohler and W.A. Hubert [eds.]. 1999. Inland Fisheries Management in North America, American Fisheries Society, Bethesda, USA.

Lange J.I. Exploring paradox in environmental collaborations. pp. 200-209. *In:* P. Brick, D. Snow and S. Van de Wetering [eds.]. 2001. Across the Great Divide: Exploration in Collaborative Conservation and the American West, Island Press, Washington, DC., USA.

Lemos, M. C. and A. Agrawal. 2006. "Environmental Governance." Annu. Rev. Environ. Resour 31: 297-325.

LVFO. 2001. The Convention for the Establishment of the Lake Victoria Fisheries Organization. Lake Victoria Fisheries Organization and the International Union for the Conservation of Nature. Jinja Uganda.

LVFO. 2005. Guidelines for Beach Management Units (BMUs) on Lake Victoria. V. Ogwang', M. Medard, E. Kilosa, J.I. Nyeko and A. Bakunda. Jinja, Uganda, Lake Victoria Fisheries Organization, East African Community.

LVFO. 2007. Guidelines for the beach management units (BMUs) on Lake Victoria. Lake Victoria Fisheries Organization. Jinja, Uganda.

Manring N.J. 2005. The politics of accountability in national forest planning. Administration & Society 37.

Manthesian C. 1998. Grassroots Charade. Governing: 38-42.

Mattes W. and J. St. Arnold, C. Otto Rasmussen, N. Kmiecik, and S. Erickson. 2005. Mazina'igan, Lake Superior fishery management supplement. Mazina'igan Fall.

McCloskey M. 2000. Problems with using collaboration to shape environmental public policy. Valparaiso University Law Review 34(2): 423-434.

Mkumbo O.C. 2006. The Fisheries of Lake Victoria: Status and Management Initiatives for Sustainable Exploitation. Paper presented at Great Lakes of the World IV. February 21, 2006, Bagamoyo, Tanzania.

Mkumbo O.C. and E. Mlaponi. 2007. Impact of the baited hook fishery on the recovering endemic fish species in Lake Victoria. Aquatic Ecosystem Health & Management 10: 458-466.

Mkumbo O.C. and P. Nsinda, C.E. Ezekiel, I.G. Cowx, and M. Aeron. 2007. Towards sustainable exploitation of Nile perch consequential to regulated fisheries in Lake Victoria. Aquatic Ecosystem Health & Management 10: 449-457.

Njiru M. and J. Kazungu, C.C. Ngugi, J. Gichuki and L. Muhoozi. 2008. An overview of the current status of Lake Victoria fishery: Opportunities, challenges and management strategies. Lakes & Reservoirs: Research and Management 13: 1-12.

Nunan, F. 2006. Empowerment and Institutions: Managing Fisheries in Uganda. World Development 34(7): 1316-1332.

Ojuok J.E. and M. Njiru, M. Ntiba, and K.M. Mavuti. 2007. The effects of overfishing on the life-history strategies of Nile talapia, *Oreochromis niloticus* (L.) in the Nyanza Gulf of Lake Victoria, Kenya. Aquatic Ecosystem Health & Management 10: 443-448.

Ostrom, E. 2009. A general framework for analyzing sustainability of social-ecological systems. Science 325: 419-422.

Pinkerton E. Introduction: Attaining better fisheries management through co-management: Prospects, problems, and propositions. pp. 3-36. *In:* E. Pinkerton [ed.] 1989. Co-operative Management of Local Fisheries: New Directions for Improved Management and Community Development, University of British Columbia Press, Vancouver, Canada.

Pinkerton E. The Skeena Watershed Partnership: Learning from success and failure in salmon management. pp. 903-919. *In:* C.C. Krueger and C.E. Zimmerman [eds.]. 2009. Pacific salmon: Ecology and Management of Western Alaska's Populations: American Fisheries Society Symposium 70, American Fisheries Society, Bethesda, MD, USA.

Pinkerton E. and L. John. 2008. Creating local management legitimacy. Marine Policy 32: 680-691.

Piper D. 1967. The International Law of the Great Lakes. Duke University Press. Durham, NC, USA.

Pomeroy R.S. and F. Berkes. 1997. Two to tango: The role of governments in fisheries co-management. Marine Policy 21: 465-480.

Rachlinski J.J. 2006. Bottom-up versus top-down lawmaking. The University of Chicago Law Review 73: 933-964.

Rettig R.B. and F. Berkes, and E. Pinkerton. The future of fisheries co-management: A multi-disciplinary assessment. pp. 273-289. *In:* E. Pinkerton [ed.] 1989. Co-operative Management of Local Fisheries: New Directions for Improved Management and

Community Development, University of British Columbia Press, Vancouver, Canada.

Ribot J.C. 2002. Democratic Decentralization of Natural Resources: Institutionalizing Popular Participation. World Resources Institute.

Schlender J.H. (n.d.). The impact of the Voigt case: On-going and upcoming issues: Great Lakes Indian Fish and Wildlife Commission, Odanah, Wi.

Sen S. and J.R. Nielsen. 1996. Fisheries co-management: A comparative analysis. Marine Policy 20: 405-418.

Stein, H. 2008. Beyond the World Bank Agenda: an institutional approach to development. Chicago, The University of Chicago Press.

Townsend R.E. and S.G. Pooley. Distributed governance in fisheries. pp. 47-58. *In:* S.Hanna and M. Munasinghe [eds.]. 1995. Property Rights and the Environment: Social and Ecological Issues, Beijer International Institute of Ecological Economics and The World Bank, Washington, DC., USA.

United States District Court. 1979. U.S. v. Michigan, 623 F2d 448: United States District Court, Western District, Michigan.

United States District Court. 2000. United States vs. Michigan Consent Decree: United States District Court, Western District, Michigan.

van der Knaap M. and W. Ligtvoet. 2006. Is Western Consumption of Nile Perch from Lake Victoria Sustainable? Paper presented at Great Lakes of the World IV; February 21, 2006, Bagamoyo, Tanzania.

van Sittert L. 2002. The tyranny of the past: Why local histories matter in South African fisheries. Ocean & Coastal Management 46: 199-219.

Weiss E.B. The emerging structure of international environmental law. pp. 98-115. *In:* N.J. Vig and R.S. Axelrod [eds.] 1999. The Global Environment, Congressional Quarterly Press, Washington, DC., USA.

Wiebe, P.D. and C.P. Dodge [eds.]. 1987. Beyond Crisis: Development Issues in Uganda. Kampala, Uganda, Makerere of Social Research, African Studies Association.

Wilson D.C. Lake Victoria Fishers' Attitudes towards Management and Co-Management. pp. 174-184. *In:* K. Geheb and M.T. Sarch [eds.]. 2002. Africa's Inland Fisheries: The Management Challenge, Fountain Publishers, Kampala, Uganda.

Witte F. and J.H. Wanink, M. Kishe-Machumu, O.C. Mkumbo, P.C. Goudswaard and O. Seehausen. 2007. Differential decline and recovery of haplochromine trophic groups in the Mwanza Gulf of Lake Victoria. Aquatic Ecosystem Health & Management 10: 416-433.

Young O.R. 1994. International Governance: Protecting the Environment in a Stateless Society. Cornell University Press. Ithaca, NY, USA.

Zorn J.E. 1989. General principles of Indian law. Great Lakes Indian Fish and Wildlife Commission. Odanah, WI, USA.

Zorn J.E. 2003. Testimony of James E. Zorn, Policy Analyst of the Great Lakes Indian Fish and Wildlife Commission. Senate Committee on Indian Affairs, June 3, 2003: 37-42.

SECTION 5

INDUSTRIAL LEADERSHIP

CHAPTER 18

Industry and Governance in the Great Lakes Basin

George H. Kuper

Council of Great Lakes Industries, 3600 Green Court, Suite 710,
Ann Arbor, Michigan 48105-1570
E-mail: ghk@cgli.org

INTRODUCTION

The Great Lakes Basin hosts a unique ecosystem with remarkable natural resources. It also operates under a unique self-imposed binational governance model in addition to the federal state/provincial constitutional structure. This model is rooted in the Canada/U.S. Boundary Waters Treaty of 1909 (Boundary Waters Treaty 1909) and propagated through additional cooperative agreements administered through the International Joint Commission (Great Lakes Water Quality Agreement, 1972, 1978, and 1987), Great Lakes Commission (Great Lakes Basin Compact, 1995) and Council of Great Lakes Governors (Creation of Council of Great Lakes Governors, 1983; The Great Lakes Charter Annex, 2001). Those who live, work and play in the basin have a strong desire to understand environmental stressors – and, more recently, alleviate those stresses – that impact the Great Lakes. The natural resources of the Great Lakes have provided a strong economic base for industry. A continuing strong economy is needed to enable the region to protect its natural resources and provide for the economic and social needs of the population. The relationship between industry and the environment can and should be mutually beneficial when governments, public interest organizations, and companies all work together (Austin 2007). For more than two decades, industry has joined forces with other stakeholders concerned about the region's sustainability and worked together within processes designed to develop and drive public policy in the region. These efforts have achieved substantial progress in ecosystem restoration and protection.

Business and industry were initially attracted to the Great Lakes region by natural resources such as timber, iron ore, other minerals, natural brine wells, a landscape suitable for agricultural development, opportunities for efficient waterborne transportation reaching far into the North American continent, and abundant supplies of fresh water. The resulting heavy concentration of industry – and the jobs industry produced – attracted people to a region where they could both work and enjoy bountiful natural resources. As in other regions, during this age of settlement and development, early industry did not understand the potential for impact on the waters. As a result, the Great Lakes suffered (Austin 2007).

In the past three decades, however, industry has worked effectively within public policy initiatives to help develop standards and practices that achieve significant environmental improvement for the region. Progress was made through the development of environmental regulatory programs and industry's heavy investment in environmental control systems.

Industry's initial efforts were focused on environmental control systems and processes that were based on municipal effluent treatment technologies. This early industry effort in the Great Lakes region preceded regulatory systems that established specific standards for each industrial category. These early pollution control systems focused on water contaminants that formed deposits on river bottoms (Total suspended solids – TSS) and depleted oxygen in streams (biochemical oxygen demand – BOD) (Eckenfelder 2009). Technologies and regulatory structures needed to address air emissions followed the implementation of water pollution controls. Concerns about chemical constituents and "toxics" came along much later and broadened the scope of policy development in water, air, solid waste and chemical exposure regulations.

BEGINNINGS OF INDUSTRY PARTICIPATION IN REGIONAL GOVERNANCE

Industry in the Great Lakes, for the most part, had not been a participant in regional policy development before the 1990s. Industry had long participated in national policy development and state and provincial legislation/regulatory affairs. But neither industry nor governments had developed capabilities for efficient and collective collaboration on regional issues in the basin. In the meantime, environmental groups in the basin were successfully presenting their views on the need for ecosystem protection to the governments. Their proactive approach included creating opportunities for environmental NGOs to get their message before government agencies and policy makers – including using measures that attracted the attention of the news media. Industry, less organized for participating in policy debates at a regional level, needed to develop vehicles to articulate their concerns about the need to insure cost effectiveness and manufacturing process compatibility as a part of the policy decision-making process. These dynamics stimulated industrial leaders to highlight the need to bring a sustainable development perspective into emerging public policy discussions (Industry Canada 2011).

To address industry's needs, the Federal Reserve Bank of Chicago, Eastman Kodak Company, and a number of other industrial organizations and companies in both the U.S. and Canada came together in 1991 to create the Council of Great Lakes Industries (CGLI). The charge given to the new organization was to collaborate with governments and stakeholders on issues specific to the binational region. The CGLI mission was established as promoting the economic growth and vitality of the region in harmony with its human and natural resources, or sustainable development. In the meantime, governments within the region were expanding or establishing venues in which collaboration activities could take place.

> Industry's initial entry into the regional conversation regarding broad issues relating to chemical substances in the environment and protection of the chemical, biological, and physical integrity of the Great Lakes ecosystem first started to take substantive form at the 1991 Biennial meeting of the International Joint Commission (IJC) in Traverse City, Michigan. At that time, the virtual elimination of selected chemical contaminants through a philosophy of zero discharge – provisions contained in the Canada/U.S. Great Lakes Water Quality Agreement – were being actively promoted through campaigns and demonstrations conducted by environmental groups. These presentations, which some observers characterized as environmental theatre[1], grabbed the attention of the press, IJC Commissioners and government representatives. Industry was stunned at what these efforts could do to their ability to operate in the Great Lakes (International Joint Commission 1978).

In response, industry – now mobilized under the banner of the binational Council of Great Lakes Industries – participated in the 1993 meeting, showcased the measures that they had been implementing as a part of their sustainability pursuits, and began to understand and explain the real impact of imposing these new philosophies on their operations. However, because industry and environmental groups rarely showed up on the same stage and had little history of working together, that initial meeting accomplished little in the way of productive collaboration or consensus building. But, the effort was worthwhile. Industry learned that it needed to be present where and when policy was decided and must make its case if it was to gain policy provisions that included elements that would allow facilities to continue to operate in the region.

The 1991 IJC meeting was a watershed event in many ways. Environmental groups and citizens made it clear that they wanted more action on reducing

[1]This "theatre" actually became quite nasty with some industry representatives actually receiving threats to their physical safety!

levels of contaminants in the Great Lakes. IJC Commissioners were listening with sympathetic ears, governments were taking notice, and industry became aware that they needed to more completely describe the pollution prevention measures they had already put in place. It also became clear that industry needed to supply information regarding the state of the ecosystem including the progress already made in clean-up efforts. Further, industry had to actively participate in the formal and informal discussions where new and additional policy needs were determined.

The early industry response took on several forms. First, industry joined the IJC's binational Virtual Elimination Task Force. This group of academic researchers, government scientists, environmental group representatives, and now industry, had been established by the IJC to interpret the Great Lakes Water Quality Agreement charge that "the discharge of any or all persistent toxic substances be virtually eliminated." Rather than simply show-up at Task Force meetings, industry representatives took an active role. They presented information describing process and feedstock changes that industry had already made to reduce persistent bio-accumulative and toxic substance (PBTs) use and generation. They provided detailed information regarding the chemistry of industrial processes, product specifications, and process safety considerations (Virtual Elimination Task Force, 1993). This enabled other Task Force members to understand – evaluate for themselves – the appropriateness and effectiveness of alternative pollution prevention measures. Second, industry became an advocate for Task Force findings. Once Task Force deliberations had been completed and the group's draft report was released for public review and comment, industry joined IJC staff and other stakeholders in describing Task Force recommendations regarding virtual elimination implementation at public meetings held throughout the Great Lakes Basin (International Joint Commission 1978).

This first experience in proactive multi-stakeholder policy development engagement turned what had been a very hostile atmosphere, in which the status of the Great Lakes ecosystem was hotly debated, to a more constructive one where plans for moving forward could be developed. In follow-up, industry went forward from this point, began its third initiative, and urged governments and environmental group representatives in both Canada and the U.S. to establish multi-stakeholder based forums on specific issues of importance to encourage the development of constructive actions.

In a fourth effort to contribute to the understanding of the status and needs for restoring and protecting the Great Lakes ecosystem, industry sponsored and developed a comprehensive assessment of the state of the Lakes ecosystem. As yet another outgrowth of the 1991 IJC Biennial Meeting, the paper industry research organization NCASI (National Council for Air and Stream Improvement) contracted with the environmental assessment and consulting firm LimnoTech of Ann Arbor, Michigan to perform an assessment of the Great Lakes ecosystem. This initiative assembled a large and distinguished panel of

Great Lakes researchers tasked with using available data and information to describe the state of Great Lakes water quality and water dependent species. Published in 1992, this assessment provided valuable information that guided Great Lakes restoration program developments rapidly taking shape in this timeframe (National Council for Air and Stream Improvement, Inc. 1993). It also assisted governments in the development of the first State of the Lakes Ecosystem Conference (SOLEC) convened by Environment Canada and U.S. EPA in 1994.

As the new multi-stakeholder forums gradually began to take shape, industry found valuable opportunities to explain how it operates and to reinforce the idea that business success in the region is essential for the good of the region. This substantial new engagement of industry was beginning to produce policy development discussions that reflected a better understanding by industry of ecosystem needs, and a better understanding by environmental groups and governments of industry's concerns and abilities (National Council for Air and Stream Improvement 1994).

Taking development of stakeholder based collaborative processes even further, the Council of Great Lakes Industries partnered with the U.S. Council for International Business, and U.S. EPA to explore implementation of these public policy development tools. This work included the involvement of people with varied interests and experience. The participants recognized their own diversity as a strength of the workshop. Participants included representatives of industry, academia, the scientific community, indigenous tribes, NGOs, and government. Coordinated by Boulder, Colorado, based CDR Associates, this work led to the development of The Boulder Principles for Effective Collaboration Processes on Sustainable Development and Environmental Policy (Woodrow and Gahis 1998). These principles have been utilized to guide discussion and consideration of process in several Great Lakes regional program development activities over the past decade.

Initially, industry had to seek out opportunities to engage in regional policy issues. Invitations were not often routinely extended. However, as the U.S. and Canadian governments began to see the value of industry's engagement and contributions, governments began to invite industry representatives into the discussion. This change resulted in much more than just participation in discussions. As noted in the next section of this chapter, it also provided the opportunity for use of expanded voluntary programs to produce additional reductions in toxic chemical releases to the Great Lakes Basin.

CHEMICALS MANAGEMENT

Several programs were in development during the period that gave rise to the use of binational multi-stakeholder processes. Remedial Action Plans (RAPs), designed to address Areas of Concern (AOCs), had been in existence since the late 1980s in response to the 1987 changes to the Great Lakes Water Quality Agreement. Lakewide Management Plans (LaMPs), also called for by this GLWQA revision (Annex 2 of the Great Lakes Water Quality Agreement of 1978, Amended 1987), had been stuck for a decade in seemingly endless rounds of drafting but were now being reconstituted with new vigor and increased stakeholder engagement. However, most significant to industry was a new effort, the Great Lakes Binational Toxics Strategy (GLBTS) (U.S. Environmental Protection Agency 2007). Created to address the virtual elimination goal of the Great Lakes Water Quality Agreement, it was designed as a multi-stakeholder effort from the ground up. In it, industry worked with the governments alongside environmental groups and tribal interests to identify toxic substance reduction measures that would achieve results for the region.

The Strategy, established in 1997, set goals and timetables for reductions of specific persistent, bio-accumulating toxins though voluntary actions. Significantly, it endorsed a rational, government defined four-step process[2] that would be applied to each of the targeted substances. The Strategy's structure provided for small workgroups focused on a specific substance to identify sources, explore potential voluntary measures, and implement actions to achieve reductions. A larger GLBTS Integration Workgroup brought stakeholders working within the individual substance groups to the table to collectively track progress, and define overarching policies and direction. A key benefit of this structure was that stakeholders got to know one another and, while actively challenging each other, developed a degree of trust not experienced before in regional policy development. Stakeholder participation and contributions still reflected parochial interests but, through the GLBTS process, participants developed the ability to work together to achieve program goals. Voluntary initiatives began to gain acceptance by governments and environmental groups while industry became a more accepted partner in dealing with toxics substances. Equally, and perhaps more significantly, the GLBTS process provided defined goals and objectives that industry could respond to, rather than abstract directives like the "virtual elimination" of toxic substances.

The targets and timetables set within the GLBTS Terms of Reference for the persistent toxic substance reductions were being met in both the U.S. and Canada. GLBTS partners celebrated its success at its 10th anniversary event in 2007 (U.S. Environmental Protection Agency 2007). As the focus of the program has shifted away from legacy contaminants to the substances now termed "chemicals of emerging concern", the specific nature of the future program is still being determined. However, as explained above, experience shows that success

[2]"1. Gather information"; 2. "Analyze current regulations, initiatives and programs which manage or control substances"; 3. "Identify cost-effective options to achieve further reductions"; 4. "Implement actions to work toward the goal of virtual elimination".

in this area is also more likely to be achieved by maintaining close working relationships with interested stakeholders, including industry. Stakeholder engagement will be an important element in whatever chemical management program replaces the GLBTS process of the past dozen years.

WATER RESOURCE MANAGEMENT

Water resource management has also been an important governance area in the Great Lakes region. The Great Lakes states and provinces put a vehicle in place to establish cooperative management of these vast water resources in 1985 when the Governors and Premiers from each of these jurisdictions signed the Great Lakes Charter – Principles for the Management of Great Lakes Water Resources (Council of Great Lakes Governors 1985). In this agreement, each of the jurisdictions pledged:

- Recognition of integrity and connectivity of Great Lakes natural resources and the ecosystem "considered as a unified whole".
- Commitment to cooperation between basin jurisdictions.
- To manage and regulate diversions and consumptive uses to avoid individual or cumulative significant adverse impacts on lake levels, in-basin uses, or the Great Lakes ecosystem.
- Jurisdictions will not approve or permit any new or increased diversion or consumptive use without notifying and seeking consent and concurrence of all affected states and provinces.
- To pursue development and maintenance of a common data base regarding use and management of basin water resources, a systematic system for the exchange of this information; create a Water Resources Management Committee; develop a Great Lakes Water Resources Management Program; and, coordinate research efforts to provide improved information for water planning and management decisions.

In June 2001, the Governors and Premiers of the Great Lakes Basin reaffirmed their commitment to the principles of the Great Lakes Charter and signed the Great Lakes Charter Annex. The Annex called for the development of an enhanced water management system for the Great Lakes. It took four years and endless rounds of discussion with multi-stakeholder input before, on December 13, 2005, the governors and premiers signed two essentially identical documents: the Great Lakes–St. Lawrence River Basin Sustainable Water Resources Agreement, among the eight states and two provinces (referred to as the Agreement); and, the Great Lakes–St. Lawrence River Basin Sustainable Water Resources Compact, a contract among the eight Great Lakes states (referred to as the Compact) (Council of Great Lakes Governors 2005). Industry was an active participant in the discussions and deliberations through the advisory committee, bringing to the negotiators attention industry's need for continued and assured access to Great Lakes water supplies and valuable expertise on how industry can contribute to sound water management. Industry called upon its experts and spent countless hours in meetings and provided reams of detailed comments. CGLI coordinated industrial stakeholder participation to enhance efficiencies and ensure that the right parties were brought into the process. Industry also participated in studies that relied on operational data typical of Great Lakes facility operations to inform decisions made by government personnel who drafted Agreement and Compact language (Council of Great Lakes Governors 2005). This participation helped assure the effectiveness of the Agreement and Compact and the ability of industry to continue to responsibly use the resource in the basin (Council of Great Lakes Industries 2003). Important issues that were addressed during the process included:

- Maintaining riparian rights as the basis for water resource management decision making in the region.
- Establishing that the primary decision making authority with regard to approval of water withdrawals would continue to rest with the states and provinces, not a regional decision making body. (And not the U.S. federal government, which technically has jurisdiction over the Great Lakes since they are international boundary waters. In Canada, federal authority is still preeminent.)
- Recognition that, though a common set of objectives and minimum standards is established within the body of Agreement and Compact language, the specific procedures and information requirements needed to comply with those objectives and standards is left to each individual state or province. Each will implement Agreement and Compact provisions through vehicles and processes that reflect their unique legal systems and sovereignty.

On the U.S. side, acceptance of an inter-state compact is a complicated process. Each state has to agree to the identical language of the compact, the U.S. Congress has to approve that same identical language, and then the President has to sign the same. The fact that this all was accomplished within seven years of the commitment by the Great Lakes Governors is remarkable testimony to the influence of multi-stakeholder action.

Collective work to completely implement and maintain the Agreement and Compact continues through proceedings of the Agreement Regional Body, Compact Council and activities of the Secretariat organization, the Council of Great Lakes Governors (CGLG). The CGLG establishes and maintains advisory panels and special working groups to gather stakeholder input and share expertise. Industry is a participant and is currently serving on these bodies (Great Lakes–St. Lawrence River Basin Water Resources Council). Another important activity is for industry to track Agreement and Compact implementation efforts within individual states and provinces, providing information regarding industry water withdrawal needs, conservation practices, and comments on other aspects of water resource management by the respective jurisdiction.

GREAT LAKES RESTORATION

In 2004, U.S. President George W. Bush issued an executive order creating the Great Lakes Interagency Task Force to improve federal coordination of Great Lakes activities and called for a "Regional Plan of Significance" to develop a strategy for restoring the Great Lakes (President Bush Executive Order, 2004). The tool used to accomplish this task was known as the Great Lakes Regional Collaboration (GLRC). Multi-stakeholder groups, known as strategy teams, were established and called together to identify needs in eight of the nine designated priority areas identified by the Great Lakes Governors. The ninth priority area was already being addressed through the activities that led to development of the Compact. The strategy teams included representatives from industry, government, environmental groups, cities and tribes. In recognition of the binational character of the region, Canadian representatives were invited to sit on each team as observers. The charge given to each team was to work together to recommend elements of a strategic plan that would guide the protection, restoration and sustainable use of the Great Lakes.

More than 1500 individuals worked on these ideologically diverse strategy teams. Viewpoints from each of the sectors represented were of value and contributed to the richness of the final product - Great Lakes Regional Collaboration Strategy to Restore and Protect the Great Lakes, December 2005 (Great Lakes Collaboration Strategy). Significantly, because this strategy was produced through the multi-stakeholder collaboration process, the resulting framework for Great Lakes restoration was unanimously supported by all participants. Development of the GLRC Strategy has proved to be of great benefit to the region. However, this benefit has not come without additional hard work - pursued with stakeholders in the lead, not governments!

The Great Lakes had a coordinated plan on the U.S. side, crafted by stakeholders, that outlined specific ecosystem needs, detailed the resources needed to meet those needs, and had the support of the region's major stakeholders. But when U.S. government agencies added up the cost of all of the elements that had been rolled into this restoration effort, the price tag was estimated at US $22 billion (now estimated at US $26 billion with inflation). The governments and stakeholders had demonstrated that it was possible to reach agreement on what needed to be done, but the cost of the plan might just prove to be a deal breaker.

Seeking a means to get past this considerable roadblock, industry and environmental groups rallied together and began to look at the economic benefits that could come from this large-scale restoration project. But, actual economic impact figures were hard to come by. The Healing Our Waters-Great Lakes Coalition (a collaboration of 90+ environmental groups), the Council of Great Lakes Industries and the Great Lakes-St. Lawrence Cities Initiative came together and developed a proposal that was presented to foundations and businesses to fund a project that would determine the economic benefit of restoration in the region. This initiative was successful, the study was conducted, and the results were two reports produced by The Brookings Institution. The first was very detailed and entitled *America's North Coast: A Benefit-Cost Analysis of a Program to Protect and Restore the Great Lakes* (Austin et al. 2007). The second, and the more concise report, was entitled *Healthy Waters, Strong Economy: The Benefits of Restoring the Great Lakes Ecosystem* (Austin et al. 2007). These reports concluded that the benefit to the economy from restoration of the Great Lakes was at least two-to-one for every restoration dollar spent. In total, implementation of the Strategy to Restore and Protect the Great Lakes could result in an economic benefit to the region of at least US $52 billion.

Though the results of the study were stunning in and of themselves, it is important to recognize that it would have been difficult, if not impossible, to produce these reports without industry involvement in the restoration planning process from the beginning. Industry participation helped other stakeholders understand the importance of the economy in restoring the ecosystem. This represented a major shift in thinking in the region – i.e. the economic impact matters in the restoration discussion[3].

The Brookings reports were widely circulated and frequently quoted, especially in Washington, D.C. where members of Congress and President Obama cited the report in support of a basin restoration effort. Collectively, the multi-stakeholder support, the economic impact study and, presidential support led to the approval by Congress of an implementation effort called the Great Lakes Restoration Initiative, based on the Collaboration Strategy. Working together again, stakeholders convinced Congress to appropriate US $475 million in 2010 to fund the beginning of the restoration process (U.S. House of Representatives 2010). President Obama has said that the US $475 million was the first payment of US $2 billion of federal restoration funding, short of the US $22 billion needed, but a substantial investment nonetheless[4]. Once again, the benefits derived from multi-stakeholder efforts, especially when industry is included, were clear and tangible.

[3]The subsequent economic down turn has reinforced that understanding.

[4]It is important to realize that a significant portion of the US $22bn cost estimate is to modernize sewerage that would come out of the State Revolving Fund, not a separate restoration initiative appropriation.

GREAT LAKES WATER QUALITY AGREEMENT

As pointed out at the beginning of this chapter, the Great Lakes Water Quality Agreement (GLWQA) has served as a guiding document for environmental action in the Great Lakes. It clearly expresses the commitment of Canada and the United States to restore and maintain the chemical, physical and biological integrity of the Great Lakes basin ecosystem. Signed in 1972, revised in 1978 and again in 1987, the GLWQA has been reviewed every six years to determine if revisions were needed. Unmodified since the 1987 amendments, it was the subject of an extensive citizen review process in 2006. Stakeholders, including industry, were again invited to participate in working groups and asked to identify issues of importance to the ecosystem. In addition, the governments held public meetings, or "listening sessions." The IJC held workshops. A final review report detailing the outcome of this very extensive collaboration effort was issued in 2007 (Binational Executive Committee 2007).

In this instance, it was not as easy to obtain consensus on what should be included in any GLWQA amendments. Opinions differ sharply between stakeholders regarding some provisions and the level of prescriptive detail that should or should not be made a part of the GLWQA (Council of Great Lakes Industries 2010). However, given the significant changes in the state of the Great Lakes ecosystem since 1987, and nature of the current environmental challenges facing the region, it was agreed that it was time for another amendment of this "guiding North Star" policy document. The governments of Canada and the U.S. are in the process of developing potential revisions to the GLWQA with a revised Agreement expected in 2012.

OUR HISTORY CAN GUIDE OUR FUTURE

These examples have fully demonstrated the value of, and success that can be attained, when multi-stakeholder processes are used to inform and aid in policy development. They illustrate the importance of maintaining the health of social, economic, and environmental factors to support a sustainable development platform. Industry must do its part to protect the ecosystem, but it also must remain economically viable so it can provide the support needed for social structure. It is hard to imagine that it would be possible to achieve that balance in the Great Lakes region, where multiple federal, provincial, state and tribal governments have jurisdiction over the region, without a multi-stakeholder centered approach.

Going forward, a revised GLWQA will presumably set the tone for environmental policy in the region for at least the next couple of decades. This revision, as envisioned by industry, must incorporate the full range of sustainable development principles for a healthy ecosystem, a robust economy and support of societal needs. It needs to be ecosystem centered, long on vision, but short on specific objectives that can quickly make the new document out of date.

Some stakeholders have often expressed exasperation or dissatisfaction over environmental policy directions articulated through Canadian or U.S. laws and regulations. They seek regional policies that replace or supersede those established through state, provincial, or federal legislative and regulatory processes (Alliance for the Great Lakes et al. 2007). In these circumstances, it is essential to remember that the GLWQA and/or other regional policy instruments cannot replace those established by the governments of the sovereign nations responsible for striking the inter-government agreements (The World Bank Group 2010).

Likewise, governments in the Great Lakes region have collectively established organizations and working groups with specific purposes, charges, and responsibilities in pursuit of cooperation and efficiency. Examples include: the International Joint Commission, the Great Lakes Commission, the Council of Great Lakes Governors, LaMP organizations, the Lake Superior Binational Program, Remedial Action Plan (RAP) organizations aimed at Areas of Concern (AOCs), the Binational Executive Committee and others. These groups have the potential to operate outside of the normal legislative or regulatory regimes as their authority derives not through legislated process but by bureaucratic interpretations of the GLWQA, terms of reference, and/or other authorizing provisions set up by the convening authority. These resulting potentially "extra-constitutional" initiatives need to be grounded in and limited to the legislative and regulatory policy directives of the individual jurisdictions. Revised GLWQA language must be carefully drafted to insure consistency with policies and legal authorities of the federal governments of Canada and the U.S. and worded clearly enough to guard against contrary interpretations.

UNITING IN WASHINGTON

The multi-stakeholder collaborative activities in the basin have not been limited to in-basin activities. For a number of years stakeholders interested in the Great Lakes Restoration agenda have been pursuing U.S. federal funds. Great Lakes groups have been coming together for Great Lakes Day in Washington, D.C. Participants include the Great Lakes Commission, Northeast-Midwest Institute, Healing Our Waters-Great Lakes Coalition, Great Lakes-St. Lawrence Cities Initiative, Great Lakes Fisheries Commission, the Great Lakes Metro Chambers Coalition, the Council of Great Lakes Industries (CGLI), and others that have come together to develop common strategies for seeking federal authorization of Great Lakes programs and funding support (Great Lakes Commission and Northeast-Midwest Institute 2010). Together, the group crafts a statement that all organizations participating can support. This combination of environmental groups, Great Lakes interest groups, industry and business has been a powerful force in Washington, once again demonstrating the power and significance of collaborative stakeholder-based activities.

Among the first legislative successes was passage of the Great Lakes Legacy Act in 2002. Industry contributed to this success through participation that helped other stakeholders understand how this program can expedite clean-up of AOCs without "letting industry off the hook" and escape their rightful liability. Provisions that enable industry to form partnerships with other parties eligible for Legacy Act funding expedite clean-up of orphan shares, settle disputes over who should cover costs not clearly belonging to responsible parties, and provide other clean-up incentives.

INDUSTRY AS A STAKEHOLDER IN CANADA

Historically, industry has participated in collaborative regulatory development processes with the Canadian federal government. Examples include deliberations over the past several decades regarding Canada Wide Standards for water and air pollution controls (Canadian Council of Ministries of the Environment 1991) and, more recently, the extensive consultation opportunities for industry and other stakeholders to address risk management provisions in Canada's Chemical Management Program Canadian Chemicals Management Plan 2006). Canadian industry has been well equipped to work in these venues and has supplied scientific information and data that governments have found useful in setting policy. As Great Lakes regional stakeholder venues and processes were developed, industry was able to use these experiences in Canada to gain comfort in multi-stakeholder processes (Industry Canada 1995). These collaborative efforts have been essential for realization of many successes that the parties to the Great Lakes Water Quality Agreement have had in the quest to restore and maintain the chemical, physical, and biological integrity of the Great Lakes (Binational Executive Committee 2007).

CHALLENGES FACING INDUSTRY AND THE REGION

Much progress has been made in ecosystem restoration and protection in the past several decades. As was detailed in the 2009 State of the Great Lakes ecosystem report, "releases of targeted bioaccumulative toxics chemicals have declined significantly from the peak period in the past decades and, for the most part, no longer limit the reproduction of fish;" "contaminants in herring gull eggs have declined 90%;" and air quality is said to be "improving on a regional scale." The report adds that "the Great Lakes provide residents with some of the finest drinking water sources found anywhere in the world." However, the report also highlights challenges remaining, such as "substances of emerging concern such as flame retardants, plasticizers, pharmaceuticals and personal care products and pesticides" found in fish and other Great Lakes environmental samples; and "the presence of invasive species can be linked" to challenges such as food web changes, "fish and waterfowl diseases, and excessive algal growth." The report warned of land use changes, and "climate in the Great Lakes region is changing" (Environment Canada and U.S. Environmental Protection Agency 2009).

These issues must be addressed through the governance provisions now being established through Great Lakes Water Quality Agreement revision efforts, the U.S. Great Lakes Restoration Initiative, pursuits of provisions of the Canadian Environmental Protection Act (CEPA), and coöperative and delegated programs such as the Canada – Ontario Agreement (COA). To be effective, these efforts must include stakeholder and collaborative based processes that have been demonstrated to be the key to success.

For industry, the challenges include increased regulation, regional policy initiatives, pressures from global competition, and weak economic environments (U.S. Environmental Protection Agency 2006). Though the restoration effort is expected to provide significant economic benefit for the region, the outcome for industry will be dependent upon these challenges (Great Lakes Restoration Initiative Action Plan 2010).

Invasive species represent a significant threat to the biological, physical and chemical integrity of the Great Lakes ecosystem. Control of biological invasives of all types needs to be a priority policy matter for Great Lakes policymakers. But, actions proposed to prevent or control aquatic invasive species impacts must reflect the needs of all sectors that rely on the Great Lakes resources. They must also be consistent with the need for continued and enhanced sustainable development within the region. Binational stakeholder involvement is necessary if policy makers are to successfully work toward solutions that maximize protection of the biological integrity and recreational value of the Great Lakes while maintaining water-related commerce, both inter-lake and to markets outside of the basin. Industry stakeholders are uniquely qualified to provide information and perspectives that will help accomplish these goals (Council of Great Lakes Industries 2009).

Climate change policies focused on emissions reductions or cap and trade proposals have been difficult issues for industry due to increased costs and competitiveness concerns. However, since the mid 1990's climate adaptation has become recognized as a necessity within many industry sectors. Most recently CGLI has worked with industry representatives to identify venues and discussions that focus on adaptation that might be required for successful operation in the Great Lakes region. CGLI collaborated with the National Wildlife Federation, the U.S. Environmental Protection Agency (EPA), and the Great Lakes Commission to sponsor a climate adaptation workshop held September 22, 2010 in Buffalo, New

York. (Proceedings are in production.) This multi-stakeholder process, designed to inform programs under consideration by U.S. EPA, provided thoughtful presentations and comments from environmentalists, industry, tribes, cities and government. Programs designed to respond to the climate change threat would be difficult to develop without the ability to tap into the long-standing trust and cooperation that can be accomplished through the multi-stakeholder model.

THE FUTURE

New issues will continue to arise in the region requiring new policies and new solutions. To ensure successful responses, all parties need to be engaged in public policy development and implementation. They need to be directed to assure that all three elements of sustainable development – environmental, social, and economic factors – be pursued in the most cost-effective and efficient manner to assure success. Industry is committed to serving as a constructive part of the process as work continues to build and support a sustainable Great Lakes.

Industry has, and will in the future, have an essential societal role to play. Industry provides the products and services that people need to maintain their quality of life. In so doing, industry provides good jobs and, importantly, creates new wealth. Part of that wealth, of course, is another resource necessary for enabling the protection of ecosystem services and the environment. In addition, it should be evident from the preceding points raised in this chapter, that industry can bring important perspectives, expertise and experience to the public policy setting process. It is important to realize that industry seldom speaks with a singular voice. Some would prefer a free market approach to dealing with ecosystem challenges; others, a regulatory approach. Indeed, if the regulatory structure that currently exists in the Great Lakes Basin was in place back in the 1930s, 40s and even the 50s, we wouldn't be having many of the discussions now needed to resolve many of the legacy problems in the basin. No doubt other problems would have surfaced because of the fact that there is no such thing as a completely risk free society. Consequently, the regulatory environment going forward will, by necessity, reflect the need to react to unexpected outcomes and the policy of the time.

Regrettably, some of the successes in the basin of late in reducing pressures from industry on the ecosystem have come through plant closures. These "successes" are not to be celebrated but viewed as a failure to create a sustainable business climate that enables maintenance of balance between social, economic, and environmental regimes. The economic future of Canada and the United States cannot rely exclusively on the growth of what is often regarded as "green industry." Industries based on the extraction of natural resources can be, and are being, operated in a sustainable mode, and have the advantage of being able to create a high level of value added wealth to our economic systems. Canada and the U.S. will continue to make products in order to capture the economic rent from the capital and intellectual property investments that have been made. Industrial manufacturing activities will be able to avoid many past mistakes relative to ecosystem pressures. But, because of the lack of complete certainty, the reality is that we must be prepared to address new ones.

However, our economic well-being requires that industrial activities advance towards the ideal state of sustainability. To do so we must be vigilant, seek to identify the real potential threats to the ecosystem, but avoid distractions from concerns over issues that represent low risk. Failure to recognize and accept the concept of risk management in the development of environmental policy would stymie future development and disrupt the ability to support all three of the sustainability elements – social well-being, economic progress, and environmental health.

ACKNOWLEDGEMENTS

Dale Phenicie and Evelyn Strader made essential contributions to the substance of this chapter; and, Janet Rieke to the mechanics of its production.

REFERENCES

Alliance for the Great Lakes, Biodiversity Project, Canadian Law Association and Great Lakes United. 2007. The Great Lakes Water Quality Agreement Promises to Keep – Challenges to meet, Perspectives for Citizens in Consultation With the Great Lakes Basin's Environmental Community (http://www.greatlakes.org/Document.Doc?id=121)

Annex 2 of the Great Lakes Water Quality Agreement of 1978 as amended by Protocol signed November 1987.

Austin, John C. 2007. The Vital Center: A Federal State Compact to Renew the Great Lakes Region. The Brookings Institution. (http://www.brookings.edu/~/media/Files/rc/speeches/2007/1001techbelt_austin/1001techbelt_austin.pdf)

Austin, John C. and Soren Anderson, Courant, Paul N. and Litan, Robert E. 2007a. Healthy Waters, Strong Economy: The Benefits of Restoring the Great Lakes Ecosystem. The Brookings Institution. (http://www.brookings.edu/~/media/Files/Projects/GLEI/20070905_GLEI.pdf)

Austin, John C. and Soren Anderson, Courant, Paul N. and Litan, Robert E. 2007b (http://www.healthylakes.org/site_upload/upload/GrtLakesCostBenefit.pdf)

Austin, John C. and Soren Anderson, Courant, Paul N. and Litan, Robert E. 2007c (http://www.healthylakes.org/site_upload/upload/Americas_North_Coast_Report_07.pdf)

Bush, George W. 2004. Presidential Executive Order 13340. (http://www.glrc.us/documents/EO13340.pdf)

Canadian Council of Ministries of the Environment. 1991. (http://www.ccme.ca/ourwork/environment.html?category_id=108)

Council of Great Lakes Governors. 1983. Creation of Council of Great Lakes Governors. (http://cglg.org/Overview/History.asp)

Council of Great Lakes Governors. 1985. http://www.cglg.org/projects/water/docs/GreatLakesCharter.pdf)

Council of Great Lakes Governors. 2001. The Great Lakes Charter Annex (Annex 2001) (http://cglg.org/projects/water/docs/GreatLakesCharterAnnex.pdf)

Council of Great Lakes Governors. 2005. (http://www.cglg.org/projects/water/index.asp)

Council of Great Lakes Governors. 2009. Water Management Initiative. (http://www.cglg.org/projects/water/links.asp)

Council of Great Lakes Industries Annex 2001 process comments and water resources policy papers (http://cgli.org/annex/annex.html)

Council of Great Lakes Industries. 2007. Building a Sustainable Great Lakes Water Quality Agreement, The Work Product from a Multi-stakeholder Review and Workshop, November 16, 2006. Final Project Report: EPA Assistance Project No. X3-83309901-0, March 2007 (http://cgli.org/sustainable/WQAConfWorkshopFinalReport3-07.pdf)

Council of Great Lakes Industries. 2009. Great Lakes Restoration Initiative Action Plan Outline Comments (http://cgli.org/comments/CGLIGLRIActionPlanOutlineComments8-09.pdf)

Council of Great Lakes Industries. 2010. Great Lakes Water Quality Agreement (GLWQA or Agreement) Revisions, Ann Arbor, Michigan (http://cgli.org/comments/CGLIGLWQAAmendment Comments7-10.pdf)

Eckenfelder, Wesley. 2009. Waste Water Extraordinaire: The Life of an Environmental Pioneer. Authorhouse Publishers (E-publication).

Environment Canada and U.S. EPA. 2007. 10 Years of Strategy Progress and Ambient Environmental Monitoring. (http://www.epa.gov/glnpo/bns/10thann/wg_for_sf_05-23-07.pdf)

Environment Canada and U.S. EPA. 2009. State of the Great Lakes 2009. State of the Lakes Ecosystem Conference 2009 Technical Report (http://www.epa.gov/solec/sogl2009/sogl2009complete.pdf)

Government of Canada Chemicals Management Plan. 2006. (http://www.chemicalsubstances chimiques.gc.ca/plan/index-eng.php)

Great Lakes Basin Compact of 2005 (http://glc.org/about/pdf/Compact.pdf)

Great Lakes Commission. 1994. Ecosystem Charter for the Great Lakes – St. Lawrence Basin (http://glc.org/ecochart/)

Great Lakes Commission and Northeast-Midwest Institute. Great Lakes Day 2010. (http://www.glc.org/greatlakesday/glday.html)

Great Lakes-St. Lawrence River Basin Water Resources Council. 2009. Chair's Proposed Invitation List for Membership on the Advisory Committee, Resource Group and Observers. http://www.glslregionalbody.org/Docs/Misc/GLSLRWRRB_Proposed_Advisory_Committee_RO_12-1-09.pdf

Great Lakes Water Quality Agreement of 1972, 1978, and 1987 (http://ijc.org/rel/agree/quality.html)

Great Lakes Water Quality Agreement Review Committee. 2007. Report to the Binational Executive Committee. Vol. I. Review of the Canada–U.S. Great Lakes Water Quality Agreement. (http://binational.net/glwqa/v1_glwqareview_en.pdf)

Industry Canada. 2010. (http://www.ic.gc.ca/eic/site/csr-rse.nsf/eng/rs00138.html)

Industry Canada. Corporate Social Responsibility, The Importance of Stakeholder Engagement. 2011. (http://www.ic.gc.ca/eic/site/csr-rse.nsf/eng/rs00138.html)

International Joint Commission 1978a. Sixth Biennial Report Under the GLWQA of 1978. (http://ijc.org/php/publications/html/6bre.html).

International Joint Commission 1978b. Seventh Biennial Report Under the GLWQA of 1978. (http://ijc.org/php/publications/html/7bre.html#appb)

International Joint Commission. 1993. A Strategy for Virtual Elimination of Persistent Toxic Substances. Vols 1 and 2, Windsor, Ontario. Canada.

International Joint Commission. 2007. Review of the Canada-U.S. Great Lakes Water Quality Agreement, Final Report to the Great Lakes Binational Executive Committee.

National Council for Air and Stream Improvement Inc. 1993. Technical Bulletin No. 662, Great Lakes Environment Assessment.

NCASI (National Council for Air and Stream Improvement). 1994. Five Great Reasons Why We Care: The Pulp and Paper Industry's Virtual Elimination Strategy, Special Report 94. (http://www.p2pays.org/ref/14/13028/refs.html)

The Boundary Waters Treaty between the United States and Great Britain (Canada) – 1909 (http://ijc.org/rel/agree/water.html#text)

The Great Lakes Regional Collaboration Strategy. 2005 (http://www.glrc.us/strategy.html)

The World Bank Group. 2010. The Worldwide Governance Indicators: Methodology and Analytical Issues. (http://info.worldbank.org/governance/wgi/resources.htm)

U.S. EPA. 2007. Canada-United States Strategy for the Virtual Elimination of Persistent Toxic Substances in the Great Lakes Basin. (http://www.epa.gov/greatlakes/p2/bnsintro.html)

U.S. Government. 2010. Great Lakes Restoration Initiative Action Plan. (http://www.greatlakesrestoration.us/pdfs/glri_actionplan.pdf)

U.S. House of Representatives. 2010. The Great Lakes Ecosystem Protection Act of 2010, H.R. 4755 (http://www.govtrack.us/congress/bill.xpd?bill=h111-4755)

Woodrow, P. and S. Gahis, CDR Associates, Boulder, CO. 1998. Effective Collaborative Processes on Sustainable Development and Environmental Policy. (http://cgli.org/positions/IPPDbig.pdf)

CHAPTER 19

Hamilton Harbour's Regeneration – An Industrial Perspective

Ian A.C. Shaw

Arcelor Mittal Dofasco, 1330 Burlington St. East,
Hamilton Ont. L8N 3J5

OVERVIEW

Two hundred years ago, the Hamilton Harbour watershed was heavily forested. Its creeks were pure and the Harbour was abundant with a diverse range of fish and wildlife. However, the effects of urbanization and the growth of the city of Hamilton took its toll on the natural habitat. By the early 1900s, the Harbour ecosystem was severely degraded as a result of direct sewage discharges, habitat loss, toxic spills, and sediment contamination. By the mid 1920s, canals and infill eliminated more than two-thirds of the original Harbour wetlands, inlets and shallows. Further deterioration occurred during the 1940s as Hamilton, now a center of heavy industry, focused on providing support to the war effort. It was not until the 1960s when there became a growing recognition that years of environmental degradation had to be rectified.

A number of forcing factors have resulted in a marked improvement in the state of the Hamilton Harbour over the past 40 years. These factors include the International Joint Commission's work regarding the control of toxic pollutants into Canada's waterways which identified the Harbour as one of 43 Areas of Concern in the Great Lakes, the establishment of environmental legislation and regulations governing water quality standards, and the Ontario Ministry of the Environment's program the Municipal/Industrial Strategy for Abatement (MISA).

Today a healthy Harbour is closer to becoming a reality as the public, industry and government recognize its value as a centerpiece for the Hamilton community. Today's Harbour serves many purposes. It is a busy working port and the location of one of Canada's largest concentrations of industrial activity. At the same time the Harbour supports environmental restoration and protection activities and provides a desirable recreational destination (also see Chapter 12 for details and case studies regarding the Hamilton Harbour RAP process).

As a large industrial site that benefits from the Harbour ArcelorMittal Dofasco (AM Dofasco) has long recognized its role as a corporate citizen and its need to actively participate in the regeneration of the Harbour. In the past 20 years, the company has reduced the discharge of all process effluent to Hamilton Harbour by 99%. This has been accomplished through the deployment of best available water treatment technologies and through the use of closed-loop water recycling systems that re-circulate nearly all of the process water used in the steel manufacturing process[1]. The company has also has been an active participant and supporter of a number of the key initiatives that have set the Harbour on a restorative track.

THE HARBOUR AND HAMILTON'S GROWTH

Hamilton Harbour lies at the western tip of Lake Ontario, and is separated naturally from the lake by a sandbar. It is the largest naturally protected Harbour on western Lake Ontario. Industry, commerce and residential areas, along with private and public open spaces share the 45 kilometer shoreline.

The Harbour's watershed covers more than 500 square kilometers and is drained by three major tributaries. The cities of Hamilton and Burlington, with a population of 650,000 people, are located in and around the watershed area.

Hamilton has long benefited from its geographical proximity to major land and water transportation routes along the Niagara Peninsula and Lake Ontario. However from the first settlement a tension has existed between maximizing economic growth and minimizing environmental impact on the local ecology. The area bounded by Burlington Bay, also known as Hamilton Harbour, and the Niagara Escarpment, has been significantly altered for residential, industrial and recreational purposes over the past 200 years.

HAMILTON, THE HARBOUR AND STEEL

The steel industry in the Hamilton area can trace its history as far back as the 1850s. Easy access to limestone from the Niagara Escarpment, coal mined in Appalachia, iron ore mined from Northern Ontario and Quebec and

[1]T.E. McGuire, J.N Lockington, D.O. Mallen, and W. Thoburn. 1993. *Environmental Legislation and the Canadian Steel Industry* - Steel Technology International

transported though the Great Lakes made Hamilton an important iron and steel producing city. During the 1940s heavy industry boomed as the Canadian and British governments' war-driven demands for steel, arms, munitions and textiles increased. During this period the shape and form of the Harbour was forever altered through infilling to increase the Harbour lands. By the end of the war the ecological cost of pollution had taken its toll on Hamilton.

Today the city of Hamilton has one of the largest concentrations of heavy industry in Canada. The steel industry is the dominant industry in the area of which the steel mills of AM Dofasco and US Steel (formerly Stelco) are located on the south shore of the Harbour. These steel plants collectively employ 10,000 people, produce 8 million tons of steel per year and occupy approximately 30% of the Harbour's waterfront. Steelmaking activities have made Hamilton one of the largest ports in Canada.

The Harbour is essential for the operation of the steel industry. It facilitates the movement of raw materials and steel products to and from the steel plants. The Harbour provides a source of water to meet a two million cubic meters per day demand by steel making processes. Finally the Harbour provides a reservoir for the discharge of the treated waters from the mills.

ARCELORMITTAL DOFASCO[2]

ArcelorMittal Dofasco Inc. is one of Canada's largest steelmakers. The company manufactures hot and cold rolled steel, galvanized steel, chromium-coated steel and tinplate. It sells its products to a wide variety of customers including those in the automotive, construction, energy, manufacturing, pipe and tube, appliance, and packaging industries.

Founded in 1912 at Hamilton Ontario, Dofasco Incorporated has a history of innovation and a reputation for a family atmosphere. At that time, known as the Dominion Steel Castings Company Ltd., the company began as a foundry making steel castings for Canada's expanding railway system.

In 1938, Dofasco introduced profit sharing for their employees. Seen as an aid to productivity, profit sharing also improved moral by breaking the divisions between blue and white collar workers, involving all employees directly in Dofasco's destiny. This was just one of the components of the "Dofasco Way", a unique employee relations program which would grow to include: a recreation park, a company pipe band, a community choir, and at one time the world's largest corporate Christmas party.

In 1954 Dofasco was the first North American company to build a basic oxygen furnace. This method for making steel became and remains today the

[2]http://www.fundinguniverse.com/company-histories/Dofasco-Inc-Company-History.html

standard for modern steel production. The year 1955 saw another first for Dofasco with the addition of Canada's first continuous galvanizing line. Capital investments continued throughout the 1960s.

Since 1970, the company has used the same corporate slogan "Our product is steel. Our strength is people" which is considered to be one of the most recognized corporate images in the country[3]. Through its successful restructuring in the 1980s, emphasis on higher-margin specialized products, and its use of the latest technology Dofasco was consistently profitable through the 1990s. In the mid 1990s Dofasco became the first steelmaker in the world to integrate lower environmental impact scrap steel based electric arc furnace technology into their traditional integrated steelmaking configuration. This technology enabled a 30% increase in steel production capacity. The technology also made Dofasco the largest steel scrap recycler in Canada.

In 1997, Dofasco signed a groundbreaking voluntary environmental agreement with the provincial and federal governments to enable accelerated environmental performance improvements beyond the existing regulatory framework. The Environmental Management Agreement (EMA) was designed to benefit Dofasco, governmental agencies and the environment in the following ways:

- Reduce business risk by clarifying future environmental expectations
- Enable focus and prioritization for the efficient allocation of capital and human resources for both the company and regulating bodies
- Ensure alignment with environment ministry goals and objectives
- Reduce regulatory administrative requirements
- Establish measurable performance based agreement targets
- Enable transparent communication and self-accountability on Dofasco's part

The EMA did not negate Dofasco's obligations to meeting existing laws and standards. A number of the provisions of the agreement were not required by law at that time.

Many onlookers initially perceived that the EMA put Dofasco at a competitive disadvantage resulting from the company's commitments to expenditure, implementation timelines and stringent performance measures beyond the laws of the day. In fact the opposite was true. The agreed upon priorities enabled proactive business planning and resource allocations within the company. The EMA also provided flexibility in implementation of the most cost effective solutions as opposed to prescriptive cost inefficient obligations. Solutions could be implemented using capital efficient approaches including new technology, modified work practices, material substitutions, or end-of-pipe technology.

[3]The 1970s Landmark Canadian Advertisements". *Marketing Magazine*. 28 September 1998. http://www.marketingmag.ca/shared/print.jsp?content=19980928_7929. Retrieved 2008-02-21.

In addition to the aggressive improvement targets, the EMA committed Dofasco to continued participation in community-focused environmental initiatives on issues including air quality, energy conservation, the reduction and elimination of toxics, the Hamilton Harbour Remedial Action Plan, Bay Area Restoration Council (BARC), Bay Area Implementation Team (BAIT) and Hamilton's sustainability based initiative VISION 2020.

The company's focus on technology, value generation, customer-focused solutions, strong environmental performance and a corporate citizen orientation has earned it several marks of distinction. For seven consecutive years Dofasco was recognized for its triple bottom line approach to business by being named a member of the Dow Jones Sustainability Group Index (DJSGI) based on financial, social and environmental performance and management approaches[4].

The company was named the Global Market Sector Leader in the Basic Materials Sector of the DJSGI index. Dofasco is the only Canadian company ranked as a sector leader. During this period of time the company also maintained the distinction of being one of the most profitable steel companies in North America.

In 2005 Dofasco became an acquisition target during a period of industry consolidation. A bidding war erupted between the French Steelmaker Arcelor and the German steelmaker Thyssen Krupp who were both vying to establish a strong North American presence. Arcelor ultimately acquired Dofasco, paying the highest price per ton of production capacity in the history of any modern steel acquisition. Soon after this event the Arcelor and Mittal steel companies merged to form the largest steelmaker in the world. Today ArcelorMittal Dofasco (AM Dofasco) is a stand alone subsidiary of the ArcelorMittal Company.

WATER USE IN THE STEEL INDUSTRY

Water is used in the steel industry for three main purposes[5]:

Heat transfer: Typically heat transfer applications account for 70% of all water consumption in integrated steel mills. In the iron and steelmaking processes it is necessary to heat raw materials in the range of 1400 to 1600 deg. Celsius. The equipment used in these processes is protected by refractory and water cooling systems. Byproduct fuel gases generated by cokemaking, ironmaking and steelmaking operations must be cooled and treated to remove air pollutants using water before they can be used as useful fuels for steelmaking processes.

Air pollution control: Water is used in wet scrubbing tower and venturis to capture particulate from gas and air streams. Usually 15% of total water consumption is associated with these pollution control systems.

[4]http://www.dofasco.ca/bins/content_page.asp?cid=2347-2349-2630

[5]Water Use in Industries of the Future: Steel Industry, U.S. Department of Energy, July 2003

Material conditioning: Water is used for dust control, mill scale removal in hot rolling operations, as a solvent for acid in pickling operations, and as a carrier for roll oil coolant and lubricant packages in cold rolling operations. Typically 15% of water consumption is associated with material conditioning operations.

AM Dofasco typically requires approximately 75 cubic meters of water per ton of steel produced. Approximately 97% of water used is pumped directly from the Harbour. The remaining 3% is purchased from the city. Roughly 30% of the water comes into direct contact with the raw materials, gases or intermediate product streams. Adequate capacity has been installed to treat all process contact water. All of these facilities meet or surpass best available technology (BAT) effluent treatment requirements. The balance of Dofasco's water requirements are for the purpose of indirect cooling. This water does not come into contact with polluting elements and is returned to the Harbour untreated. Only 2% of Dofasco's total water intake is consumed by evaporation.

WATER CONSERVATION AT DOFASCO

While AM Dofasco benefits by being located adjacent to the Great Lakes and 20% of the earth's accessible freshwater, conservation practices go hand in hand with maintaining water quality.

There are eight major water conservation techniques employed at Dofasco. The installation of these technologies plant-wide has reduced the need for baywater withdrawals by over 100 million cubic meters per year. The conservation techniques that have been successfully deployed include:

1. *Closed loop cooling systems*: AM Dofasco utilizes over a dozen cooling towers that utilize evaporative cooling systems to extract heat from steelmaking processes.
2. *Contact process water conditioning*: There are a number of treatment processes that are designed to treat specific parameters found in the process water. Examples include deep bed filtration for the removal of suspended solids and ion exchange to capture dissolved metals while returning deionized water back to the manufacturing process.
3. *Condensation of water vapor*: Steam is condensed to extract as much energy as is possible and to recover the hot condensate to be recycled as boiler feedwater.
4. *Direct contact evaporative cooling*: Slab reheat furnaces are equipped with evaporative skid cooling systems to manage the heat load within the furnace slab walking beam system and to generate steam for plant use. On extraction of the useful energy the steam is condensed and used for boiler feedwater makeup.

5. *Water cascading*: Directing water from processes requiring higher water quality to lower quality demands is a common water reuse practice at Dofasco. The blast furnaces cascade water used for cooling the furnace into the gas cleaning system. The water is then processed to remove solids picked up from the gas and recirculated back to the furnace.
6. *Counterflow rinsing*: A conservation technique to reduce the required water volume is to stage rinse tanks in a manner that they flow counter to the direction of the steel that is being rinsed. This technique is utilized at several of the finishing mill operations.
7. *Thermostatic control*: The flow of cooling water is controlled at the caster mold machine to maintain constant temperature at the mold.
8. *Preventative maintenance*: Regular maintenance of pumping and steam systems eliminates water loss due to leaks that are found in valve packing, flanges and steam traps. Water waste can be reduced by as much as 15% through preventative maintenance practices.

THE HAMILTON REMEDIAL ACTION PLAN (RAP)

The goal of the Hamilton RAP is to bring about sustainable natural ecosystems in Hamilton Harbour and its entire watershed, and to improve the potential for more extensive recreational uses while maintaining the Harbour's and the watershed's essential economic function. Stage 1 – Problem Identification was initiated in 1985. Stage 2 – Selection of Remedial Measures was completed in 1992 and updated in 2002. Stage 3 – Delisting of AOC is currently proposed for 2015.

A stakeholder group representing over 40 agencies from industry, environment and government developed the RAP Stage 1 and 2 reports. Implementation is mandated to the Bay Area Implementation Team (BAIT) representing 18 key government and industrial stakeholders, co-chaired by Environment Canada and the Ontario Ministry of the Environment. Overseeing public participation to both scrutinize and encourage remedial actions is the Bay Area Restoration Council (BARC) with its own elected president and board. A "who does what best approach" to implementation is used by the various RAP stakeholders.

It is estimated that prior to 1990, CDN $ 600 million was spent by industry and government; between 1990 and 2000, CDN $ 200 million was spent; and from 2000 to 2015, CDN $650 million will be required.

HAMILTON RAP FINDINGS

The types of problems which were covered in the IJC's Water Quality Board included: elevated levels of conventional pollutants, elevated heavy metal and complex organic compound concentrations, contaminated sediments, diminished aesthetic quality, and eutrophication.

The IJC Water Quality Board identified the following major sources of the Harbours problems to include: municipal sewage plants, industrial point sources, combined sewer overflows, upstream tributaries, urban non-point sources, and in-place pollutants.

RAP GOALS, PRINCIPLES AND THE STEEL INDUSTRY

Nine use goals for the Harbour were identified by the stakeholders. These goals include: recreational boating, water sports, shipping and navigation, industrial uses for water intake, water receiving body, fisheries, wildlife and habitat appreciation, swimming and an educational resource. All of these goals have the overriding requirement of the improvement of Harbour water quality or a reduction in negative impacts on water quality.

The stakeholders developed two primary principles and five secondary principles which have served the basis of reasoning and a guide to the development of recommended actions. The two primary principles have had a direct impact on the approaches to remedial action taken by the steel industry.

1. The Ecosystem Principle states that the RAP must embody a comprehensive ecosystem approach to restoring and protecting beneficial uses to the Harbour in that economic, social and environmental goals must be viewed in an integrated manner.
2. The Philosophy of Zero Discharge principle states that "inputs of persistent toxic substances shall be zero discharge, and therefore, the discharge of any or all persistent substances shall be virtually eliminated."

The second principle was concurrent and re-enforced by the Ontario Ministry of the Environments toxic reduction program, Municipal/Industrial Strategy for Abatement (MISA).

MUNICIPAL/INDUSTRIAL STRATEGY FOR ABATEMENT (MISA[6])

The purpose of the MISA program was to reduce water pollution systematically at its source with enforceable regulations that become more stringent as abatement technology improves. The ultimate goal is the virtual elimination of

[6]http://www.ene.gov.on.ca/environment/en/industry/standards/spills/index.htm

T.E. McGuire, J.N Lockington, D.O. Mallen, and W. Thoburn. 1993. *Environmental Legislation and the Canadian Steel Industry*-Steel Technology International

persistent toxic contaminants for all discharges into Ontario waterways echoing the Great Lake Water Quality Agreement's objective.

The program initially dealt with all direct dischargers and divided the group into two categories, municipalities and nine major industrial sectors. Each sector was dealt with separately by the Ministry of the Environment resulting in sector-specific regulations.

The MISA program consisted of four phases. The first involved the development of an Effluent Monitoring Regulation. This regulation was undertaken to develop a list of parameters to be monitored and the monitoring frequency. In the case of the iron and steel sector 154 parameters were selected for monitoring.

On completion of the Effluent Monitoring Regulation the second phase of monitoring data collection began. In the case of AM Dofasco four types of effluent were monitored at 12 discharge points including five process effluents, two cooling water effluents, two storm water effluents and three final effluents – a combination of process, cooling water and storm water.

The third phase of the MISA program was the development of an Effluent Limits Regulation for each sector. This involved:

- Evaluation of the effluent monitoring data
- Selection of parameter for limit setting
- Selection of Best Available Technology (BAT)
- Assessment of economic viability
- Establishment of plant limits.

The final phase involved the implementation and regulation including a phasing in period to enable companies to install the technologies to enable compliance with the limits.

While AM Dofasco was not specifically selected as a BAT plant for any process category a number of installations were already in place that met or exceeded effluent limits.

One of these systems included the ironmaking gas washing recirculation system. The gas cleaning system scrubs particulate from the top gases of the blast furnaces using countercurrent water scrubbing towers. The blast furnaces produce approximately 100,000 cubic meters per day of scrubber water that is processed through a physical/chemical treatment plant before being recirculated back to the gas scrubbers. The recycle system has resulted in a 99% reduction in water discharges.

A second system involved a hot rolling mill filtration plant that treats more than 260,000 cubic meters per day of water from the AM Dofasco hot mill and continuous caster. The system has also reduced water discharge by 99% with the remaining 1% discharge containing only 2.2 parts per million (ppm) particulate. This constitutes only 10% of the average particulate concentration of water drawn from the bay.

In 1998 at a cost of CDN $ 34 million AM Dofasco completed its water recycle and processing systems. As a result process water particulate loadings to Hamilton Harbour have been reduced by over 98% since 1989.

LOOKING TO THE FUTURE

In today's world water is becoming increasingly scarce. Macro trends including population growth, climate change, expanding urbanization and industrialization are challenges that will strongly affect water availability and consumption. Looking to the future it is expected that three major factors will influence water availability and cost:

- ***Supply and demand imbalance***: The highlighted macro trends will have a significant effect on the supply and demand imbalance for fresh water. These trends show no sign of abating, making this imbalance a growing and sustained change that will drive a requirement for new water technologies and better water conservation practices.
- ***Higher input costs***: The cost of delivering water to all end users has risen dramatically due to increasing demand, but also due to underlying factors such as the increase in the price of energy, chemicals and basic raw materials. The issue has been further compounded by increased regulation and higher quality standards which have led to an increasing total cost of managing water processes. These factors are driving the need for increased efficiency and the implementation of innovative water technologies.
- ***Increasing value of water***: The value of water has two components: the market price and its intrinsic value. The fundamental value of water extends beyond its price. Without water, power generation plants cannot operate, most industrial processes would be constrained, and our built environment could not be heated or cooled.

Water constraints pose a potential multi-billion dollar risk for businesses worldwide. Globally, business related water consumption accounts for approximately 20% of all water use[7]. The diminishing quality of water supplies, increasing water purchase costs and tightening environmental standards are forcing industries towards increased water efficiency and conservation. These factors and an estimated five-fold increase in worldwide manufacturing water use by the year 2030 will continue to drive increased water costs[8].

A tangible business case exists for water management within corporations. Dimensions of the business case include:

[7]WBCSD Water Facts & Trends". http://www.wbcsd.org/includes/getTarget.asp?type=d&id=MTYyNTA

[8]M. Ellis. 2000. *Industrial Water Use and Its Energy Implications*, US Department of Energy Energy-Office of Industrial Technologies

1. *Increased production and treatment costs*: These costs will be borne by business as a result of decreasing availability, reliability of supply and declining quality. Many municipalities now charge financial penalties for the introduction of non-municipal water loadings into the municipal treatment system.
2. *Environmental regulation*: Not only are company effluent discharges being more closely regulated but environmental agencies in many jurisdictions are now placing limitation on the volume of water withdrawals.
3. *Loss of license to operate*: In localities where business water use competes with meeting the basic needs of local communities there is potential for conflict and the loss of access to water resources.

Future water constraints are not only about business risk but also provide opportunities for business to contribute with sustainable solutions. Companies will have to respond by employing a suite of solutions leveraging both technology and best management practices. Some of these solutions will include:

- Advanced water treatment systems that are capable of cost effective high pollutant removal efficiencies
- Separation of process and non-contact water streams to provide focused effective treatment
- Instituting pollution prevention philosophies as opposed to end-of-pipe treatment systems
- Closed loop water recycle systems
- Storm water capture, retention and discharge
- Adoption of manufacturing processes that are designed to be less water intensive
- Movement to dry type cooling and pollution control systems
- Deployment of technologies to extract low grade heat for process stream preheating or conversion into electricity.

CONCLUSION

AM Dofasco's pursuit of sustainability, the alignment of business, environmental and social objectives, has enhanced its ability to compete globally and have positioned the company to thrive in its second century of existence. This strategy has guided the organization's decision making and approach in the day-to-day business as well as its outward multi-stakeholder view to include customers, government and the community. The strategy also provided a fertile environment that fostered the innovative approach embodied in the AM

Dofasco Environmental Management Agreement. The EMA and the benefits that resulted further reinforce the ability of companies to generate cost effective, practical solutions that strengthen the company's competitive position while making tangible environmental improvement.

Engaging and leveraging the pooled knowledge, ideas and resources of a diverse group of stakeholders have resulted in a practical revitalization plan for the Harbour. While there is still outstanding work required before the Hamilton RAP can meet its ultimate objective of being delisted, there are many tangible signposts indicating the Harbour is on a path of recovery. This is evidenced by improvements in a broad spectrum of water quality measures, increasing ecological diversity and improved aesthetics. These improvements can be related directly to actions instituted by the RAP process. AM Dofasco has been a strong supporter and active participant in the clean-up and restoration of Hamilton Harbour. Since 1989 AM Dofasco has reduced total process effluent discharges to Hamilton Harbour by 99%. In the future it is expected that water will become an even more critical issue for industry driving the need for further innovation in water efficiency and conservation.

ACKNOWLEDGEMENTS

The author wishes to thank Mr. Thomas E. McGuire, Environment Manager, AM Dofasco for his significant contributions to the writing of this chapter. Tom's distinguished 30-year career has been singularly focused on ensuring AM Dofasco acts as an exemplary corporate citizen and to striving towards the ideal of making AM Dofasco a sustainable corporation. Tom represented AM Dofasco from the outset of the Hamilton RAP, as well as playing a significant role in shaping the corporation's approach towards constructively addressing the RAP objectives. Tom's knowledge and perspectives were indispensible in the development of the chapter's content. In addition to providing exemplary corporate environmental leadership he has patiently mentored several generations of environmental engineers including myself.

REFERENCES

Bay Area Restoration Council. www.hamiltonHarbour.ca. October 2010

Ellis, M. 2000. Industrial Water Use and Its Energy Implications, US Department of Energy Energy – Office of Industrial Technologies.

Environment Canada. Hamilton Harbour Area of Concern. www.ec.gc.ca/raps-pas/default.asp?lang=En&n=18AD73C1-1, October 2010

History of Hamilton. www.wapedia.mobi/en/History_of_Hamilton, October, 2010.

International Joint Commission. www.ijc.org, October 2010

Johnson, R. 2003. Water Use in Industries of the Future: Steel Industry – US Department of Energy – Office of Energy Efficiency and Renewable Energy. July 2003

McGuire, T.E. 1998. An Industrial Approach to Water Conservation

McGuire, T.E. and M.S. Greenfield, and J.N. Lockington. 1990 The Steel Industry and Hamilton Harbour – A Remedial Action Plan.

McGuire, T.E. and J.N Lockington, D.O. Mallen, and W. Thoburn. 1993. Environmental Legislation and the Canadian Steel Industry – Steel Technology International

McGuire, T.E. and M.S. Greenfield, and J.N. Lockington. 1990 The Steel Industry and Hamilton Harbour – A Remedial Action Plan.

SECTION 6

CONCLUSIONS

CHAPTER 20

Conclusion – Lessons Learnt

Velma I. Grover

916 – 981 Main St West, Hamilton, ON, L8S1A8
E-mail: Vgrover@rogers.com

The Great Lakes are an important part of the physical landscape and cultural heritage of the United States of America and Canada. They are important for both the Canadian and US economy and also form a substantial network of trade between the two countries. The eco-system services provided by the Great Lakes Basin include: water supply for residential, commercial, industrial and agricultural use; commercial and sport fishing; hunting; boating; swimming; hydropower production; nutrient cycling; biodiversity, and responding to climate change in the region. Despite their immense size the Great Lakes are very sensitive to chemical and biological stresses, which threaten the resilience of any restoration activities.

Boundary Waters Treaty is a historic accord signed between the US and Britain (on behalf of Canada) in 1909 that calls for cooperative management of the many lakes and rivers that flow along and across the common boundary. The Treaty created the International Joint Commission (IJC) to resolve any transboundary disputes (described in detail by the authors – Crane, Muldoon, Krantzberg). The increase in industrialization and commercial agriculture caused concern over quality of the Great Lakes. A study commenced by the IJC in the 1960s led to the signing of Great Lakes Water Quality Agreement in 1972. One of the main goals of the Great Lakes Water Quality Agreement is to “restore and maintain the physical, chemical, and biological integrity of the waters of the Great Lakes Basin Ecosystem". To deal with the significant issues of toxic substances, agricultural and urban run-off threatening water quality and aquatic wildlife, amendments to the Great Lakes Water Quality Agreement in a 1987 Protocol designated, under Annex 2 of the Agreement, 42 severely degraded areas in the nearshore environment along the Great Lakes as Areas of Concern (AOCs) (described in detail by several authors – Crane, Muldoon, Krantzberg). Degradation in environmental quality in these AOCs resulted in measurable "impairment" of numerous chemical, physical, or biological aspects of the water body termed "beneficial uses" (discussed by Krantzberg, Hall, Lawrence, Boughton and Sommer). To deal with the issues in these AOCs, Remedial Action Plans (RAPs) were formed for each AOC.

Annex 2 of the Great Lakes Water Quality Agreement stipulated that RAPs are to "embody a systematic and comprehensive ecosystem approach to restoring and protecting beneficial uses" in AOCs. This was an ambitious forward-looking plan since the ecosystem approach (as described by Krantzberg) was not the traditional way to deal with the issues at that time especially when the original agreement was signed in 1972. It is quite interesting to note that at the time of signing, the then Canadian Prime Minister, Pierre Trudeau, said that the Agreement "promises to restore to a wholesome condition an immense area which, through greed and indifference, has been permitted to deteriorate disgracefully." As discussed by McLaughlin and Krantzberg (2006) this statement is remarkable for two reasons: Trudeau refers to the "wholesome condition" where he was probably looking at the issues more holistically while the term "ecosystem approach" (incorporated in the Agreement in 1978) was not conceived at that time; secondly his tone and mention of "greed and indifference" suggests that he had more of a utilitarian view or approach to nature while in reality decision-makers are still building it into their thinking.

Jones and Taylor (1999) mention that the eco-system approach is a holistic way of solving problems in the Great Lakes and quote Ferreri et al. (1999) to describe the ecosystem management approach as a philosophy that emphasizes: "(1) the integration of the physical, chemical, biological, and social components of the resource; (2) the interactions of these four components (as mentioned in point one); and (3) how these components relate to ecosystem productivity and resource sustainability." They further quote Christensen et al. (1996) to describe the eight principles that should build the foundation of the eco-system management as mentioned by an ad hoc committee of the Ecological Society of America:

1. long-term sustainability as a fundamental value;
2. clear, operational goals;
3. sound ecological models and understanding;
4. understanding complexity and interconnectedness;
5. recognition of the dynamic character of ecosystems;
6. attention to context and scale;

7. acknowledgement of humans as ecosystem components;
8. commitment to adaptability and accountability.

As described by the authors in the case study section (to name a few: Krantzberg, Hall, Lawrence, Boughton), the RAP process consists of a few basic steps: identification of problems in a particular area of concern, list causes of the problems identified, options and preferred actions to solve these problems, identification of the persons or agencies responsible for implementation of remedial measures, a schedule for implementation, and an evaluation process for monitoring and to track progress (that can eventually confirm that the problem(s) has been resolved. The end goal of all the RAP exercises is to get "delisted", meaning that all BUIs (beneficial use impairments) identified have been restored according to defined criteria (Botts and Muldoon 2005). RAPs are based on some principles such as: requirement to include multiple stakeholders (including non-technical stakeholders), the accountability involved in specifying responsibilities for action, and involvement of the IJC to review and evaluate progress of each objective identified by AOCs (Hartig and Vallentyne 1989).

Although, a lot of chapters in the case study section demonstrate that progress has been made in AOC regions, only a few of them have been delisted (one of the delisted successes in Collingwood is described by Krantzberg). Places such as Hamilton have made progress in rehabilitation of the eco-system, however, they are still far from being delisted mainly because of chemicals present in certain areas of the harbour, which are expensive and difficult to get rid off (Hall). Muldoon rightfully mentions that the ecological state of the Great Lakes presents a mixed picture. Although, some argue that significant progress has been made to improve the state of Great Lakes since the early 1970s, few of the threats identified in the early 1970s have been resolved and new ones seem to constantly appear (Muldoon).

The IJC concluded in its 2006 13th Biennial Report (and discussed by Muldoon) that the future of the Great Lakes remains "uncertain." A number of findings reinforce that, despite past progress, there are continuing threats to the health of the Great Lakes:

- There are on-going threats to wetlands considering that over 90% of pre-settlement wetlands along Lake Huron/Lake Erie corridor have been lost.
- There are various reports which outline that in the Areas of Concern in the Great Lakes, there are reproductive problems, reduced stress response, disrupted thyroid function, suppressed immune function, among other impairments in wildlife (snapping turtles, herring gulls) and reproductive failures in various species of fish (including lake trout, sturgeon, lake herring).

- "General water quality degradation, associated algal blooms, Type E botulism in fish and waterfowl, and contamination of drinking water".

This essentially leads to a conclusion that there are problems in the current institutional and management structure and thinking (including the RAP program), and we need a new governance regime to manage the Great Lakes.

Some of the key ingredients for the success of RAP program include: involving all the key stakeholders from the very beginning in a meaningful way, have a well-defined work plan and right partners, a right mix of funding, innovation and leadership (facilitator) to lead the process; accountability, transparency, good communication plan and celebrating every small success along the way to delisting. Although, there was a lot of enthusiasm at the beginning of RAP programmes the momentum often stalled for various reasons. For example, the ecosystem approach was a new concept, and it posed implementation challenges, since insufficient thinking or time was devoted to the process in the beginning. Further, the partial or complete withdrawal of funding created a resources gap in the implementation of RAPs, a result of a lack of a strong leadership and probably not enough oversight from the IJC (Botts and Muldoon 2005). Gaps in continued monitoring and accountability (IJC 2003) also impinged on implementation progress. The failure of many RAPs can be explained by the use of a traditional technocratic approach to solve environmental problems. That is, designing solutions on analytical understanding based on technical/scientific information in a process that establishes primacy for technological expertise (Miller 1985) and results in piece-meal solutions that does not account for holistic solutions and can result in unintended undesired consequences.

Even though some progress has been made in RAPs implementation, it has been slow and in the meantime some of the Agreement's provisions have become outdated. One of the biggest challenges will be the need to change the governance structure. According to Botts and Muldoon (2005), the Great Lakes regime lacked a plan for institutional development for the RAPs programme and depended more on the science and technical experts leading to weak institutional structure for implementation. As discussed by several authors in the book, the change required was not just a scientific solution but needed changes at societal level in terms of changing behavior of people on the use of water, setting levels for industrial effluents, and setting standards for agricultural pesticide use, this means involving community in the RAP process, initiating educational and awareness programs in addition to technical solutions and more effective institutional/governance/management structure.

Lessons learnt from the success and failure of RAPs (discussed by Fox et al., Zimmer et al., Krantzberg, Hall, Lawrence, Boughton) can be translated to other places for lake restoration. These include:

- Identifying problems from science, social and policy/institutional perspective.
- Selection of a good leader, involvement of all stakeholders (preferably from the very beginning of the process, involvement should be more meaningful and effective with a clear goal/aim so as to keep people engaged for a long time with the aim of achieving certain goals).
- A framework that links larger lake basin objectives for restoration and development to local priorities that would also engage local partners and citizens.
- Involvement (and leadership) of local industries is also important (see Kuper, Shaw).
- Making a plan before implementation (understand the process to have an implementable plan but be adaptable to the changes occurring in the environment and learn from failures along the journey).
- There will need to be a strategic shift in current land use practices towards low impact development principles in both developing new areas and retrofitting urban areas.
- Establish a monitoring plan with a feedback loop to change the plan when needed.
- A transparent and accountable system.
- Do cost/benefit analysis of the restoration costs (see Kubursi) which is important for long term sustainability.
- Long-term funding.
- Good communication plan to keep communities, stakeholders and funding agencies aware of the progress and successes to keep the momentum going.
- Part of the solutions to the issues along nearshore basins will also involve changing the behavior of people. This means raising awareness and developing right educational programs for both children and adults. Zimmer et al. in the book discuss how marketing and outreach initiatives, informed by sound market research and a fulsome understanding of the target demographic(s), is an effective approach to securing uptake and desired long-term behavioral change within the intended population. Behavioral change can only happen only through initiatives that reach the intended people on an emotional level and tap entrenched, intrinsically-held beliefs, will the constraints to at-source controls be effectively addressed?
- Political will for implementation (a strong community support can often lead to strong political support as well).

REFERENCES

Botts L. and P. Muldoon. 2005. The Evolution of the Great Lakes Water Quality Agreement. Michigan State University Press. East Lansing, MI.

Christensen, N.L. and A.M. Bartuska, J.H. Brown, S. Carpenter, C.D. D'Antonio, R. Francis, J.F. Franklin, J.A. MacMahon, R.F Noss, D.J. Parsons, C.H. Peterson, M.C. Turner and R.G. Woodmansee. 1996. The report of the Ecological Society of America Committee on the Scientific Basis for Ecosystem Management. Ecological Applications 6: 665-691.

Ferreri, C.P. and W.W. Taylor and J.M. Robertson. 1999. Great Lakes fisheries futures: Balancing the demands of a multi-jurisdictional resource. *In:* W.W. Taylor and C.P. Ferreri (eds.). Great Lakes Fisheries Policy and Management: A Binational Perspective, Michigan State University Press, East Lansing, MI.

Hartig, J.H. and J.R. Vallentyne. 1989. Use of an ecosystem approach to restore degraded areas of the Great Lakes. Ambio, 18: 423-428.

IJC [International Joint Commission], 2003. Priorities 2001–2003: Priorities and progress under the Great Lakes Water Quality Agreement. Science Advisory Board, Chapter 5 (Emerging Great Lakes Issues in the 21st Century). Accessed on June 26, 2011 from http://www.ijc.org/php/publications/pdf/ID1568.pdf

Jones, M.L. and W.W. Taylor. 1999. Challenges to the implementation of the ecosystem approach in the Great Lakes basin. Aquatic Ecosystem Health and Management 2: 249–254.

McLaughlin, C. and G. Krantzberg. 2006. Toward a 'better understanding' of the Great Lakes Basin Ecosystem. Journal of Great Lakes Research 32: 197-199.

Miller, A. 1985. Ideology and environmental risk management. The Environmentalist, 5: 21-30.

Index

R

S

About the Editors

Velma I. Grover is an Adjunct at the Arcelor Mittal DOFASCO School of Engineering and Public Policy at McMaster University in Hamilton, Canada, and an Adjunct Professor at the Faculty of Environmental Studies at York University in Toronto. Velma has worked with nonprofit organizations, consulting companies and International Organizations mainly in the field of water and waste management. Some of the research appointments include: visiting fellow at Fudan University, China; visiting fellow at Smith School of Enterprise and the Environment, Oxford University and Kalmar University, Sweden.

She has a Masters in Environmental Management and a Ph.D. from University of London, UK. Velma has published on various environmental management issues, including seven edited volumes.

Gail Krantzberg is Professor and Director of the Dofasco Centre for Engineering and Public Policy in the School of Engineering at McMaster University offering Canada's first Master's Degree in Engineering and Public Policy. Gail completed her M.Sc. and Ph.D. at the University of Toronto in environmental science and freshwaters. She worked for the Ontario Ministry of Environment from 1988 to 2001, as Coordinator of Remedial Action Plan and Great Lakes Programs, and Senior Policy Advisor on Great Lakes. She is a past president of the International Association of Great Lakes Research. Dr. Krantzberg was the Director of the Great Lakes Regional Office of the International Joint Commission from 2001 to 2005. She has authored more than 100 scientific and policy articles on issues pertaining to ecosystem quality and sustainability and is a frequent speaker to media and the public.

Color Plate Section

CHAPTER-8

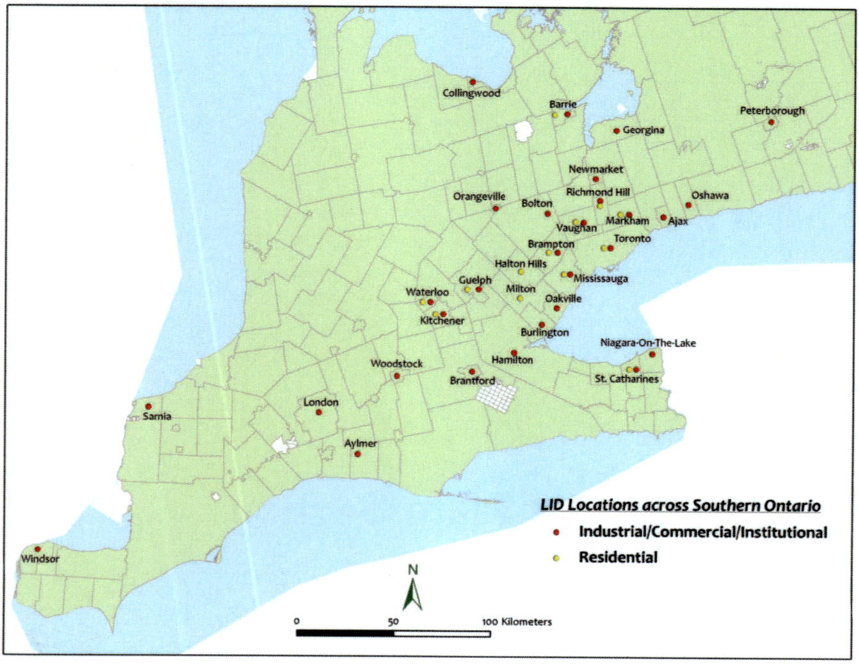

Fig. 8.1 LID implementation across southern Ontario

Fig. 8.5 Respondent's drawings of an ideal landscape and sustainable landscape

CHAPTER-10

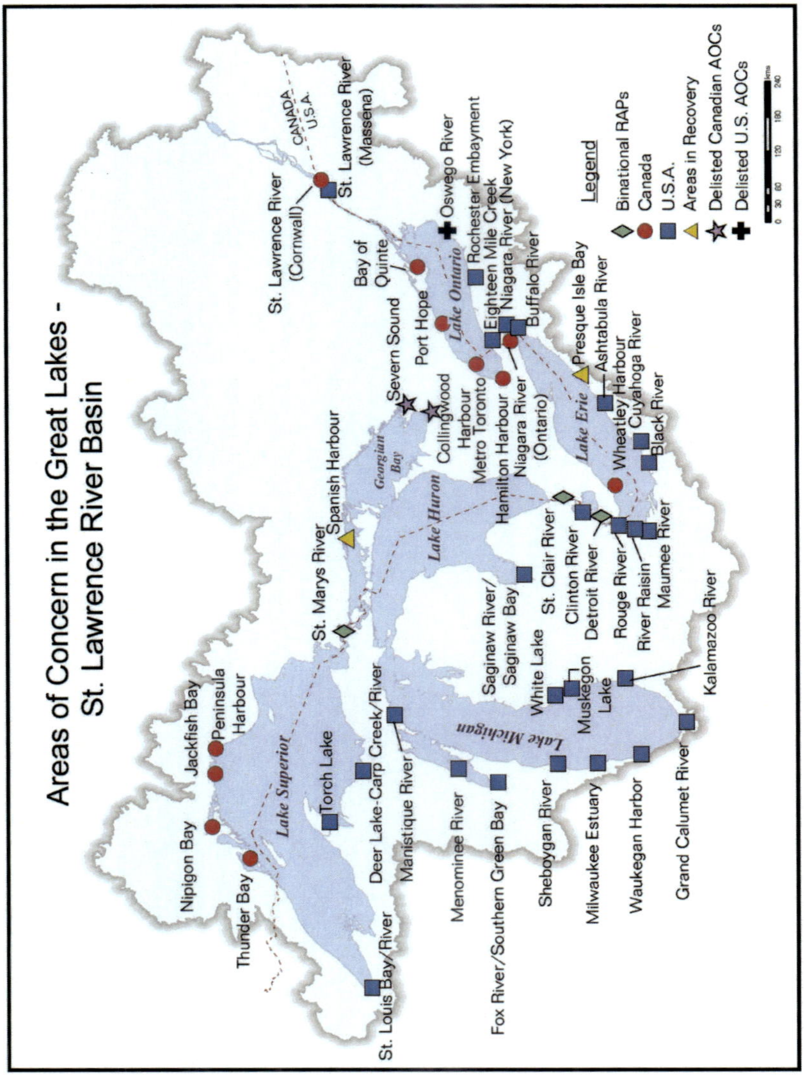

Fig. 10.1 The location of the geographic Areas of Concern for which Remedial Action Plans are underway or completed

CHAPTER-11

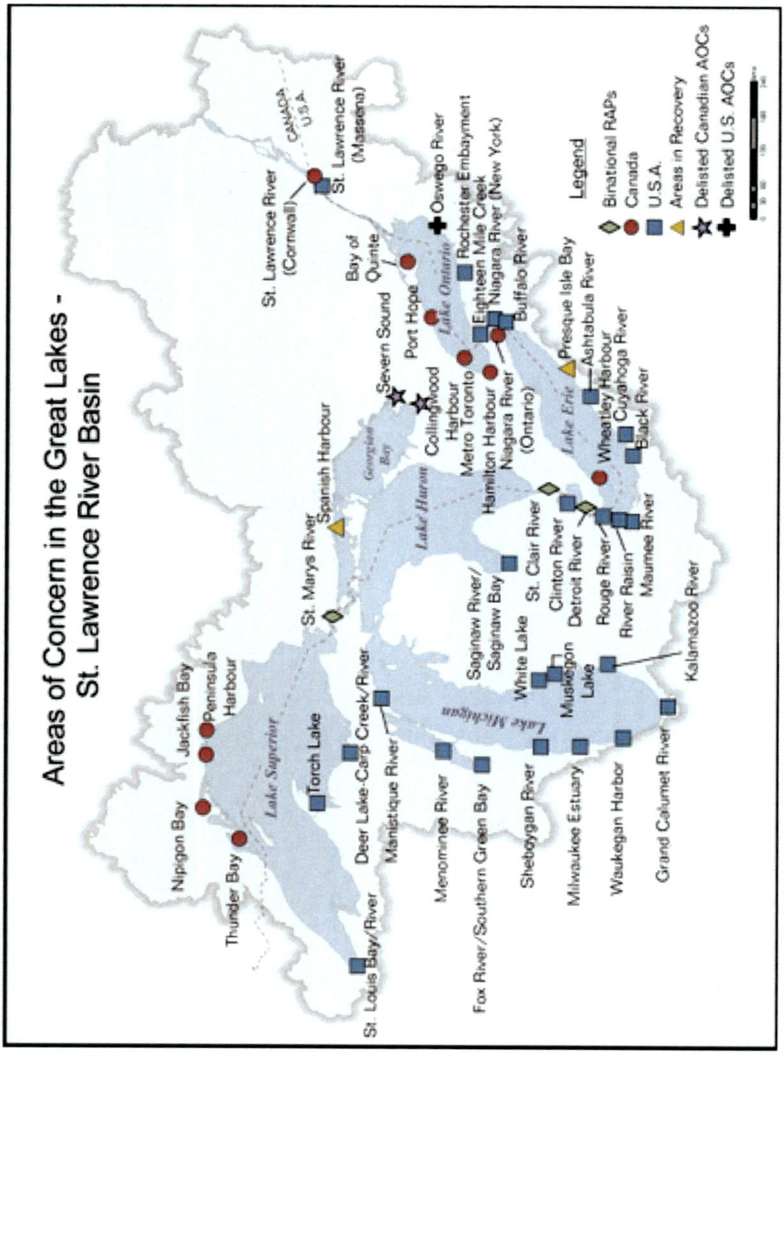

Fig. 11.2 Location of the Great Lakes Areas of Concern
source: http://www.epa.gov/glnpo/aoc/

Fig. 11.4 The shipyards proposed development plan.

CHAPTER-13

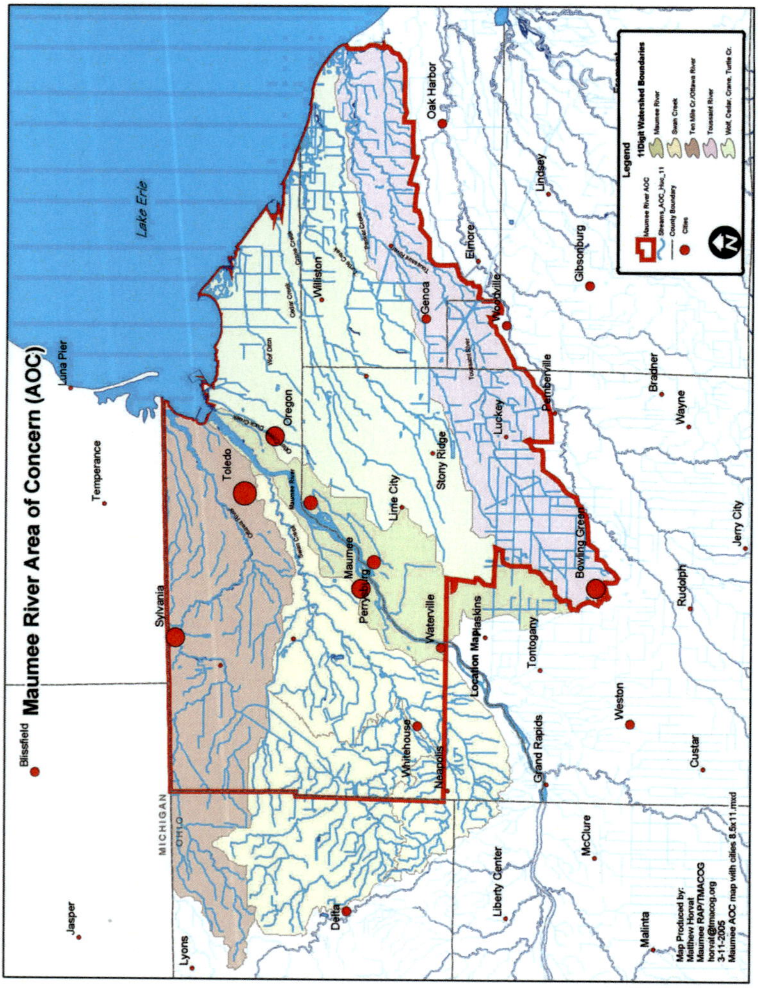

Fig. 13.1 Maumee Area of Concern, NW Ohio

CHAPTER-14

Fig. 14.1 Aerial view of Presque Isle Bay

Fig. 14.4 The City of Erie's Bay Front Today

CHAPTER-16

Fig. 16.1 Map of Lake Victoria and its catchment

Fig. 16.3 Water hyacinth in Lake Victoria (March 2006)